MW00846144

WOMEN'S HEALTH AND MENOPAUSE

RISK REDUCTION STRATEGIES – IMPROVED QUALITY OF HEALTH

Medical Science Symposia Series

Volume 13

The titles published in this series are listed at the end of this volume.

Women's Health and Menopause

Risk Reduction Strategies – Improved Quality of Health

Edited by

R. Paoletti

Institute of Pharmacological Sciences, University of Milan, Italy

P.G. Crosignani

Department of Obstetrics and Gynaecology, University of Milan, Milan, Italy

P. Kenemans

Department of Obstetrics and Gynaecology, University Hospital Vrije Universiteit, Amsterdam, The Netherlands

N.K. Wenger

Department of Medicine, Division of Cardiology, Emory University School of Medicine, Atlanta, Geogia, U.S.A.

and

A.S. Jackson

Giovanni Lorenzini Medical Foundation, Houston, Texas, U.S.A.

KLUWER ACADEMIC PUBLISHERS

DORDRECHT / BOSTON / LONDON

Fondazione Giovanni Lorenzini, Medical Science Foundation, Milan, Italy

Giovanni Lorenzini Medical Foundation, Houston, Texas, U.S.A.

A C.I.P. Catalogue record for this book is available from the Library of Congress.

ISBN 0-7923-5906-2

Published by Kluwer Academic Publishers,
P.O. Box 17, 3300 AA Dordrecht, The Netherlands.

Sold and distributed in North, Central and South America
by Kluwer Academic Publishers,
101 Philip Drive, Norwell, MA 02061, U.S.A.

In all other countries, sold and distributed
by Kluwer Academic Publishers,
P.O. Box 322, 3300 AH Dordrecht, The Netherlands.

Printed on acid-free paper

Printed in the Netherlands.

In Memoriam

MAURIZIO R. SOMA

Dr. Maurizio R. Soma, an investigator at the Institute of Pharmacological Sciences, University of Milan, Milan, Italy, died suddenly on May 22, 1998 at the age of 38.

His untimely death interrupted a brilliant career in the area of atherosclerosis, devoted to the investigation of mechanisms for the progression and regression of arterial lesions and drugs acting directly on the arterial wall, and cardiovascular disease and menopause.

Dr. Soma was responsible for the laboratory of experimental atherosclerosis and was establishing the laboratory of Nuclear Magnetic Resonance at the Institute of Pharmacological Diseases, University of Milan, Milan, Italy. He was also very active as scientific secretary of significant international meetings on lipids, menopause, and cardiovascular diseases.

After graduating Magna cum laude in chemistry and technology of drugs from the University of Milan (1984) and receiving a Ph.D. in Experimental Medicine, Dr. Soma spent from 1984-1989 at the Baylor College of Medicine in Houston conducting research under the direction of Dr. Antonio M. Gotto, Jr.. During this time, he began publishing a significant body of work which continued throughout his life.

More recently, he helped establish a European network directed by Dr. J. Martin of London, on arterial pathophysiology and gene therapy in vascular diseases which was rated one of the outstanding projects by the Committee of Biomed 2 of the European Union (1995-1998).

His interest and expertise in the field of women's health had been evidenced in his role as Scientific Secretary for the series of international symposia on Women's Health and Menopause (1993, 1996, and 1998) and as editor of the Proceedings resulting from these meetings. He was also the Scientific Secretary for the development and publication of the European position paper on Hormone Replacement Therapy and the Menopause, published in 1996 and updated in 1997. In Italy, Dr. Soma played an instrumental role as Scientific Secretary to the Italian National Campaign on Women's Health and Menopause.

His intense research activity and teaching did not detract him from tutoring on an individual basis many young and capable collaborators, who are now continuing his work. His many friends in Europe and the United States shall remember him as an extremely creative scientist with a very rich personality full of humor and insight, always interested in culture, in learning, and in sharing his knowledge. They will ensure that his work, interrupted by his untimely death, will continue.

P.G. Crosignani *Rodolfo Paoletti* *Peter Kenemans* *Nanette Wenger*

CONTENTS

VIII. ONCOLOGY

IX. MENOPAUSE IN ITALY: EPIDEMIOLOGICAL STUDIES AND MEDICAL PERCEPTION

PREFACE

The population structure in the world is rapidly changing. In 1975 people aged 65 and older represented only 5-6% of the population; today the percentage has increased to 7% and is expected to increase to 10% by the year 2025. By 2050 this age group will constitute 15-18% of the total population.

In less than a century—only 75 years—we will face a tripling of the elderly population. The phenomenon is particularly striking in Europe. Today the population aged 65 and over represents 15% of the total European population; by 2025 this percentage will increase to 20-25%, while the number of centenarians will double every decade. In fact, according to World Bank estimates, within 50 years, there will be 2.2 million of people in the world over a hundred years old. Therefore, in the next few decades, age-related biological degenerative processes and their sequelae will be the most important issues facing social medicine and forcing operating costs to rise for all levels of governmental health agencies.

Although women are favored in terms of life expectancy, they live with a longer period of disability, approximately twice that of aging men. Cardiovascular disease, dementia, and cancer are potential enemies of all the elderly. Menopause is the endocrine event that overlaps with aging, potentially worsening both the quality of life and the risks of disease in women. This volume of *Proceedings* of the 3rd International Symposium on WOMEN'S HEALTH AND MENOPAUSE: RISK REDUCTION STRATEGIES—REVISED QUALITY OF HEALTH analyzes many of these complex issues facing both clinicians and researchers treating this population.

The role assigned to ovarian deficiency as part of the global calculation of risks for aging women, for example, is not easily determined for a number of methodological reasons. Thus, it is not surprising that hormone replacement therapy (HRT) efficacy with regard to CHD risk reduction is evaluated in several different ways. While the effect of HRT on menopausal symptoms is generally viewed as rapid and consistent, and is thereby accepted by the scientific community, its relationship to CHD risks is considered variable and controversial.

In analyzing these complex issues, this volume of *Proceedings* of the 3rd International Symposium on WOMEN'S HEALTH AND MENOPAUSE: RISK REDUCTION STRATEGIES—IMPROVED QUALITY OF LIFE yields new and significant insights both to the study of menopause-related disorders and to their treatment, by illustrating the most recent information on mechanisms of action of new estrogen receptors and on the use of sophisticated techniques of statistical analysis for population-based studies. Although we realize that the content of this volume does not and can not answer all the questions pending while awaiting the results from ongoing clinical trials, we do hope that this book will help the many scientists and clinicians involved in the exciting and growing area of women's health and menopause.

The Editors

LIST OF CONTRIBUTORS

Pietro Affinito, *Via Fiume 65, 81032 Carinaro, Italy*

Daniele Agostinelli, *Presidio Hospital, M. Sarcone Ambulatorio per la Menopausa, Via P. Fiore 135, 27038 Terlizzi, Italy*

Cesare Albanese, *Associazione Studi della Menopausa, Via Cardinal Pacca 16, 00165 Rome, Italy*

Louis V. Avioli, *Schoenberg Professor of Medicine, Professor of Orthopedic Surgery, Director, Division of Bone and Mineral Diseases, Washington University School of Medicine, 216 S. Kings Highway, St. Louis, Missouri 63110, USA*

Jole Baldaro Verde, *President, Ricerche in Sessuologia, Via Angelo Ceppi 1 int. 16, 16126 Genoa, Italy*

Fulvio Bartiromo, *Monaldi Hospital, Division of Gynecology and Obstetrics, Centro Menopausa, Via L. Bianchi, 80131 Naples, Italy*

Angela Maria Becorpi, *Diagnostica Medica, P.za Stazione 2, 50123 Florence, Italy*

Francesca Bernardi, *Department of Reproductive Medicine and Child Development, Division of Obstetrics and Gynecology, University of Pisa, Via Roma 35, Pisa, Italy*

Nicoletta Biglia, *Department of Gynecological Oncology, University of Turin, Mauriziano Umberto I Hospital and IRCC of Turin, Largo Turati, 62, 10128 Turin, Italy*

Stanley J. Birge, *Washington University School of Medicine, Division of Geriatrics and Gerontology, 4488 Forrest Park Blvd., St. Louis, Missouri 63108, USA*

Adriano Bompiani, *President, Bambino Gesu Hospital, Piazza S. Onofrio 4, 00165 Rome, Italy*

Gloria Bonaccorsi, *Centro di Servizio e Ricerca per lo Studio della Menopausa e Osteoporosi, Via del Pozzo 71, 41100 Modena, Italy*

Matteo Bottari, *Nuovo Ospedale Papardo, Contrada Sperone, 98100 Messina, Italy*

Maria Luisa Brandi, *Endocrine Unit, Department of Clinical Physiopathology, University of Florence, Viale Pieraccini, 6, 50132 Florence, Italy*

Curt W. Burger, *Academic Hospital Vrije Universiteit, Department of Obstetrics and Gynaecology, De Boelelaan 1117, 1081 HV Amsterdam, The Netherlands*

Angelo Cagnacci, *Policlinico Clinica Ostetrica Ginecologica, Centro Menopausa, Via del Pozzo 71, 41100 Modena, Italy*

Elisa Cevasco, *AGICOL (Assoc. Med. e Progresso), P.zza Palermo 5, 16129 Genoa, Italy*

Raffaella Chionna, *Centro della Menopausa, Hospital San Raffaele, Via Olgettina, 60, 20132 Milan, Italy*

Claus Christiansen, *Center for Clinical and Basic Research, Ballerup Byvej 222, DK-2750 Ballerup, Denmark*

Mary T. Chunko, *Office of Research on Women's Health, National Institutes of Health, Office of Research on Women's Health, National Institutes of Health, Building 1, Room 201, 9000 Rockville Pike, Bethesda, Maryland 20892, USA*

Massimo Ciammella, *Centro per lo Studio e la Terapia del Climaterio, Clinica Mangiagalli, University of Milan, Via Commenda, 12, 20129 Milan, Italy*

Arcangelo Cordopatri, *Division of Obstetrics and Gynecology, Papardo Hospital, 98100 Messina, Italy*

Marilena Cozzarella, *Department of Gynecological Oncology, University of Turin, Mauriziano Umberto I Hospital and IRCC of Turin, Largo Turati, 62, 10128 Turin, Italy*

Francesco d'Agostino, *Presidente Comitato Nazionale per la Bioetica, Via Veneto 56, 00186 Rome, Italy*

Giovanni de Gaetano, *Department of Vascular Medicine and Pharmacology, Istituto di Ricerche Farmacologiche, Consorzio Mario Negri Sud, Via Nazionale, 66030 S. Maria Imbaro, Italy*

Vincenzo De Leo, *Centro della Menopausa, University of Siena, Via Mascagni, 46, 53100 Siena, Italy*

Giovanni De Luigi, *Centro di Fisiopatologia del Climaterio, Department of Obstetrics and Gynecology, Via Ventimiglia, 3, University of Turin, 10126 Turin, Italy*

Michael Dey, *ESI Lederle, 555 Lancaster Avenue, St. David's, Pennsylvania 19087, USA*

Egon Diczfalusy, *Karolinska Institute, Rönningevägen 21, SE-14461 Rönninge, Sweden*

Maria Benedetta Donati, *Consorzio Mario Negri, Via Nazionale, 66030 Santa Maria Imbaro, Italy*

Alberto Falchetti, *Endocrine Unit, Department of Clinical Physiopathology, University of Florence, Viale Pieraccini, 6, 50132 Florence, Italy*

Marie L. Foegh, *Department of Surgery, Georgetown University Medical Center, Washington, D.C. and Solvay Pharmaceuticals, Marietta, Georgia 30062, USA*

Lynda Frassetto, *University of California, San Francisco, Box 0126, San Francisco, California 94143, USA*

Mario Gallo, *Centro di Fisiopatologia del Climaterio, Department of Obstetrics and Gynecology, University of Turin, Via Ventimiglia, 3, 10126 Turin, Italy*

Morrie M. Gelfand, *Department of Obstetrics and Gynecology, McGill University, The Sir Mortimer B. Davis Jewish General Hospital, 5750 Côte des Neiges, Suite 600-A, Montreal, Quebec H3S 1Y9 Canada*

Andrea Riccardo Genazzani, *Department of Reproductive Medicine and Child Development, Division of Obstetrics and Gynecology, University of Pisa, Via Roma 35, Pisa, Italy*

Luigi Gennari, *Endocrine Unit, Department of Clinical Physiopathology, University of Florence, Viale Pieraccini, 6, 50132 Florence, Italy*

Lorenzo Ghiadoni, *Department of Internal Medicine, University of Pisa, Via Roma, 67, Pisa, Italy*

Maurizia Giai, *Department of Gynecological Oncology, University of Turin, Mauriziano Umberto I Hospital and IRCC of Turin, Largo Turati, 62, 10128 Turin, Italy*

Stefano Govoni, *Institute of Pharmacology, University of Pavia, Viale Taramelli, 14, 27100 Pavia, Italy*

Alessandra Graziottin, *Via San Secondo 19, 10128 Torino, Italy*

Antonio Guaita, *Istituto Geriatrico C. Golgi, P.zza Golgi, 20081 Abbiategrasso, Italy*

Secondo Guaschino, *Clinica Ostetrica Ginecologica, Centro Menopausa, University of Trieste, Viale dell'Istria 65/1, 34100 Trieste, Italy*

Jan-Åke Gustafsson, *Department of Medical Nutrition and Dept of Biosciences at Novum, Huddinge University Hospital, S-141 86 Huddinge, Sweden*

Uriel Halbreich, *BioBehavioral Program, School of Medicine and Biomedical Sciences, SUNY Clincial Center, BB170, 462 Grider Street, Buffalo, New York 14215-3098, USA*

Gerard Hall, *St. Peters Hospital, Guildford Road, Chertsey, Surrey KT16 0PZ, UK*

Menno V. Huisman, *Department of General Internal Medicine, Leiden University Medical Center, Postbox 9600, 2300 RC Leiden, The Netherlands*

Licia Iacoviello, *Unit of Genetics of Vascular Risk Factors, Consorzio Mario Negri Sud, Via Nazionale, 66030 Santa Maria Imbaro, Italy*

Linda S. Kahn, *BioBehavioral Program, School of Medicine and Biomedical Sciences, SUNY Clincial Center, BB170, 462 Grider Street, Buffalo, New York 14215-3098, USA*

Peter Kenemans, *Academic Hospital Vrije Universiteit, Department of Obstetrics and Gynaecology, De Boelelaan 1117, 1081 HV Amsterdam, The Netherlands*

Christopher Lademacher, *Merck KgaA, Frankfurther Str. 250, D-64271 Darmstadt, Germany*

Antonio La Marca, *Centro della Menopausa, University of Siena, Via Mascagni, 46, 53100 Siena, Italy*

Carlo La Vecchia, *Istituto di Ricerche Farmacologiche "Mario Negri", Via Eritrea, 62, 20157 Milan, Italy and Istituto di Statistica Medica e Biometria, Università degli Studi di Milano, 20133 Milan, Italy*

Marianne J. Legato, *Columbia University College of Physicians and Surgeons, 622 West 168th St., New York, New York 10032, USA*

Filippo Leonardo, *Department of Cardiology, Istituto H San Raffaele, Via Elio Chianesi 33, 00144 Roma, Italy*

Remo Lombardo, *Hospital of Soriano, Via dei Basiliani, 8, 88018 Vibo Valentia, Italy*

Douglas W. Losordo, *Associate Professor of Medicine, Tufts University School of Medicine, Division of Cardiovascular Medicine, St. Elizabeth's Medical Center, 736 Cambridge St., Boston, Massachusetts 02135, USA*

C. Richard Lyttle, *Wyeth-Ayerst Research, Women's Health Research Institute, 145 King of Prussia Road, Radnor, Pennsylvania 19087, USA*

Francesco Mangino, *Clinica Ostetrica Ginecologica, Centro Menopausa, University of Trieste, Viale dell'Istria 65/1, 34100 Trieste, Italy*

Teri Manolio, *Director, Epidemiology and Biometry, Division of Epidemiology and Clinical Applications, National Heart, Lung, and Blood Institute, Two Rockledge Center, MSC 7934, Bethesda, Maryland 20892-7934, USA*

Patrick Marquis, *Mapi Values Limited, Adelphi Mill, Bollington, Macclesfield, Cheshire SK10 5JB UK*

Laura Masi, *Endocrine Unit, Department of Clinical Physiopathology, University of Study of Florence, Viale Pieraccini, 6, 50132 Florence, Italy*

Patrizia Masi, *Poliambulatorio Mazza Corati, Via Toscana 19, 40141 Bologna, Italy*

Marco Massobrio, *Centro di Fisiopatologia del Climaterio, Department of Obstetrics and Gynecology, Via Ventimiglia, 3, University of Turin, 10126 Turin, Italy*

Maurizio Mauloni, *Clinica Ostetrica Ginecologica III, Hospital San Orsola, Via Massarenti, 1, 40138 Bologna, Italy*

Bruno Mauro, *Monaldi Hospital, Division of Gynecology and Obstetrics, Centro Menopausa, Via L. Bianchi, 80131 Naples, Italy*

Massimo Milani, *Via A. Nota, 18, 20126 Milan, Italy*

Bruno Molteni, *Ospedale Trabattoni Borella, Via Milano 65, 20038 Seregno, Italy*

Daniela Morano, *Centro di Servizio e Ricerca per lo Studio della Menopausa e Osteoporosi, Via Boschetta*

29/31, 44100 Ferrara, Italy

R. Curtis Morris, *University of California, San Francisco, Box 0126, San Francisco, California 94143, USA*

Carmine Nappi, *President, Italian Society of Menopause, Direttore Cattedra di Ginecologia e Ostetricia, Universita Federico II di Napoli, Via Pansini 5, 80131 Naples, Italy*

Polly A. Newcomb, *Fred Hutchinson Cancer Research Center, 1100 Fairview Ave N. MP702, Seattle, Washington 98109, USA*

Morris Notelovitz, *Women's Medical & Diagnostic Center, 2801 NW 58th Boulevard, Gainesville, Florida 32605, USA*

Umberto Omodei, *Clinica Ostetrica Ginecologica, The Civil Hospital, P. le Spedali Civili, 1, 25100 Brescia, Italy*

Silva Ottanelli, *1 Clinica Ostetrica Ginecologica, University of Florence, Via G.B. Morgagni 39, 50134 Florence, Italy*

Gaia Panina, *Department of Cardiology, Istituto H San Raffaele, Via Elio Chianesi 33, 00144 Roma, Italy*

Socrates E. Papapoulos, *Department of Endocrinology and Metabolic Diseases, Building 1, C4-R, Leiden University Medical Center, Albinusdreef 2, 2333 ZA Leiden, The Netherlands*

Vivian W. Pinn, *Office of Research on Women's Health, National Institutes of Health, Building 1, Room 201, 9000 Rockville Pike, Bethesda, Maryland 20892, USA*

Stefania Pinto, *Department of Internal Medicine, University of Pisa, Via Roma, 67, Pisa, Italy*

Riccardo Ponzone, *Department of Gynecological Oncology, University of Turin, Mauriziano Umberto I Hospital and IRCC of Turin, Largo Turati, 62, 10128 Turin, Italy*

Marco Racchi, *IRCCS San Giovanni di Dio FBF Brescia, Via Pilastroni 4, 25135 Brescia, Italy*

Gian Luigi Radici, *Clinica Ostetrica Ginecologica Macedonio Melloni, Via M. Melloni 52, 20129 Milan, Italy*

Giorgio Ranchet, *San Antonio Abate Hospital, Division of Obstetrics and Gynecology, Via Pastori 4, 21013 Gallarate, Italy*

Riccardo Roagna, *Department of Gynecological Oncology, University of Turin, Mauriziano Umberto I Hospital and IRCC of Turin, Largo Turati, 62, 10128 Turin, Italy*

Giuseppe M.C. Rosano, *Department of Cardiology, Istituto H San Raffaele, Via Elio Chianesi 33, 00144 Roma, Italy*

Henri Rozenbaum, *15, rue Daru, 75008 Paris, France*

Antonio Salvetti, *Department of Internal Medicine, University of Pisa, Via Roma 67, 56100 Pisa, Italy*

Göran Samsioe, *Department of Obstetrics and Gynecology, Lund University Hospital, S-221 85 Lund, Sweden*

Fredde Scheele, *St. Lucas Andreas Ziekenhuis, Jan Tooropstraat 164, 1061 AE Amsterdam, The Netherlands*

Giansalvo Sciacchitano, *Clinica Ostetrica Ginecologica, Hospital V. Emanuele, Via Plebiscito, 95100 Catania, Italy*

Anthony Sebastian, *University of California, San Francisco, Box 0126, San Francisco, California 94143, USA*

Giovan Battista Serra, *Division of Obstetrics and Gynecology, Hospital Cristo Re, Via delle Calasanziane 25, 00167 Rome, Italy*

Luca G. Sgro, *Department of Gynecological Oncology, University of Turin, Mauriziano Umberto I Hospital and IRCC of Turin, Largo Turati, 62, 10128 Turin, Italy*

Piero Sismondi, *Department of Gynecological Oncology, University of Turin, Mauriziano Umberto I Hospital and IRCC of Turin, Largo Turati, 62, 10128 Turin, Italy*

Daniela Solano, *Institute of Pharmacological Sciences, University of Milan, Via Balzaretti 9, 20133 Milan, Italy*

Bruno S. Solerte, *Department of Internal Medicine, Geriatrics and Gerontologic Clinic, P.zza Borromeo, 27100 Pavia, Italy*

Antonio G. Spagnolo, *Institute of Bioethics, Catholic University of the Sacred Heart, School of Medicine, Largo F. Vito 1, 00168 Rome, Italy*

Angela Spinelli, *Istituto Superiore di Sanita, Department of Epidemiology and Biostatistics, viale regina Elena 299, 00161 Rome, Italy*

John C. Stevenson, *Rosen Laboratories of the Wynn Institute, Endocrinology and Metabolic Medicine, Imperial College School of Medicine, St. Mary's Hospital Medical School, 3rd Floor, Norfolk Place, London, W2 1PG, UK*

Isabella Sudano, *Department of Internal Medicine, University of Pisa, Via Roma, 67, Pisa, Italy*

Jay M. Sullivan, *Professor of Medicine, Chief, Division of Cardiovascular Diseases, The University of Tennessee, Memphis, 951 Court Avenue, Room 353D, Memphis, Tennessee 38163, USA*

Stefano Taddei, *Department of Internal Medicine, University of Pisa, Via Roma, 67, Pisa, Italy*

Jos H.H. Thijssen, *Department of Endocrinology, University Hospital G02.625, PO Box 85500, 3508 GA Utrecht, The Netherlands*

Karen Todd, *University of California, San Francisco, Box 0126, San Francisco, California 94143, USA*

Giorgio Tresoldi, *Italian Society of General Practitioners, Viale Europa, 190, 20062 Cassano d'Adda (Milan), Italy*

Wulf H. Utian, *Department of Reproductive Biology, Case Western Reserve University, Department of Obstetrics and Gynecology, University MacDonald Women's Hospital, 11100 Euclid Avenue, Cleveland, Ohio 44106, USA*

Alessandra Valerio, *Centro di Servizio e Ricerca per lo Studio della Menopausa e Osteoporosi, Via Boschetta 29/31, 44100 Ferrara, Italy*

Martin P. Vessey, *University Department of Public Health, Institute of Health Sciences, Old Road Headington, Oxford, OX3 7LF, UK*

Sandro Viglino, *AGICOL (Assoc. Med e Progresso), P.zza Palermo 5, 16129 Genoa, Italy*

Agostino Virdis, *Department of Internal Medicine, University of Pisa, Via Roma, 67, Pisa, Italy*

Nanette Kass Wenger, *Department of Medicine, Division of Cardiology, Emory University School of Medicine, Thomas K. Glenn Memorial Building, 69 Butler Street, S.E., Atlanta, Georgia 30303, USA*

Callum S. Wingrove, *Rosen Laboratories of the Wynn Institute, Endocrinology and Metabolic Medicine, Imperial College School of Medicine, St. Mary's Hospital Medical School, 3rd Floor, Norfolk Place, London, W2 1PG, UK*

Lucio Zichella, *I Clinica Ostetrica e Ginecologica, Universita La Sapienza Policlinico Umberto I, Viale Policlinico 155, 00161 Rome, Italy*

AN AGING HUMANKIND: NEW REALITIES

Egon Diczfalusy

Alles Leben ist Problemlösen
(Life is solution of problems)
Sir Karl R. Popper
(Title of book)

Sometimes there is foolishness in wisdom and wisdom in foolishness and there is a great deal of wisdom in the praise of foolishness (*Encomium moriae*) as was published in 1509 by Erasmus of Rotterdam: "All things in life are so multifaceted, contradictory, and obscure that we can never be sure about the truth." Some 150 years later, Blaise Pascal reflecting on this in his *Pensées sur la religion* (1669) said "Yes, but it is not certain that everything is uncertain." There are some realities. There are old realities and new realities. This paper will discuss some of the new realities.

In 1920, the global population was less than 2 billion people; next year it will exceed 6 billion. Hence my generation has seen the birth of another two worlds equal in numbers, needs, aspirations, hopes, and dreams. The twentieth century has been characterized by the steepest rise in population ever seen, which is still continuing, albeit at a slower pace, and is projected to reach 9 billion shortly after 2050.

It is thus understandable that our generation was mesmerized by population growth and its likely consequences, so mesmerized by it that the simultaneous change in population structure hardly received the necessary attention. In fact, there is no longer such a thing as a population pyramid in the developed countries and the population pyramid of the developing world is also rapidly changing shape. By 2025 it will be a population globe rather than a population pyramid and by the end of the twenty-first century its population structure might look like a mushroom. Humankind has never grown so rapidly and then aged so rapidly. It is an entirely new experience for *Homo sapiens* and he has to learn how to deal with the new realities.

There is a gender difference in all regions of the world in terms of life expectancy at birth. Even today life expectancy for women in Africa is more than 20 years less than in Europe. The most pronounced difference is seen in North America and Europe, where during the past 100 years, life expectancy at birth for women has increased by more than 50%, an increase of 30 years from 47 to 77 years of age. It is projected by the United Nations [1] that by 2050 it will be 83, an increase of 6 years from the current 77; in Italy

R. Paoletti et al. (eds.), Women's Health and Menopause, 1–4.

it is projected to be 86.2 which will be the highest in the world.

In 1996, there were few countries where 20% or more of the population was aged 60 and over, mainly European countries and Japan. In 2025, it is expected that huge areas of the world will be inhabited by populations in which 20% or more are aged 60 and over. The increase in life expectancy is not limited to that at birth. There is an increase at every age group, even in the oldest old (those aged 80 and over). It is projected that the population aged 75 and over will increase between 1996 and 2035 by almost 160% in China and 127% in India.

In the year 2025, the population aged 75 or over will be almost 15% in Japan, almost 13% in Italy, and 12% in Greece and Sweden [2]. Moreover, the number of centenarians currently grows the fastest—at an annual rate of 8%. Their number in the world, which at present is at 60,000, will double every decade [3] .

Using the UN definition that the elderly are those aged 65 years and over, it is projected that between 1950 and 2050 the elderly population of the world is going to increase from 5.2% to 15.1%. By the year 2050, the elderly population in Europe will be more than 25% of the total with the elderly population of Italy more than 35%, again another world record [1].

Rapidly increasing elderly populations represent a social phenomenon with no historical precedent, with the results that governments are uncertain as to what to do and the scientific community unable as to how to advise them. The problem is compounded by another issue of fundamental importance: while the elderly population is increasing, the population of the children is decreasing dramatically. In 1975, almost 37% of the world population consisted of children aged 15 years or less. By the year 2050, it is projected to be around 20%, in Europe 16%, in Italy 12.4% [1]. If a simple ratio is calculated, in 1975 the ratio of children to the elderly in the world was 6.6%; it is projected to decrease to 1.4% in the year 2050. By that time, the ratio in Europe will be 0.6% and in Italy it may reach 0.3%, or 3 elderly for every child.

The most important factor underlying these changes is a worldwide dramatic decline in total fertility rate per woman. The concept of total fertility rate can be approximated by the number of children a woman will have in her lifetime. In 1950, it was five around the world. Between 1975 and 1995 the fertility rate per woman in the world declined from 4.48 to 2.96. In 1975, in Europe and in Italy the rate was still above replacement level, around 2.1 children per woman, which would be expected to keep a population in balance. Only five years later, in 1980, it was below replacement level both in Europe and in Italy. In 1995, it was very low in Europe (1.57) and in Italy, the lowest in the world (1.24). Only the former East Germany had lower levels than that. In 1995 the fertility of 51 countries representing 45% of the global population was already at or below replacement levels. It is projected that by the year 2025 there will be 102 such countries with 76% of the world population [4]. Never before in history have birth rates fallen so far, so fast, so low, for so long all around the world. The implications for good and for ill are as yet unclear. There is no theory as yet that would explain why, when, or how long-term below-replacement fertility rates would ever go back up to replacement levels.

Thus the population growth is slowly coming to an end. In 1975 the world population increased with an annual rate of 2%. It is expected that by the year 2050, the growth will be less than half a percent. Population growth in Europe and Italy is virtually nonexistent today and is expected to turn increasingly negative. If the most likely medium-variant projection materializes, by the year 2050 the population of Europe is expected to decrease yearly by half a percent and that of Italy by one percent. In the case of the low-variant projection, the decline will be more dramatic. As an example, the population of Europe in the year 2000 is expected to be 729 million people. By the year 2050 it is projected to decline to 637 million according to the medium projection, and to 537 million according to the low projection; a decline by almost 200 million people. Let us season these figures with just a bit of so-called geopolitical reflection: In 1950 the population of Europe was 547 million and that of Africa 224 million people. In the year 2000, Europe will have a population of 729 million and Africa 820 million. According to the medium-variant projection of the UN, in 2050 Europe might have a population of 637 million and Africa 2,046 million people [1].

We are living in complicated times and complicated times are always golden opportunities for the prophets of gloom and doom. In the December (1997) issue of *The Economist* there was an amusing statement that the prophets of gloom and doom appear to think that having been invariably wrong in the past makes them more likely to be right in the future. And the prophets of gloom and doom are, of course, again wrong. They are concerned about economic incompatibility and intergenerational conflict in our aging society. In fact, there is little, if any, evidence of this in the aging societies, for instance in Europe and Japan. Moreover, in Europe, longevity is increasing more rapidly than the cost of healthcare. At a recent International Congress on Longevity and Quality of Life (UNESCO, Paris, 18-20 May 1998), the former German Minister of Social Affairs, Professor Ursula Hahn, presented interesting data indicating that today in Germany 79% of men and 74% of women aged 80 or older can still manage their daily lives.

A few years ago, some African leaders reflected upon this state of affairs and they concluded with a somewhat rhetorical question: “Longevity has been our quest for ages; now that we have found it, can we afford it?” [5]. Of course the question is meant to be provocative; after all, what is the alternative? At any rate, the answer is definitely yes, provided one remembers what Edmund Burke said in 1796, almost 200 years ago: “You can never plan the future by the past.”

There is a need for urgent institutional reform. Why? Because most of our institutions, the admirable brain-children of our great-grandfathers, are catering to a population structure (with many children and few elderly) that does not exist any longer. In fact, there are major efforts in progress in Europe to redress the imbalance between the increasing number of beneficiaries and the decreasing number of contributors and to create new pension schemes for the twenty-first century [6]. The task will be much more complicated in Eastern Europe and in some other countries where the implicit pension debt as a percentage of gross domestic product is between 100-200% [7]. This is not only a problem for the developed world, because today more than 150 countries around the world have public old age, disability, and survivor programs of different kinds [8].

Other institutions catering to the needs of yesterday include social security, healthcare, education, housing, and last, but not least, marketing, which predominantly addresses the youth and not the elderly. The danger is that addressing problems that no longer exist may prevent us from focusing on the real issues of the present and thus also on those of the future. If there is such a need, what will be the major impediment to the implementation of the necessary reforms? Most probably political inertia.

The second half of the twentieth century was the epoch of revolutions. Among others, we have witnessed revolutions in science, technology, the generation revolution and that of transmission of information, globalization, preservation of our environment, the universal acceptance of contraception, reproductive health and gender equity, and last, but not least, the most overwhelming of all of them, the demographic revolution. What will then be the most important revolution of the twenty-first century? I am convinced that it will be the universal achievement of human dignity. A world will be created, in which every human individual has sufficient food, water, shelter, sanitation, health services, a healthy environment, education, employment, and personal security. This will be a world ruled by the 30 articles of the Universal Declaration of Human Rights [9]. Yes, it is a dream today but it may be reality tomorrow. It can happen, if we let it happen. As Khalil Gibran said, "Trust your dreams, for in them is hidden the gate to eternity" [10]. And as another great humanist and believer in the future of the human race, Eleanor Roosevelt used to say, "The future belongs to those who believe in the beauty of their dreams."

References

1. United Nations Department of Economic and Social Affairs Population Division. World population prospects: The 1996 revision. New York: United Nations, 1997.
2. U.S. Department of Commerce Economics and Statistics Administration, Bureau of the Census. Global aging into the 21st century. Bethesda: U.S. National Institute on Aging (NIA), December 1996.
3. Vaupel JW. Demographic analysis of aging and longevity. Plenary lecture presented at the XXIII International Population Conference, Beijing, China, 11-17 October, 1997.
4. World Health Organization. The world health report 1998. Life in the 21st century. A vision for all. Geneva: World Health Organization, 1998.
5. Tulanian 57, No.1., 1986. In: Coordinating action on aging. Report of the first NGO/WHO Roundtable. World Health Organization, Regional Office for Africa. Annex 4. 1988:26.
6. Organization for Economic Co-Operation and Development (OECD). Maintaining prosperity in an ageing society. Paris: OECD, 1998.
7. The World Bank. World development report 1997. The state in a changing world. Oxford: Oxford University Press, 1997.
8. Diczfalusy E. An aging humankind revisited. The Aging Male 1998;1:89-99.
9. Brownlie I. Basic documents on human rights. III. Universal declaration of human rights, 1948. Oxford: Clarendon Press, 1992.
10. Gibran K. The Prophet. New York: Alfred A. Knopf, Publisher, 1923.

Does Menopause Increase the Risk of Coronary Heart Disease?

Menno V. Huisman

Introduction

It has been a constant finding for more than a century that there is a marked sex difference in risk of coronary heart disease (CHD). There is however little understanding in explaining the basis of this difference. The often-held assumption of the ovarian estrogen as a principal endogenous cardioprotective factor to explain this difference has been based on the finding of a decline in the ratio of male-to-female deaths from CHD that begins in the fifth decade. This decline in the male/female ratio is thought to be related to an increase in mortality among women following the menopause. If this were true an increase in the rate of female mortality rates should be visible following the age at which the menopause usually occurs. This chapter reviews both epidemiologic studies, which had as its objective to find evidence for this increase, and mechanistic studies.

Parallel to these epidemiologic studies, several lines of research have been performed to explore further the role of estrogen in coronary risk in women as a main cause for the presumed increase in coronary risk after the menopause. These include studies in women with artificial or early menopause, studies of changes in lipids, and observational studies on the effect of postmenopausal hormone replacement therapy. Because the results of these studies often are used as indirect evidence for the menopause as a causal factor in the presumed increase in coronary risk these studies are reviewed as well.

Studies in Vital Statistics

Quite contrary to the general opinion it is relatively unknown that in the literature over the years several studies on mortality data can be found that have all been remarkably consistent in demonstrating a lack of any increase in risk after menopause.

Furman, studying United States of America mortality data from the year 1955, found that although the mortality rate from CHD increased exponentially in both men and women with age, it was always higher in men [1]. Confirming the data of Tracy (see below) he found that the acceleration of the mortality rate in men began to diminish during the fifth decade, while for women a steadily accelerating mortality rate was observed.

R. Paoletti et al. (eds.), Women's Health and Menopause, 5–13.

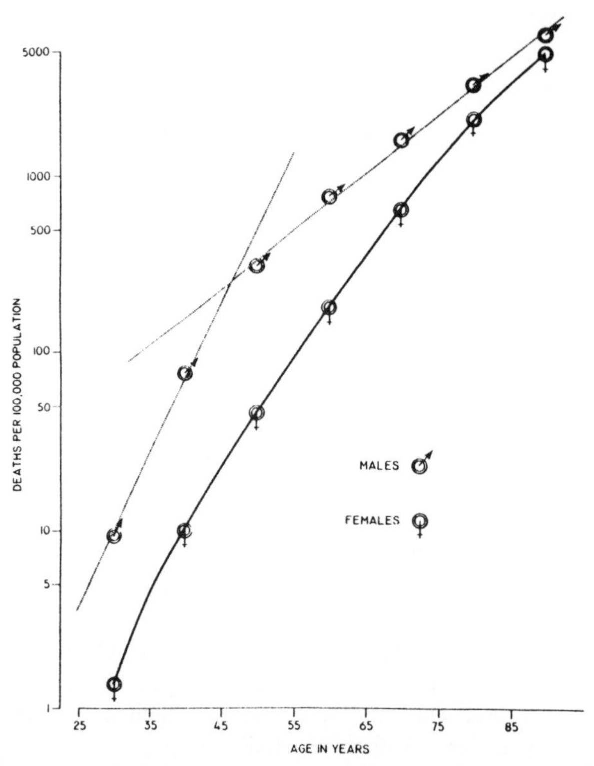

Figure 1. Male and female mortality rates, U.S. white population, 1955, plotted as a function of age (reprinted with permission from Ann NY Acad Sci).

He concluded that it is this diminishing acceleration in men beginning in the fifth decade that is responsible for the declining male/female ratios of deaths from CHD. Tracy in 1966 [2] showed from mortality data of the U.S.A National Office of Vital Statistics from the years 1962 and 1963, that, while there was a deflection in the rate of increase in coronary mortality with increasing age among men, this could not be observed for women, in whom the increase in mortality followed a straight line. The argument against this observation that the inability to see a change in risk at menopause was based on a broad range for age at which menopause begins or that a weak correlation between the last menstrual period and the decline in estrogen is masking any increase was elegantly denied by the same author who showed a sharp alteration in the death rate pattern of breast cancer at menopause, a disease strongly correlated to endogenous estrogen levels.

Heller and Jacobs examined the differences between men and women in mortality from CHD from England and Wales over the years 1970-1974 and likewise found that after the menopause there was no acceleration in the increase of CHD death rates after the menopause and that after the age of 50 the rate of increase in men was much lower than it was both in younger groups of men and in women [3]. He concluded that a change occurred around the time of the menopause, but rather in men than women and that the data could not support the idea that women lost protection from CHD after the menopause. In the discussion an alternative hypothesis of lowering unbound testosterone, which occurs in men after the age of 50, as an explanation for the findings was put forward.

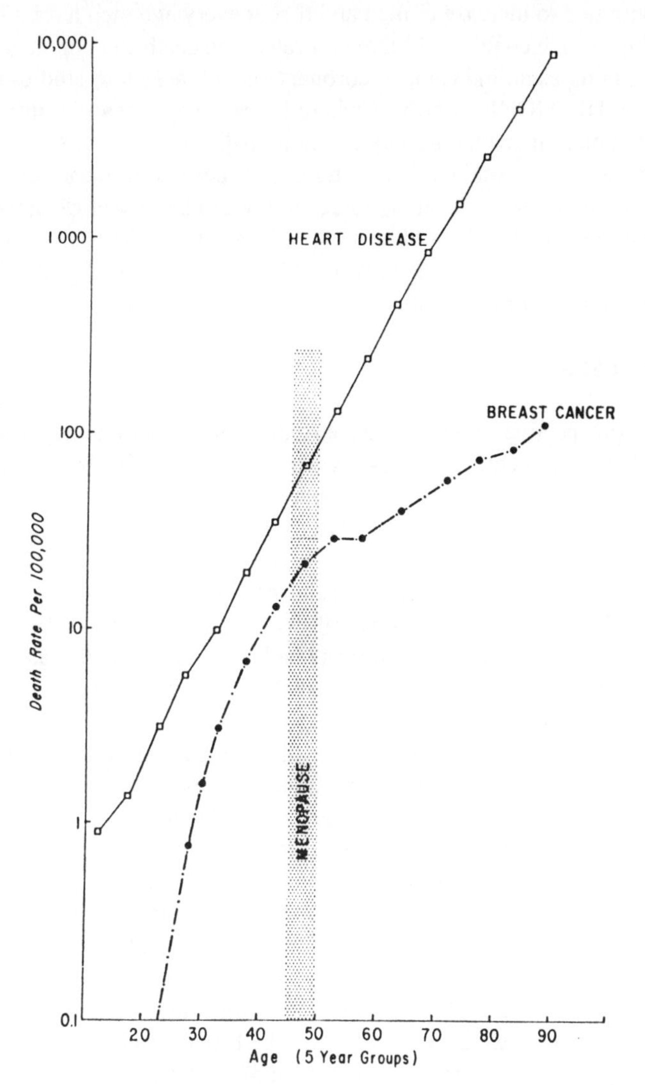

Figure 2. Semi-logarithmic plots of age-specific death rates versus age (Gompertz plots) in U.S.A females 1952 (Reprinted from Tracy RE. Sex difference in coronary disease: Two opposing views. J Chron Dis 1996;19:1245-51 with permission from Elsevier Science)

Finally, Tunstall-Pedoe published mortality statistics from coronary heart disease from England and Wales over the years 1989-1993 [4]. For women a constant acceleration of coronary risk in middle life was found, while for men a decline in the rate of acceleration around the age of the menopause was shown. He also demonstrated that the sex ratios in rates peaked at age 35-39 years and then declined steadily, without a clear inflection at the average age of menopause. Finally, it was found that the absolute difference in risk of coronary death continued to increase in men and that at every age men have a higher rate than women, findings which could be observed repeatedly in death and population data [5]. That the use of mortality as an indicator of coronary risk in these four studies is justified was shown in the WHO MONICA project where the risk of myocardial infarction was correlated across populations with coronary mortality [6].

Taken together, these studies demonstrate that there is no clear acceleration in coronary risk by the menopause itself, If there is any hidden effect of menopause on coronary risk, it is either small and could be masked by the variation of time at which the menopause is occurring, or the effect is prevented by yet unidentified protective factors associated with the ending of menstruation.

Prospective Cohort Studies

There are two large prospective studies that have studied the effect of menopause on CHD. These two studies have found conflicting results. The Framingham Heart Study studied the effect of natural menopause on the risk of coronary disease in 1,934 women who had a natural menopause during follow-up [7]. They were examined every two years for up to 24 years. In the initial report of this study group of women with a natural menopause there were 18 cases of major coronary disease (myocardial infarction, coronary insufficiency, and coronary death) and five cases in premenopausal women. In the original article a relative risk of 2.7 ($p < 0.01$) was quoted and was only attained by including angina as an endpoint. Recalculation has confirmed the relative risk of 2.7 but with borderline significance ($p=0.07$) [8]. For major coronary disease excluding angina, the relative risk was 2.0 (95% confidence interval 0.6-6.3, $p= 0.26$). Importantly, although the analyses were age-adjusted by 5-year age groups, within each such group the premenopausal women were younger and hence at lower risk by virtue of age. A third caveat was that no adjustment for cigarette smoking was made, while it is known that smokers tend to have an earlier menopause [9]. These observations have been followed by an update of the Framingham study in 1986, where natural menopause was associated with a 4.1-fold higher 10-year incidence of CHD compared with rates in premenopausal women, but no adjustments for age and cigarette smoking was made [10].

The largest prospective study, the Nurses' Health study, studied 116,258 female nurses aged 33-55 years, who were free of diagnosed CHD at baseline were followed for 6 years by biennial questionnaires for the occurrence of nonfatal and fatal coronary heart disease [11]. Coronary endpoints were documented by medical records and this resulted in very high follow-up (98.3%for fatal endpoints, 95.4%for nonfatal endpoints). There were 105 cases of major coronary disease (84 nonfatal myocardial infarction and 21 coronary deaths) that occurred in 102,114 person-years of follow-up among women with a natural menopause and 88 cases in 436,003 person-years in premenopausal women. This rate of CHD was significantly elevated in the postmenopausal women with a relative risk of 1.7 (95% confidence interval 1.1-2.8). Further analysis include control for age and cigarette smoking was performed in this study. When the results were analyzed using 1-year age groups instead of 5-year age groups, the relative risk was reduced to 1.2 (95% confidence interval 0.8-1.9), demonstrating the need for close control of age. Control for cigarette smoking resulted in no difference between premenopausal and naturally postmenopausal women with a relative risk of 1.0 (95% confidence interval 0.8-1.3).

It can be concluded that these large prospective studies have not refuted the lack of association of increased coronary risk by menopause that was found in studies in vital statistics.

Hormone Replacement Studies

Numerous studies have been performed to evaluate the effect of postmenopausal estrogen therapy, however, all but a very small trial performed in the 1970s [12] have been of observational design. In a meta-analysis the observed risk reduction based on the postulated cardioprotective effect of estrogen has been calculated to be 46% (95% confidence interval 39-50%) [13]. The results of this and other analyses have been challenged by authors who have postulated and proved that some sort of selection bias on health might have operated [14-17]. It is apparent that randomized studies would give the ultimate evidence for the protective effect of estrogen. If such studies demonstrate an unequivocal protection by exogenous estrogen, it then would also support any effect of the menopause on CHD risk.

Studies in Early Menopause

If endogenous estrogen do indeed protect against cardiovascular disease an early menopause might be associated with a higher risk. Evidence for this hypothesis have mostly come from small studies, used angina pectoris as an endpoint, and had as a main problem a short follow-up period, with resulting insufficient cardiovascular endpoints for statistical analysis.

One of the larger studies has been carried out by van der Schouw et al. who studied 12,115 postmenopausal women with a follow-up period up to 20 years (median 10 years) [18]. At enrollment the women were 50-65 years. The cohort was initially created for a breast screening project. During follow-up all women attended screening visits during which questions on menopausal status, age at menopause, medication use, cardiovascular risk factors, and ovarian function were asked. Deaths were ascertained from the family physicians. The endpoint of analysis was total cardiovascular mortality. The association between age at menopause and cardiovascular mortality was assessed by life-table analysis and Cox regression analysis and all analyses were adjusted for biological age.

Results show that the risk of cardiovascular mortality was higher for women with early menopause than for late menopause. The age-adjusted hazard ratio of age at menopause was 0.982 (95% confidence interval 0.968-0.996, p=0.01). This means that each year the menopause is delayed the annual hazard of cardiovascular death is diminished by 2%. The effect of age at menopause on cardiovascular death was shown to be fairly linear (p=0.35), i.e. a menopause delay of 1 year at age 45 had the same effect as a 1-year delay at age 55.

Adjustment for potential confounders, such as hypertension, body-fat distribution, and presence of diabetes, did not change the results for the estimated hazard ratios. Importantly, in the study by van der Schouw, the effect of natural menopause on cardiovascular death was almost neutral in smoking women [18]. The effect of smoking on cardiovascular endpoints seems to be so strong that it overwhelms the effect of an early menopause. It is of interest that a recently published study in 25 postmenopausal women showed that smoking significantly diminished the HDL cholesterol increase by estrogen replacement in postmenopausal women [19]. In the group of nonsmokers a HDL-cholesterol increase of 0.18 $\pm$ 0.06 mmol/l ($p < 0.05$) was seen after treatment with 0.2 mg estradiol for 8 weeks, while in the smokers the increase was only 0.07 $\pm$ 0.06 mmol/l (p=ns). In the same study a possible explanation for this difference was provided by the observation that plasma concentrations of estrone and estradiol were approximately twice as high in nonsmoking women on estrogen than in smokers. This observation has potential important implications since it has been shown that HDL-cholesterol is the main risk factor for CHD in women [20].

Studies with Intermediate Endpoints

Numerous studies have been performed over the last ten years to lend support to the concept of the menopause as an endocrinopathy, in which estrogen deficiency plays a central role. These studies have included studies on the metabolism of lipoproteins and vasoreactivity. The following part concentrates on studies with lipid measurements as endpoints, since the lipid profile seems to be a logical candidate to explain any effect of estrogen.

Lipids

Studies of risk factors of CHD suggest that postmenopausal women have an atherogenic risk factor profile. They have higher plasma levels of cholesterol, triglycerides, very low density lipoprotein (VLDL) cholesterol, and low density lipoprotein (LDL) cholesterol than their premenopausal counterparts in cross sectional studies. In a large study Matthews and colleague prospectively followed 541 healthy, initially premenopausal women 42 to 50 years of age for 5 years [21]. After 2.5 years follow-up, in the 69 women who had a natural menopause and were not receiving HRT, serum levels of high density lipoprotein (HDL) cholesterol declined as compared with those of premenopausal controls (-0.09 versus 0.00 mmol/l; p=0.01) and levels of LDL cholesterol increased (+0.31 versus +0.14 mmol/l; p=0.04).

In another study Lindquist found an increase in total cholesterol of 0.43 mmol/l among 325 ($p < 0.01$) women who were premenopausal at the start of their study and became postmenopausal during the 5-year follow-up. Triglycerides in the same group of women increased 0.24 mmol/l after menopause [22].

Estrogen replacement therapy has been shown to reverse some of the above mentioned adverse changes in postmenopausal women with normal serum lipid levels. In a randomized, double-blind crossover study in 31 healthy postmenopausal women who were all nonsmokers and had normal blood lipid levels (total cholesterol and triglyceride concentrations below the 95th percentile for age) Walsh et al. demonstrated that a low dose of conjugated estrogens of 0.625 mg daily or micronized estradiol of 2 mg lowered LDL cholesterol levels and raised HDL cholesterol levels both by 14 to 16% and that a higher dose of estrogen (1.25 mg) did not substantially enhance this effect. Triglyceride concentrations rose stepwise by 24% with 0.625 mg estrogen per day and by 38% with 1.25 mg per day [23].

While it seems attractive to have these lipid parameters ameliorated by estrogens, the pharmacologic mechanism by which estrogens protect against atherosclerosis is not fully resolved and if the effect is indeed mediated largely by an improvement in the lipid profile a more direct attack on lipid abnormalities, e.g. by HMG-CoA reductase inhibitors, would seem a logical and strong alternative. Few studies of a direct comparison between statins and estrogens have been performed. Darling compared the effect of postmenopausal therapy up to 1.25 mg of conjugated equine estrogens daily, along with 5 mg of medroxyprogesterone acetate versus simvastatin 10 mg for eight weeks in a randomized cross over design in 58 hypercholesterolemic postmenopausal women who had an intact uterus and were thus to be prescribed a progestin for protection of the uterus [24]. At baseline the mean total cholesterol was 305 ± 39 mg/dl, HDL-cholesterol 62 ± 19 mg/dl, and LDL-cholesterol 217 ± 39 mg/dl. Total cholesterol with hormone therapy decreased with a mean 14% (95% confidence interval [CI] 11-16%), versus a mean decrease of 26% (95% CI 23-29%) with simvastatin. For LDL cholesterol the mean decrease was 24% (95% CI 20-28%) with hormone therapy and 36% (95% CI 32-40%) with simvastatin ($p < 0.001$). HDL cholesterol increased similarly with hormone therapy (mean 7%, 95% CI 2-12%) and simvastatin (mean increase 7%, 95% CI 4-10%). Finally triglycerides levels increased with hormone therapy (mean increase 29%, 95% CI 15-42%), while it decreased with simvastatin therapy (mean decrease 14%, 95% CI 8-20%). The results of this study clearly show the problem with which one is confronted when facing a treatment decision in postmenopausal women. Although treatment with hormone therapy resulted in amelioration of several lipid parameters (total cholesterol, LDL cholesterol, HDL cholesterol) statin treatment was significantly more effective in the reduction of LDL cholesterol and decreased triglycerides that went up by hormone therapy. In view of the uncertainty as to whether reducing LDL cholesterol is more important than increasing HDL and in spite of the uncertainty of the clinical importance of estrogen-induced hypertriglyceridemia with respect to CHD it might be to early to conclude, as the authors did, that individualized hormone therapy should be strongly indicated as pharmacotherapy in postmenopausal women with hyper-cholesterolemia.

It is concluded that although the menopause is accompanied with a more atherogenic lipoprotein profile and estrogens can reverse part of these changes, the "lipid-hypothesis" for estrogens is not fully proven and only prospective data linking clinical endpoints of CHD to both plasma lipoproteins and endogenous hormone levels will resolve these uncertainties.

Conclusions

In spite of numerous publications on the effects of menopause on ischemic cardiovascular endpoints the results of epidemiologic studies demonstrate that there is no clear acceleration of coronary risk after the menopause. It could be concluded from this that the influence of risk factors such as lipids, vasoreactivity, and time of menopause is therefore not present or it is too small to show up against the effects of age. Alternatively there could be unknown coronary protective factors associated with the end of menstruation.

The absence of a change in slope of the CHD incidence curve at the time of the menopause does not support the concept that there is a role for postmenopausal hormone replacement therapy as a primary prevention against coronary heart disease.

Epidemiological evidence, derived from meta-analyses of studies in estrogen replacement therapy have indicated a reduction in cardiovascular disease of up to 50%. These studies however all have been observational in design and ultimately only the results of randomized controlled trials with hormone replacement therapy will give the ultimate evidence of any protective effect of such therapy.

If such trials would indicate a protective effect it would then challenge the above-mentioned results of the epidemiologic studies that have not found an increase in CHD risk as a result of the menopause.

References

1. Furman RH. Are gonadal hormones(estrogens and androgens) of significance in the development of ischemic heart disease? Ann NY Acad Sci 1968;149:822-33.
2. Tracy RC. Sex difference in coronary disease: Two opposing views. J Chron Dis 1966;19:1245-51.
3. Heller RF, Jacobs HS. Coronary heart disease in relation to age, sex, and the menopause. BMJ 1978;i:472-74.
4. Tunstall-Pedoe H. Value of Dundee coronary risk-disk. BMJ 1992;305:1096.
5. Tunstall-Pedoe H. Viewpoint: Myth and paradox of coronary risk and the menopause. Lancet 1998;351:1425-27.
6. Office of Populations Censuses and Surveys. Mortality statistics:cause 1989-1992 England and Wales. London: Her Majesty's Stationary Office (HMSO), 1991-1993.
7. Tunstall-Pedoe H, Kuulasmaa K, Amouyel P, et al. Myocardial infarction and coronary deaths in the World Health Organisation MONICA project. Registration procedures, event rates, and case-fatality rates in 38 populations from 21 countries in four continents. Circulation 1994;90:583-612.
8. Gordon T, Kannel WB, Hjortland MC, et al. Menopause and coronary heart disease. The Framingham study. Ann Int Med 1978;89:157-61.

9. Stampfer MJ, Colditz GA, Willett WC. Menopause and heart disease. A review. Ann NY Acad Sci 1990;529:193-203.
10. Kaufman DW, Slone D, Rosenberg L, et al. Cigarette smoking and age at natural menopause. Am J Publ Health 1980;70:420-22.
11. Lerner DJ, Kannel WB. Patterns of coronary heart disease morbidity and mortality in the sexes: A 26 year follow-up of the Framingham population. Am Heart J 1986;111:383-90.
12. Colditz GA, Willett WC, Stampfer MJ. Menopause and the risk of coronary heart disease in women. N Engl J Med 1987;316:1105-10.
13. Nachtigall LE, Nachtigall RH, Nachtigall RD, et al. Estrogen replacement therapy II: A prospective study in the relationship to carcinoma and cardiovascular and metabolic problems. Obst Gynecol 1979;54:74-79.
14. Stampfer MJ, Colditz GA. Estrogen replacement therapy and coronary heart disease: A quantitative assessment of the epidemiologic evidence. Prev Med 1991;20:47-63.
15. Vandenbroucke JP. Postmenopausal estrogen and cardioprotection. Editorial. Lancet 1991;337:833-34.
16. Barrett-Connor E, Bush TL. Estrogen and coronary heart disease in women. JAMA 1991;265:1861-67.
17. Petitti D. Coronary heart disease and estrogen replacement therapy: Can compliance bias explain the results of observational studies? Ann Epidemiol 1994;4:115-18.
18. van der Schouw YT, van der Graaf Y, Steyerberg EW, et al. Age at menopause as a risk factor for cardiovascular mortality. Lancet 1996;347:714-18.
19. Hoogerbrugge N, Zillikens MC, Jansen H, et al. Estrogen replacement decreases the level of antibodies against oxidised low-density lipoprotein in postmenopausal women with coronary heart disease. Metabolism 1998;47:1-7.
20. Bass KM, Newshaffer CJ, Klag MJ, et al. Plasma lipoprotein levels as predictors of cardiovascular death in women. Arch Int Med 1993;153:2209-16.
21. Matthews KA, Meilahn E, et al. Menopause and risk factors for coronary heart disease. N Engl J Med 1989;321:641-46.
22. Lindquist O. Intra individual changes of blood pressure, serum lipids, and body weight in relation to menstrual status: Results from a prospective population study of women in Goeteborg, Sweden. Prev Med 1980;11:162-72.
23. Walsh BW, Schiff I, Rosner B, et al. Effects of postmenopausal estrogen replacement on the concentrations and metabolism of plasma proteins. New Engl J 1991;325:1196-1204.
24. Darling GM, JA Johns, Mc Cloud PI, Davies SR. Estrogen and progestin compared with simvastatin for hypercholesterolemia in postmenopausal women. New Engl J Med 1997;337:595-601.

Chronic Low-Grade Metabolic Acidosis in Normal Adult Humans: Pathophysiology and Consequences

Lynda Frassetto, R. Curtis Morris, Jr., Karen Todd, and Anthony Sebastian

Introduction

Normal adult humans eating modern-day diets have a chronic low-grade metabolic acidosis whose severity is determined in part by the net rate of endogenous acid production (NEAP). NEAP varies mainly with diet composition. The greater the quantity of organic and sulfuric acids produced from metabolism of animal foods, and the lower the amounts of potassium salts metabolizable to bicarbonate, which come mainly from fruits and vegetables, the greater the production rate of acid. It had previously been thought that "healthy" kidneys were capable of excreting any excess acid produced by the body's metabolism [1]; the authors' research suggests that the normally occurring slow decline in renal function with age allows the kidneys to merely mitigate the degree of severity of the acidosis, and with increasing age, the steady-state levels of acidity in the body slowly rise [2-3].

This rise in blood acidity levels may be pathogenic in the age-related decline in bone mass that leads to osteoporosis and increased bone fractures. Epidemiologic studies of hip fracture incidence worldwide show an increasing incidence of hip fractures in populations with higher intakes of animal protein, a surrogate marker for dietary acid [4]. In short-term General Clinical Research Center (GCRC) studies of postmenopausal women (weeks to months), neutralizing the acid in the diet with dietary supplements of potassium bicarbonate improves calcium and phosphate balance and improves biochemical markers of bone breakdown [5]. It increases the average 24-hour blood concentration of growth hormone, another factor that declines with age. And, it decreases the amount of nitrogen excreted in the urine, which may reflect a decrease in muscle breakdown. Decreased muscle mass and strength with increasing age are risk factors for falls, and therefore for increased fractures. A simple method for predicting diet acid load from diet protein and potassium content developed by the authors is described. Since future studies will be necessary to show whether long-term neutralization of dietary acid reduces fracture incidence; a simple method to evaluate diet acid load may make this type of study more widely usable to the nonacid-base physiologist.

R. Paoletti et al. (eds.), Women's Health and Menopause, 15–23.

Description of Chronic Low-Grade Metabolic Acidosis

In normal adult humans eating ordinary American diets, systemic acid-base equilibrium is maintained within narrow limits [6-7]. That occurs despite a continuing load of net acid to the systemic circulation generated as end products of metabolism of neutral precursors in the diet (e.g. sulfuric acid generated from the metabolism of sulfur-containing amino acids) [6,8]. Stability of systemic acid-base equilibrium is critically dependent on excretion of acid in urine [9], the rate of which the kidney adjusts in keeping with diet-induced variations in the net endogenous acid load. Since renal function progressively decreases with age [10], for a given net endogenous acid load blood acidity might increase with age, and plasma bicarbonate concentration might decrease. Such metabolic acidosis, though mild, might over time contribute to the pathogenesis of the physiologic disturbances and degenerative diseases characteristic of aging.

Endogenous acid production can be considered as comprising three components: (1) organic acids produced during metabolism that escape complete combustion to carbon dioxide and water; (2) sulfuric acid (H_2SO_4) produced from the catabolism of methionine and cystine, the sulfur-containing amino acids in dietary proteins; and (3) potassium bicarbonate ($KHCO_3$) produced from the metabolism of the potassium salts of organic anions in the vegetable foods of the diet, for example potassium citrate and potassium malate. The potassium bicarbonate so produced can neutralize an equal amount of sulfuric acid produced from protein metabolism, and thereby reduce net endogenous acid production.

The endogenous acid production rate then is computed as the sum of organic acid production and sulfuric acid production minus the intestinally absorbed potassium salts of organic anions that are metabolized to potassium bicarbonate. Potassium bicarbonate production is subtracted because potassium bicarbonate neutralizes acid, thereby reducing endogenous acid production.

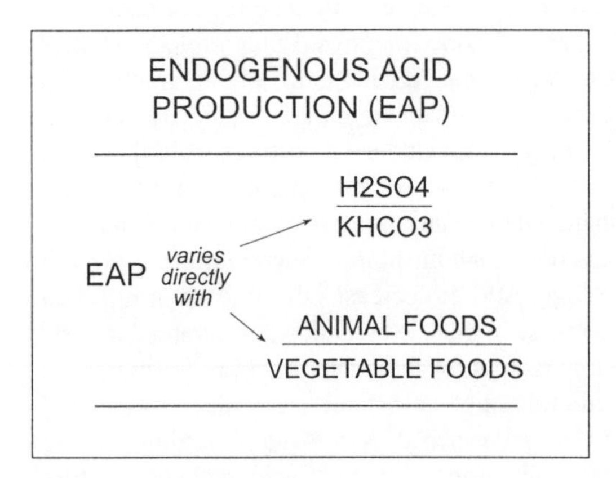

Figure 1. Acid production from foods.

Of these three components, organic acid production is the least variable. The largest differences in endogenous acid production among differing diets are due to differences in sulfuric acid production and potassium bicarbonate production. All foods contain sulphur-containing amino acids, although fruits in general contain less; animal products contain very little potential base–this comes mainly from fruits and vegetables. The greater the quantity of organic and sulfuric acids produced from metabolism of animal foods, and the lower the amounts of potassium salts metabolizable to bicarbonate, which come mainly from fruits and vegetables, the greater the production rate of acid (see Figure 1).

INCREASING THE ACID LOAD IN THE DIET AND THE BODY'S ABILITY TO MAINTAIN BLOOD ACIDITY AND PLASMA BICARBONATE LEVELS

Chronic metabolic balance studies were carried out in 64 normal subjects (39 men, 25 women), whose ages ranged from 17 to 74 years. Each of the data points in Figure 2 represents steady-state data on a normal individual admitted to the University of California San Francisco (UCSF) GCRC, and placed on one of a variety of typical American diets that yielded different net acid excretion (NAE), where NAE is used as a measure of the acid load in the diet (Figure 2 A on left). Those subjects whose diet acid loads are in the upper ranges have higher steady-state levels of blood acidity ($[H^+]$) and lower levels of steady-state plasma bicarbonate ($[HCO_3^-]$) than those whose diet acid loads are in the lower range. In these same normal healthy people, the older they were, the more likely they were to have higher levels of blood acidity and lower levels of plasma bicarbonate, independent of the diet they were eating (Figure 2 B on right). Age and renal function (glomerular filtration rate [GFR] measured by 24-hour creatinine clearance) were highly correlated in this analysis, and were not independent predictors of blood acidity or plasma bicarbonate. One explanation may be that renal function tends to decline with increasing age [10-11].

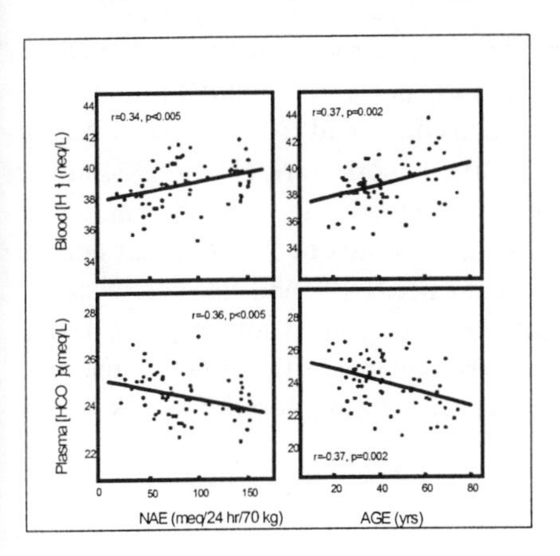

Figure 2 A and B. Changes in acid-base equilibrium with age and NAE.

Pathophysiologic Consequences Of Diet- And Age-Related Metabolic Acidosis

What are the pathophysiologic consequences of this progressively worsening metabolic acidosis? In this article, three possibly affected systems will be discussed; bone, muscle and growth hormone.

BONE

With increasing age, bone mineral density (BMD) decreases and the chances of fracturing bones increases. The decrease in BMD is considered to be multifactoral, and is partially but not completely explained by hormonal status, vitamin and mineral balance, and exercise. Other factors, such blood acidity levels, may also have roles in the age-related decline in bone mass that leads to osteoporosis and increased bone fractures. This was suggested in the 1968 Lancet article by Wachman and Bernstein, "The increased incidence of osteoporosis with aging reflects in part the life-long utilization of basic salts of bone to buffer the constant assault on pH homeostasis ... by the fixed acid load imposed by eating an 'acid-ash' diet" [12]. What evidence is there to support this idea?

Epidemiologic studies of hip fracture incidence worldwide show an increasing incidence of hip fractures in populations with higher intakes of animal protein, a surrogate marker for dietary acid. In 1992, Abelow and coworkers reported that the incidence of fractures of the hip in women correlates positively with the amount of animal protein consumed [4]. They correlated fracture incidence among 34 surveys in 16 countries worldwide with estimates of each countries' per capita animal protein consumption obtained from the United Nations Food and Agricultural Organization. Abelow et al. interpreted the correlation of hip fracture rate and animal protein intake as consistent with the hypothesis that hip fractures result in part from deleterious effects on bone of prolonged exposure to dietary acid.

If that interpretation is correct, however, the decisive risk factor for hip fracture would not be the rate of production of fixed acid (sulfuric acid) from animal protein, but would be the net rate of endogenous acid production, considering all sources of dietary acid and base. Using the method published by Abelow, the authors were able to extend the analysis to 33 countries, as well as analyze the effects of vegetable foods on the incidence of hip fractures (see Table 1) [13]. This confirmed the findings of Abelow that animal protein intake correlates positively with hip fracture incidence (HFI), and also demonstrates that hip fractures rates correlate negatively with vegetable protein intake (VP). Using vegetable protein intake as a surrogate marker for bicarbonate intake, this is consistent with the idea that the lower the net amount of acid in the system, the lower the risk of hip fractures.

Table 1. Regression Equations for Diet and Hip Fracture Incidence

Y variable	X variable	a	b	r	R^2	p
HFI	TP	-97.9	2.1	+0.67	0.45	< 0.001
HFI	AP	-26.6	2.4	+0.82	0.67	< 0.001
HFI	VP	167.3	-2.1	-0.37	0.14	< 0.04

TP: Total protein intake; AP: Animal protein intake

These however were epidemiologic studies, which can only show correlations. What direct evidence is there that neutralizing endogenous acid production actually improves bone health? In short-term (weeks to months) studies on the UCSF GCRC by Sebastian et al., postmenopausal women were fed a typical high protein, high acid-producing diet every day for 2 months, and then for part of time, the acid in the diet was neutralized with dietary supplements of potassium bicarbonate [5].

In these studies, calcium and phosphate balance increased (+56 ± 76 mg/d/60 kg and +47 ± 64 mg/d/60 kg respectively, $p < 0.01$), consistent with the idea that bone is being deposited, not broken down. The rates of bone resorption, measured as 24-hour urinary hydroxyproline, declined from 28.9 ± 12.3 to 26.7 ± 10.8 mg/day ($p < 0.05$), and the rate of bone formation, measured by mean serum osteocalcin levels, increased from 5.5 ± 2.8 to 6.1 ± 2.8 ng/ml ($p < 0.001$). Those findings are consistent with the reports of bone physiologists which indicate that an acidic milieu *in vitro* stimulates osteoclasts, the cells responsible for resorbing bone, and inhibits osteoblasts, the cells that form bone [14].

MUSCLE

Muscle mass also declines with increasing age, and decreased muscle function has been implicated as a risk factor for falls in the elderly, thereby increasing the risk for hip fractures. In humans and experimental animals with renal failure and more severe degrees of metabolic acidosis, there is increased skeletal muscle breakdown, protein turnover, and nitrogen wasting or negative nitrogen balance. Correction of the acidosis in these states with exogenous alkali therapy has been shown to reverse the nitrogen wasting [15]. What might such treatment do in so-called normal humans with age- and diet-dependent acidosis?

In these same postmenopausal women from the previous study the authors also analyzed urinary nitrogen excretion [16]. During the time that their diet was supplemented with KHCO3, the amount of nitrogen excreted in the urine decreased; both ammonium excretion and urea excretion (Figure 3). Stopping the alkali therapy caused a prompt increase in urinary nitrogen losses.

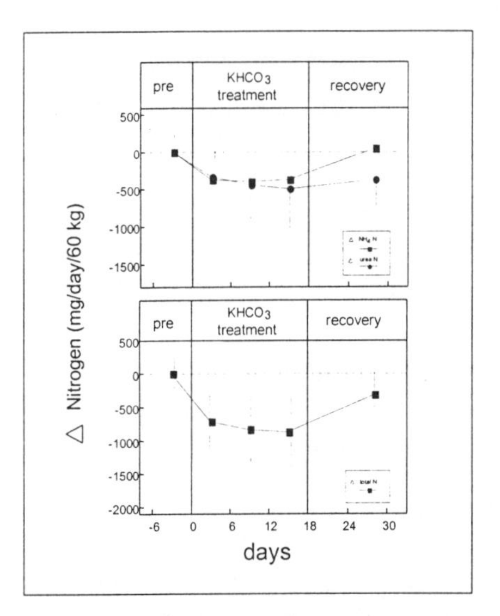

Figure 3. Decline in nitrogen excretion with $KHCO_3$ treatment.

GROWTH HORMONE SECRETION

The authors also evaluated the effect of alkali therapy on growth hormone (GH) secretion [17]. Growth hormone is an important factor in bone and muscle growth during childhood. With increasing age, GH levels slowly decline. The reason for this is unknown; the pathophysiologic consequences are unclear. Twelve postmenopausal women had mean 24-hour circadian GH levels measured at baseline. Then they were treated with $KHCO_3$ approximately 1 meq/kg body weight daily for one month, after which they had repeat 24-hour mean GH levels measured. GH levels increased from 826 $\pm$ 548 to 915 $\pm$ 631 pg/ml ($p < 0.05$), an 11 % increase over baseline. Whether this small but statistically significant increase in 24-hour mean growth hormone concentration is biologically significant remains to be determined. Because serum GH levels decline by as much as 50% or more from young adulthood to old age, this small $KHCO_3$-induced increase might seem relatively minor. On the other hand, treatment with $KHCO_3$ for only four weeks may not be enough time to reverse suppressive effect on GH of a low-grade metabolic acidosis that has been acting for decades.

Predicting Diet Net Acid Load From Dietary Protein And Potassium Contents

So what does the average person do to determine the amount of acid generated from metabolism of the food in the diet? One way is to eat just one item of food and do 24-hour urine collections to evaluate NAE. This was exactly the approach used by an Englishman named Blatherwick about 80 years ago [18]. Volunteers ate large amounts of one particular food item for approximately one week, while doing sequential 24-hour urine collections, which were then analyzed for ammonia, titratable acids, and total carbon dioxide–the constituents of NAE, analyses of which are usually only available in a research lab. This approach has a number of drawbacks; not only is it tedious and time-consuming, but as Blatherwick wrote in his article in the Archives of Internal Medicine in 1914, discussing the effects of a boiled cauliflower diet, "It became very distasteful after the third day, so that the experiment was discontinued."

The authors have developed a more practical approach, using an algorithm to predict diet acid load from only two diet constituents; diet protein and potassium content [19]. Using data from healthy subjects at steady-state, eating one of 20 whole food diets as part of metabolic balance studies that measured NAE, and analyzing these diets for their constituent parts, both dietary protein intake and dietary potassium intake were demonstrated to be independent predictors of NAE, when evaluated by multiple regression analysis.

Because protein and potassium were not correlated to each other, the ratio of dietary protein to potassium was evaluated. This ratio correlated significantly with the difference between the sulfur (i.e. potential acid) and potential base contents of the diets (Figure 4 A and B). This simple algorithm, using readily available components of diet composition analyses, accounts for 70-75% of the variation in NAE of the diets studied.

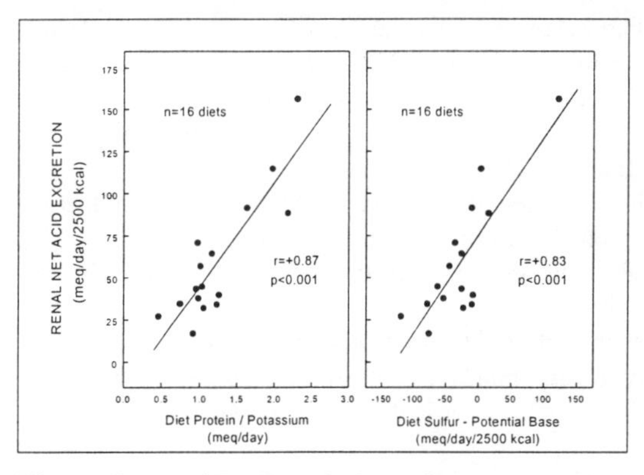

Figure 4 A and B. Correlation of NAE with dietary protein/potassium and acid-base.

Stone Age Diets for the 21st Century?

It is been inferred that the diet to which we are genetically adapted–similar perhaps to the diet of our stone age ancestors–contained far greater amounts of vegetable foods, capable of producing potassium bicarbonate in considerably greater amounts than we administered to subjects in our studies. By comparison, the average diet of people living in highly industrialized countries is low in potassium and bicarbonate-generating organic anions. In effect, they are suffering from a state of chronic potassium bicarbonate deficiency which is manifested by a chronic low grade metabolic acidosis. The data presented here suggest that the acid-producing nature of this modern diet is a factor in diseases which are now considered to degenerative diseases associated with aging. It further suggests that neutralization of the acids in the diet can help reverse the breakdown in these systems.

Future studies will be necessary to show whether long-term neutralization of dietary acid reduces fracture incidence, improves muscle mass, increases GH secretion which could then further improve bone and muscle mass, and affects other systems that as yet have not been analyzed. The simple method of evaluating diet acid load proposed in this paper may make this type of study more widely usable to the nonacid-base physiologist.

References

1. Shock NW, Yiengst MJ. Age changes in the acid-base equilibrium of the blood of males. J Gerontology 1950;5:1-4.
2. Frassetto L, Sebastian A. Age and systemic acid-base equilibrium: Analysis of published data. J Gerontol 1996;51A:B91-B99
3. Frassetto L, Morris RC, Jr., Sebastian A. Effect of age on blood acid-base composition in adult humans: Role of age-related renal functional decline. Am J Physiol 1996;271:1114-22.
4. Abelow B. Dietary protein and acid production. Calcif Tissue Int 1990;46:348-49.
5. Sebastian A, Harris ST, Ottaway JH, Todd KM, Morris RC, Jr. Improved mineral balance and skeletal metabolism in postmenopausal women treated with potassium bicarbonate. N Engl J Med 1994;330:1776-81.
6. Kurtz I, Maher T, Hulter HN, Schambelan M, Sebastian A. Effect of diet on plasma acid-base composition in normal humans. Kidney Int 1983;24:670-80.
7. Madias NE, Adrogue HJ, Horowitz GL, Cohen JJ, Schwartz WB. A redefinition of normal acid-base equilibrium in man: Carbon dioxide as a key determinant of normal plasma bicarbonate concentration. Kidney Int 1979;16:612-18.
8. Lennon EJ, Lemann J Jr, Litzow JR. The effect of diet and stool composition on the net external acid balance of normal subjects. J Clin Invest 1966;45:1601-7.
9. Schwartz WR, Hall PW, Hays RM, Relman AS. On the mechanism of acidosis in chronic renal disease. J Clin Invest 1959;38:39-52.
10. Davies DF, Shock NW. Age changes in glomerular filtration rate, effective renal plasma flow, and tubular excretory capacity in adult males. J Clin Invest 1950;29:496-507.
11. Lindeman RD. Anatomic and physiologic age changes in the kidney. Exp Gerontol 1986;21: 379-406.
12. Wachman A, Bernstein DS. Diet and osteoporosis. Lancet 1968;1:958-59.
13. Frassetto L, Todd KM, Morris RC, Jr., Sebastian A. Role of diet net acid load on hip fracture incidence worldwide. J Am Soc Nephrol 1997;551A.
14. Krieger NS, Sessler NE, Bushinsky DA. Acidosis inhibits osteoblastic and stimulates osteoclastic activity in vitro. Am J Physiol 1992;262:F442-F448
15. Papadoyannakis NJ, Stefanidis CJ, McGeown M. The effect of the correction of metabolic acidosis on nitrogen and potassium balance of patients with chronic renal failure. Am J Clin Nutr 1984;40:423-627.
16. Frassetto L, Morris RC, Jr., Sebastian A. Potassium bicarbonate reduces urinary nitrogen excretion in postmenopausal women. J Clin Endocrinol Metab 1997;82:254-59.
17. Frassetto L, Morris RC, Jr., Sebastian A. Potassium bicarbonate increases serum growth hormone concentrations in postmenopausal women. J Am Soc Nephrol 1996;1349A.

18. Blatherwick NR. The specific role of foods in relation to the composition of the urine. Arch Int Med 1914;14:409-50.
19. Frassetto L, Todd KM, Morris RC, Jr., Sebastian A. Estimation of net endogenous noncarbonic acid production in humans from diet potassium and protein content. Am J Clin Nutr 1998; in press.

THE MENOPAUSE, SEX HORMONES, AND RHEUMATIC DISEASE

Gerard Hall

Introduction

The potential association between menopause, sex hormones, and rheumatic disorders has long been alluded to. In the early nineteenth century Haygarth noted the "nodosities" of joints that occurred at the menopause and Weil, in 1928, pointed to the inevitability of arthritis following female castration. Many of the comprehensive studies of symptoms seen in perimenopausal women cite joint pains as one of the most frequently encountered complaints [1].

But what are these rheumatic complaints? Are they any different than well-recognized rheumatic diseases such as osteoarthritis or is menopausal arthralgia a specific disease entity in itself? How might endogenous sex hormones effect the rheumatic disorders and could exogenous hormones be helpful in their treatment?

Menopausal Joint Pains

It is most likely that joint and muscle symptoms seen in the climacteric woman represent either the onset of a common and well-recognized rheumatic disorder or an exacerbation of it. A study from a menopause clinic in London showed that 28 of 42 presenting women were experiencing new rheumatic symptoms. Detailed questioning, examination, and investigations showed these to be typical rheumatic disorders such as fibromyalgia, carpal tunnel syndrome, osteoarthritis, and tendinitis (see Table 1) [2]. Hormone replacement therapy (HRT) resulted in considerable improvement in those patients with recent onset carpal tunnel syndrome and fibromyalgia but not of other disorders. Carpal tunnel syndrome is probably more common at the time of the menopause due to the alterations in forearm fat that occurs at the menopause [3]. Poor sleep patterns are an important part of the fibromyalgia syndrome and sleep disturbance is commonly experienced during the menopause; its rectification will probably lead to an improvement in myalgic symptoms.

R. Paoletti et al. (eds.), Women's Health and Menopause, 25–31.

Table 1. Rheumatic diagnoses seen in 28 out of 42 patients attending a menopause clinic who complained of recent onset musculoskeltal pain. Three patients had more than one diagnosis.

Rheumatic Disorder	Frequency
Osteoarthritis	28%
Carpal tunnel syndrome	25%
Fibromyalgia	21%
Nonspecific arthragia	18%
Spondylosis	11%
Tendinitis	7%

Menopause and Rheumatoid Arthritis

Other rheumatic disorders have a higher incidence at the menopause, most noticeably rheumatoid arthritis (RA). Figure 1 shows a frequency histogram of the age of onset of first symptoms of RA in relation to menopausal transition, a classic Gaussian distribution with the mean coinciding almost exactly with the menopausal transition [4]. Other features of RA clearly suggest that female hormones may play an important role in this condition. RA is three times more common in women, it is a more aggressive condition after the menopause, it remits in 75% of pregnancies only to flare postpartum, and there may be a protective effect of the oral contraceptive pill against RA [5]. So, how might sex hormones exert an effect on an inflammatory disorder such as RA?

Mechanisms of Action of Sex Hormones

Androgens and estrogens have potential effects on both inflammatory and immune pathways with androgens generally being more immunosuppressive and estrogens being immuno-stimulatory. It is clear that the female immune response is more powerful than the male. They show higher levels of the immunoglobulins IgG and IgM and have a more sustained antibody production following immunization and infection [6,7]. They reject allografts more rapidly and show a greater resistance to a number of different infective organisms [8,9]. Estrogens have been found to inhibit T suppressor cell function and facilitate T-helper cell function [10,11]. In the male there is a reduced production of antibodies compared with the female and this coincides with the surge of androgen production at puberty suggesting that androgens may suppress immune mechanisms [12]. This is supported by the observation that orchidectomy increases protection against infection and also results in more rapid rejection of allograft [13].

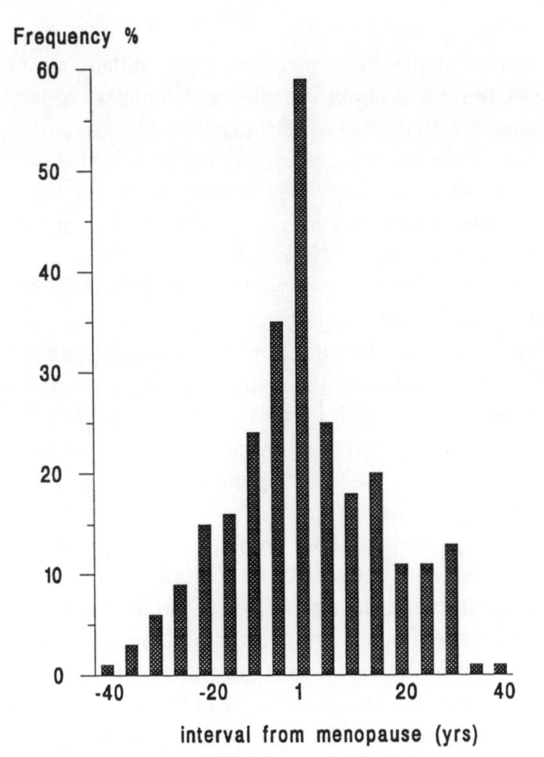

Figure 1. Frequency histogram showing the age of onset of first symptoms of rheumatoid arthritis in relation to menopausal transition.

Cytokines are important mediators of inflammation in most models of inflammation. Interleukin-1 (IL-1) secretion has been demonstrated to rise at the time of the menopause and may be important in a number of different menopausal conditions, including osteoporosis. It is also likely to play an important role in inflammatory rheumatic disorders and the perimenopausal rise is likely to be clinically significant. It is interesting that estrogen replacement leads to suppression of this IL-1 rise [14].

Estrogen levels are generally normal in RA but most studies have consistently shown reductions in some androgens, notably DHEA and DHEAS [15]. It is not known whether this is a primary disorder of androgen secretion that results in a predisposition to RA or whether it is a secondary effect of the disease process itself. Interestingly, a recent study by Masi et al. showed that normal premenopausal women who later developed RA had lower DHEA levels 12 years before the onset of their disease compared with the women who did not develop RA [16].

A more direct action of androgens and estrogens has been suggested following the discovery of the presence of estrogen and androgen receptors within synovial tissues [17,18].

Possible Treatments

Given these observations, it has been postulated that inflammatory arthritis could be manipulated using sex hormones, both in males and females. Animal experiments have shown promise, particularly with the use of estrogens. A series of studies from Scandinavia showed that oophorectomy in the collagen type II-induced mouse model of arthritis leads to an exacerbation of arthritis [19]. Pretreatment with estrogen will prevent this exacerbation. Using the same model, they also showed that the postpartum flare of arthritis could also be prevented with estrogen [20].

In humans there have been three randomized controlled trials of HRT in the postmenopausal women with RA. The largest study of 200 patients randomized 100 patients each to transdermal estrogen or low dose calcium [21]. Overall there was no effect of HRT on a number of different measures of rheumatoid disease activity. However, on further examination of serum estradiol levels it was noted that a number of patients allocated to HRT failed to achieve appropriate estradiol levels. When patients were divided into those with E_2 levels > 100 pmol/l (termed "compliers") and those with E_2 levels below 100 pmol/l (termed "noncompliers") there were significant improvements in some disease activity measures in the complier group compared with the noncomplier group. The two other studies did not show a significant effect of HRT on RA but these studies were much smaller and did not measure treatment E_2 levels [22,23]. Overall, it would seem that HRT at standard doses does not have a major effect on modulating disease activity in RA. It is possible that the levels of estradiol seen with conventional HRT are not sufficient to allow any detectable clinical changes in disease activity and that higher doses may be necessary. This is supported by the dynamic study of Lobo et al. showing that the hypophyseal-pituitary-adrenal axis is much more likely to be effected by 2.5 mg of conjugate equine estrogens compared with the more commonly used 0.0625 mg per day [24].

Can estrogen treatment protect against the onset of rheumatic disorders? The European epidemiological studies have pointed towards a protective effect of the oral contraceptive pill (OCP) but this has not been confirmed by U.S. studies [5]. The first study looking at the effect of HRT use on the later development of RA showed that estrogen replacement therapy may indeed protect against the subsequent development of RA [25]. Further studies, probably of better design, have not confirmed these initial findings [26].

Osteoarthritis

A radiographic study of osteoarthritis (OA) by Spector et al. looked at X-ray changes in the hands and knees of 1,000 women [27]. X-rays were scored for the presence of OA and HRT exposure documented. There was a tendency towards a protective effect of HRT in current users with odds ratios of 0.31 and 0.48 for early knee and distal interphalangeal joint OA respectively. A study of symptomatic OA did not show any protective effect [28].

Systemic Lupus Erythematosus

Systemic lupus erythematosus (SLE) is another condition in which sex hormones are likely to have important roles. It is nine times more common in the female, tends to remit after menopause, and is exacerbated by pregnancy. The male most likely to develop SLE is the Klinefelter male who has a relatively lower androgen:estrogen ratio. There have been attempts to manipulate the androgen:estrogen ratio in women with SLE and an open study of 10 patients using DHEA showed possible improvements in disease activity [29]. Given the potential negative effects of pregnancy and the oral contraceptive pill (OCP) on SLE, concern has been expressed over the possible effect of HRT in women with SLE. A controlled study of 60 women with SLE showed that HRT does not have any deleterious effect on measured disease parameters autoantibody profiles or history of thromboembolism [30]. HRT is, therefore, generally safe in SLE but it would seem prudent to avoid both HRT and the OCP in patients who have the anticardiolipin antibody.

Summary

Although rheumatic symptoms are common at the menopause a distinct menopausal arthritis does not exist, most symptoms representing commonly encountered rheumatic disorders. Several disorders coincide with the menopause suggesting important roles for sex hormones in their etiopathogenesis and possible roles for hormonal manipulation in their treatment. Laboratory work has shown numerous different direct and indirect mechanisms by which sex hormones, androgens especially, may influence immune and inflammatory pathways. Despite these findings, our attempts to date at hormonal manipulation of rheumatic disorders have failed to produce any clear positive findings although studies have been few.

References

1. Neugarten BL, Kraines RJ. Menopausal symptoms in women of various ages. Psychom Med 1965;27:266-73.
2. Hall GM, Spector TD, Studd JWW. Carpal tunnel syndrome and hormone replacement therapy. Br Med J 1992;304:382.
3. Hassager C, Christiansen C. Estrogen/gestagen therapy changes soft tissue body composition in postmenopausal women. Metabolism 1989;38:662-65.
4. Goemare S, Ackerman C, Goethals K, et al. Onset of symptoms of rheumatoid arthritis in relation to age, sex and menopausal transition. J Rheumatol 1990;17:1620.
5. Spector TD, Hochberg MC. The protective effect of the oral contraceptive pill on rheumatoid arthritis: An overview of the analytic epidemiological studies using meta-analysis. J Clin Epidemiology 1990;43:1221.
6. Rhodes K, Markham RL, Maxwell PM, Monk-Jones NE. Immunoglobulins and the X chromosome. Br Med J 1969;3:439.
7. Buckley CE, Dorsey FC. Serum immunoglobulin levels throughout the lifespan of healthy man. Ann Intern Med 1971;75:673.
8. Graff RJ, Lappe MA, Snell GD. The influence of the gonads and adrenal glands on the

immune response to skin grafts. Transplantation 1969;7:105.
9. Goble FC, Konopka EA. Sex as a factor in infectious disease. Trans NY Acad Sci 1971;35: 325.
10. Ansar Ahmed S, Dauphinee MJ, Talal N. Effects of short-term administration on normal and autoimmune mice. J Immunol 1985;134:204.
11. Paavonen T, Anderson LC, Adlercreutz H. Sex hormone regulation of in vitro immune response. Estradiol enhances B-cell maturation via inhibition of suppressor T-cells in pokeweed-mitogen stimulated cultures. J Exp Med 1985;154:1935.
12. Blazkovec AA, Orsini MW. Ontogenetic aspects of sexual dimorphism and the primary immune response to sheep erythrocytes in hamster from pre-puberty through senescence. International Archives of Allergy and Applied Immunology 1976;50:55.
13. Grossman CJ. Regulation of the immune system by sex steroids. Endocr Rev 1984;5:435.
14. Pacifici R, Rifas L, McCracken R. Ovarian steroid treatment blocks a postmenopausal increase in blood monocyte interleukin-1 release. Proc Natl Acad Sci 1989;86:2938.
15. Hall G, Perry L, Spector T. Depressed levels of dehydroepiandrosterone sulphate in postmenopausal rheumatoid arthritis but no relationship with axial bone density. Ann Rheum Dis 1993;52:211.
16. Masi AT, Catterton RT, Comstock GW, Malamet RL, Hochberg MC. Decreased serum dehydroepiandrosterone sulfate levels before onset of RA in younger premenopausal women: A controlled prospective study. Arthritis Rheum 1994;37(Suppl.):S315.
17. Cutolo M, Accardo S, Villagio B, et al. Evidence for the presence of androgen receptors in the synovial tissue of rheumatoid arthritis patients and healthy controls. Arthritis Rheum 1992;35: 345.
18. Cutolo M, Accerdo S, Villagio B, et al. Presence of estrogen binding sites on macrophage-like synoviocytes and CD8+, CD29+, CD45RO+ T-lymphocytes in normal and rheumatoid synovium. Arthritis Rheum 1993;36:1087.
19. Holmdahl R, Jansson L, Anderson M. Female sex hormones suppress development of collagen induced arthritis in mice. Arthritis Rheum 1986;29:1501.
20. Mattson R, Mattson A, Holmdahl R, Whyte A, Rook GA. Maintained pregnancy levels of oestrogen afford complete protection from post-partum exacerbation of collagen induced arthritis. Clin Exp Immunol 1991;85:41.
21. Hall GM, Daniels M, Huskisson EC, Spector TD. A randomized controlled trial of hormone replacement therapy in postmenopausal rheumatoid arthritis. Ann Rheum Dis 1994;.
22. MacDonald AG, Murphy EA, Capell HA, Bankowska UZ, Ralston SH. Effects of hormone replacement therapy in rheumatoid arthritis: A double blind placebo-controlled study. Ann Rheum Dis 1994;53:54.
23. van den Brink HR, van Everdingen A, van Wijk MJG, Jacobs JWG, Bijlsma JWJ. Adjuvant estrogen therapy does not improve disease activity in postmenopausal patients with rheumatoid arthritis. Ann Rheum Dis 1993;52:862.
24. Lobo RA, Goebelsmann U, Brenner P, Mischell D. The effects of estrogen on adrenal androgens in oophorectomized women. Am J Obstet Gynecol 1982;142:471.
25. Vandenbroucke JP, Witteman JCM, Valkenburg HA, et al. Noncontraceptive hormones and rheumatoid arthritis in perimenopausal and postmenopausal women. JAMA 1986;255:1299-1303.
26. Spector T, Brennan P, Harris P, Studd J, Silman A. Does oestrogen replacement therapy protect against rheumatoid arthritis? J Rheumatol 1991;18(10):1473.
27. Spector T, Nandra D, Hart D, Doyle D. Is hormone replacement therapy protective for hand

and knee osteoarthritis in women? Ann Rheum Dis 1997;56:432.

28. Oliveria S, Felson D, Klein R, Reed J, Walker A. Estrogen replacement therapy and the development of osteoarthritis. Epidemiology 1996;7:415-19.

29. van Vollenhoven R, Engleman E, McGuire J. An open study of dehydroepiandrosterone in systemic lupus erythematosus. Arthritis Rheum 1994;37:1305.

30. Arden NK, Lloyd ME, Spector TD, Hughes GR. Safety of hormone replacement therapy in systemic lupus erythematosus. Lupus 1994; 3:11-13.

NOVEL MECHANISMS OF ESTROGEN ACTION

Jan-Åke Gustafsson

Estrogen action has long been thought to be mediated via one unique estrogen receptor. This receptor was originally discovered by Elwood Jensen who managed to synthesize estrogens with sufficiently high specific radioactivity to be used as tracers for the receptor [1]. Utilizing these radioligands Jensen could first show that estrogen target tissues like the uterus and vagina displayed a specific retention of radioactivity following administration of radiolabelled estrogens to whole animals. Later on it was demonstrated that this retention was due to the presence in these target tissues of a specific protein binding the radioligands with high affinity and low capacity. After the introduction of the technique of density gradient centrifugation to monitor the behavior of the estrogen receptor, these original contributions opened a new science on the characteristics of the estrogen receptor as well as somewhat later other members of the steroid receptor family.

Solely on the basis of biochemical and immunological techniques well in advance of receptor cloning our own laboratory could show that the glucocorticoid receptor consisted of three domains: a ligand binding domain, a DNA binding domain, as well as an "immunodominant" or "immunomodulatory" domain to which most of our antibodies bound [2-4]. This three-domain structure was later confirmed following the cloning of the glucocorticoid as well as several other steroid receptors during the eighties. Another important event was the demonstration that purified receptor preparations could bind specifically to DNA from genes under the control of the steroid receptor ligand in question [5]. The first experiment in this regard demonstrated specific binding of purified glucocorticoid receptor to MMTV DNA, a well-known glucocorticoid-regulated gene. The concept of hormone responsive elements was thus founded when this experiment was complemented with the demonstration that the receptor interacting DNA fragment was capable of conferring hormonal responsiveness onto normally hormonally unresponsive genes when placed in front of such genes [6].

Over the recent decade we have seen yet another fascinating aspect of nuclear receptors coming into focus. This concerns the expansive nature of the family; using molecular biological techniques a large number of novel members of the nuclear receptor supergene family have been discovered. Since most of these new receptors lack known ligands they have been termed orphan receptors and much effort has been devoted to finding ligands for these novel family members. It was during investigations aimed at identifying novel nuclear receptors in the prostate that our group succeeded in cloning a

R. Paoletti et al. (eds.), Women's Health and Menopause, 33–38.

second estrogen receptor which we termed estrogen receptor β (ERβ) [7]. The classical estrogen receptor was accordingly named estrogen receptor α (ERα). This new estrogen receptor has turned out to be of great interest and, somewhat surprisingly, it appears to be the quantitatively dominating estrogen receptor. The reason why it has remained undiscovered throughout so many years is probably that the classical receptor seems to predominate in the estrogen target tissues most often studied, namely the uterus and the mammary gland. Other tissues where estrogen receptor β expression predominates over estrogen receptor α expression such as bone, cardiovascular system, urogenital tract, central nervous system, immune system, as well as kidney and lung, have not until recent years come into focus as important estrogen target tissues. When Jensen and others purified the estrogen receptor for generation of antibodies they used calf uterus as source and these antibodies were used to clone the estrogen receptor. Several attempts were made to use molecular biological techniques to find other estrogen receptors but these efforts instead resulted in cloning of two other interesting receptors namely estrogen receptor related receptor 1 and estrogen receptor related receptor 2 (ERR1 and ERR2, respectively) [8]. Although these two orphan receptors are clearly related to the estrogen receptors they have no known ligands and they appear to be much less widespread than the estrogen receptors.

Estrogen receptor α and estrogen receptor β share an almost identical DNA binding domain; only three amino acids differ, probably indicating that they interact with similar response elements on DNA. On the other hand it is quite clear, based on studies on several nuclear receptors that also regions flanking the DNA binding domain may be of importance in determining nuclear receptor DNA binding specificity and it may be anticipated that DNA sequences may well occur in promoters of estrogen responsive genes which may differentiate between ERα and ERβ, respectively. The ligand binding domains of ERα and ERβ show approximately 58% homology, a figure which is relatively low considering that this degree of similarity also characterizes the glucocorticoid receptor versus the mineralocorticoid receptor and the glucocorticoid receptor versus the progesterone receptor. Thus, in a sense, ERα and ERβ represent two quite different steroid receptors which should be expected to have different biology. On the other hand one should also expect several similarities in ligand binding in much the same way as the glucocorticoid receptor and the mineralocorticoid receptor share several ligands, and as is also the case for the glucocorticoid receptor and the progesterone receptor. The relatively limited homology between ERα and ERβ LBDs should make it possible to develop quite specific ligands for ERα and ERβ respectively.

The N-terminal domain of the ERs differs in length between the two isoforms. The N-terminal domain of ERα is about 10K shorter than that of ERβ and although a longer form of ERβ seems to exist ERα is still longer than the long form of ERβ. Furthermore the sequence homology between the two ERs in the N-terminal domain is very low (approximately 20%).

As indicated above ERα and ERβ display quite dissimilar distribution in the body. Whereas ERα seems to predominate in the uterus and in the mammary gland ERβ appears to be the dominating isoform in lung, kidney, urogenital tract, cardiovascular system, central nervous system, bone, and immune system. In certain tissues and cells ERα and ERβ are obviously colocalized. This has been studied in most detail in the central nervous system where discrete neurons have been shown to contain both ERα and ERβ, even though the most common situation appears to be that ERα and ERβ are expressed separately in different cells. Coexpression of ERα and ERβ in one and the same cell may be of physiological significance since the two receptors are known to heterodimerize and it cannot be excluded that the heterodimer has other functional properties than the respective two homodimers.

The expression of ERβ in so many different tissues not involved in reproductive phenomena indicates that estrogens are indeed of much wider physiological significance than only or mainly in the context of regulation of reproductive events. Even though some ERβ has been found to be expressed in both uterus and mammary gland the main impression is that these tissues predominantly contain ERα. It may perhaps be suggested that, overall, ERα is mainly involved in regulation of reproduction whereas ERβ mainly controls estrogen-regulated nonreproductive phenomena. Obviously, epidemiological and other data have indicated that tissues like bone, cardiovascular system, urogenital tract, central nervous system, and immune system are indeed under the influence of estrogens. Furthermore, hormone replacement therapy given to women after the menopause is to a large extent aimed at preventing estrogen insufficiency expressing itself in the form of osteoporosis, cardiovascular disease, urinary incontinence, and dry membranes, as well as mood swings and neurogenerative diseases.

One of the biggest surprises is the widespread occurrence of ERβ in the male reproductive system. It is not immediately apparent which is the ligand for ERβ in these male tissues. Even though estrogens definitely do occur in the male, they are normally present at levels equal to or lower than those of females after the menopause. Since the decline in estrogens in the female after the menopause leads to estrogen insufficiency one may ask what makes the male resistant to this type of problem. One possibility is that ERβ can be activated by other ligands than classical estrogens. Such candidates are 5-androstene-3β,17β-diol and 5α-androstane-3β,17β-diol which have both been shown to be weak estrogens. Possibly they interact somewhat better with ERβ than with ERα. 5-Androstene-3β,17β-diol is a metabolite of dehydroepiandrosterone (DHEA), a steroid which is secreted from the adrenal cortex in the form of DHEA sulphate and which is one of the quantitatively predominant steroid hormones in the human. 5α-Androstane-3β,17β-diol is a commonly occurring metabolite of testosterone. Further studies are necessary to show whether these steroid metabolites might function as ERβ agonists in the male.

Even though estradiol-17β appears to bind approximately equally well to ERα and ERβ, differences exist with which other ligands interact with the two receptors. For instance, phytoestrogens like genistein and daidzein appear to interact better with ERβ than ERα [9] and it cannot be excluded that the dietary phytoestrogens constitute important physiological ligands for ERβ. This is obviously of particular interest in view the various beneficial effects ascribed to the intake of phytoestrogens. These compounds which are particularly abundant in Japanese and Chinese diets and are found in tofu and soya-derived foodstuffs allegedly decrease the risk for development of cancer in tissues like breast and colon. The mechanism is unclear but has been proposed to be related to the character of phytoestrogens of weak estrogens which might block estrogen receptors from more efficient endogenous estrogens otherwise contributing to increased cancer incidence in reproductive tissues. However other mechanisms are also plausible to explain the supposed anticarcinogenicity of phytoestrogens.

Phytoestrogens and also antiestrogens like tamoxifen are sometimes described as antioxidants. Such compounds are today often recommended as dietary supplements in order to diminish the risk for development of diseases related to overproduction of free radicals and reactive oxygen, notably cancer. In this context it is of interest that an enzyme, quinone reductase, implicated in the protection against free radicals and reactive oxygen in the cell has recently been shown to be induced by tamoxifen acting through ERβ via an antioxidant response element in the 5'-flank of the quinone reductase gene [10]. Antioxidant response elements are known to have a certain resemblance to AP1 elements which bind the transcription factor complex fos-jun. We have previously shown that tamoxifen acts as an agonist when signaling through ERβ via an AP1 response element whereas classical agonists like estradiol-17β act as antagonists under the same conditions [11]. Thus in both these cases tamoxifen behaves as an agonist and, in the case of quinone reductase, also as an antioxidant turning on an enzyme of importance in protection against oxidative stress. It may be speculated that quinone reductase is merely one example of a whole cascade of genes which might be turned on by tamoxifen also encompassing superoxide dismutase, glutathione peroxidase, glutathione transferase, etc. Such a mechanism might explain the presence of ERβ in rapidly replicating cells such as granulosa cells in developing ovarian follicles, colonic mucosal cells as well as proliferating smooth muscle cells in rat carotids following experimental denudation. Presumably, proliferating cells need to be protected against free radicals to an extra large extent in order to avoid mutational events. Indeed such a protective role of ERβ is also in consonance with the fact that ERβ is downregulated in male urogenital tissues following neonatal treatment with estrogen, an imprinting phenomenon known to lead to increased susceptibility to estrogen-induced carcinogenesis in the urogenital tract of the adult animals [12]. Obviously such a putative protective effect of ERβ against free radicals needs to be supported by additional further experimentation and, furthermore, it presumably only represents part of the function of ERβ.

Recently our possibilities to understand ERβ function have significantly increased by the development of mice with the ERβ gene deleted [13]. It is apparent that these mice show several interesting phenotypic characteristics which probably will significantly deepen our understanding of estrogen action. The most conspicuous phenotype in young animals is an infertility problem in female mice owing to follicular arrest with anovulatory cycles. This phenotype is not complete and certain fraction of -/- mice may ovulate. However their fertility is much lower than that of wild type animals and it is quite apparent that expression of ERβ is necessary for normal ovarian function. It will be interesting to see whether related diseases in the human, such as polycystic ovarian syndrome, may be associated with a defect ERβ. We are currently further investigating the following tissues in the ERβ knockout mice which appear to be malfunctioning or show a pathological histological appearance: bone, prostate, abdominal fat, and urinary bladder. We are also attempting to generate double knock-out mice with both ERα and Erβ deleted. It will be of particular interest to study such animals in view of a quite widespread opinion that life might not be possible without estrogen action.

Acknowledgments

This work was supported by grants from the Swedish Cancer Society and from KaroBio AB.

References

1. Jensen EV, Jacobson HI. Basic guides to the mechanism of estrogen action. Recent Progr Horm Res 1962;18:387-414.
2. Wrange Ö, Gustafsson J-Å. Separation of the hormone- and DNA-binding sites of the hepatic glucocorticoid receptor by means of proteolysis. J Biol Chem 1978;253:856-65.
3. Wrange Ö, Carlstedt-Duke J, Gustafsson J-Å. Purification of the glucocorticoid receptor from rat liver cytosol. J Biol Chem 1979;254:9284-90.
4. Carlstedt-Duke J, Okret S, Wrange Ö, Gustafsson J-Å. Immunochemical analysis of the glucocorticoid receptor: Identification of a third domain separate from the steroid-binding and DNA-binding domains. Proc Natl Acad Sci USA 1982;79:4260-64.
5. Payvar F, Wrange Ö, Carlstedt-Duke J, Okret S, Gustafsson J-Å, Yamamoto KR. Purified glucocorticoid receptors bind selectively *in vitro* to a cloned DNA fragment whose transcription is regulated by glucocorticoids *in vivo*. Proc Natl Acad Sci 1981;78:6628-32.
6. Yamamoto KR, Payvar F, Firestone GL, et al. Biological activity of cloned mammary tumor virus DNA fragments that bind purified glucocorticoid receptor protein *in vitro*. In: Cold Spring Harbor Symposia on Quantitative Biology, Vol. XLVII, Cold Spring Harbor Laboratory, 1983:977-84.
7. Kuiper GGJM, Enmark E, Pelto-Huikko M, Nilsson S, Gustafsson J-Å. Cloning of a novel estrogen receptor expressed in rat prostate and ovary. Proc Natl Acad Sci 1996;93:5925-30.
8. Giguere V, Yang N, Segui P, Evans RM. Identification of a new class of steroid hormone receptors. Nature 1988;331:91-94.
9. Kuiper GGJM, Lemmen JG, Carlsson B, et al. Interaction of estrogenic chemicals and phytoestrogens with estrogen receptor β. Endocrinol 1998;139:4252-63.
10. Montano MM, Jaiswal AK, Katzenellenbogen BS. Transcriptional regulation of the human quinone reductase gene by antiestrogen-liganded estrogen receptor-α and estrogen receptor-β. J Biol Chem 1998;273:25443-449.
11. Paech K, Webb P, Kuiper GGJM, et al. Differential transactivation properties of the estrogen receptor isotypes (α, β): Estrogen-like effects with antiestrogens and antiestrogen effects with estrogen. Science 1997;277:1508-10.
12. Prins GS, Marmer M, Woodham C, et al. Estrogen receptor beta mRNA ontogeny in the rat prostate and effects of neonatal estrogen exposure on the expression pattern.

Endocrinol 1998;139:874-83.

13. Krege JH, Hodgin JB, Couse JF, et al. Generation and reproductive phenotypes of mice lacking estrogen receptor β. Proc Natl Acad Sci 1998;95:15677-82.

The Action of Ovarian Steroid Hormones on Tissues and Organs

Marianne J. Legato

Introduction

The gap between the physician's conviction that gonadal steroids mitigate the consequences of aging and the postmenopausal patient's willingness to use hormonal replacement therapy (HRT) is widening. An explosion of information over the past decade indicates that estrogen is a growth and maintenance hormone which acts on virtually every organ and tissue in the body; among other things, it maintains the central nervous system's plasticity, safeguards the stability of the coronary vasculature, and ensures the optimal mass and architecture of bone. Nevertheless, this information has had little impact on patients: only 15-20% of postmenopausal women use HRT [1]. Over 1/3 of patients who begin therapy discontinue it within a matter of months. Berman's group surveyed over 2,000 women for two years and found that 20% had stopped their medication at 6 months, 38% by a year, 51% at 18 months and 59% by two years. Roughly 2.5-3.0 of women quit each month, in a steady rate of decline [2]. Reasons for this include a reluctance to continue menstrual periods, fear of cancer, and a reaction to the side effects of HRT, usually related to progestin rather than to estrogen.

Molecular Basis of Action

The molecular mechanism of ovarian steroids' action is still unfolding [3,4]. Estrogen, carried in plasma on a binding globulin, diffuses through the lipid layer of the cell membrane, leaving its carrier behind. It reacts with a steroid receptor molecule in the cell nucleus where it induces or represses transcription of selected genes. Such genes have discrete consensus regulatory sequences called EREs. Estrogen binding induces conformational changes in the receptor, enhancing access to EREs that in turn promotes subsequent activation or inhibition of transcription [5].

One of the most intriguing subjects in ovarian steroid biology is the nature of the estrogen receptor itself. It is becoming evident that there are whole families of estrogen receptors and it is probably their different characteristics and distribution in tissues and organs that will eventually explain paradoxical effects of estrogen and its analogues in the body. Two major categories of estrogen receptors, alpha and beta, have now been identified and it is becoming apparent that there are probably families of receptors within each of

R. Paoletti et al. (eds.), Women's Health and Menopause, 39–45.

them [6]. Some tissues contain only one receptor, others the other. To make the story more intriguing, other players are involved in the process: Melzer and his colleagues, for example, have identified a nuclear protein produced by a gene named "amplified in breast cancer" (AIBI) which facilitates estrogen receptors in turning on genes [7]. Many factors, then, help explain the broad spectrum of the hormone's activity and its different impact in individual organs and tissues: the existence of different receptors, their heterogeneous distribution throughout the body, as well as the influence of other modulators and repressors on the effect of estrogen. The unique roles of these factors are the basis for the variation in activity of estrogen-like anticancer drugs and analogues of estrogen.

Other mechanisms of hormonal action exist: they can combine with specific receptors in target cell membranes and influence the activity of ionic channels. Naturally occurring metabolites of progesterone, for example, bind to the alpha amino butyric acid (GABA) receptor which is coupled to the membrane chloride channel [8]. This mechanism explains the sedative effect of progesterone; it is this receptor which also binds to benzodiazepam and barbiturates.

Estrogen and the Brain

The gonadal hormones increases dendritic spine density on central nervous system neurons [9] and can do so very quickly: in rats, synaptic plasticity and differences in dendrite density happen within a single ovarian cycle [10]. Dendritic spine density is related to the plasticity of the central nervous system defined as the ability to learn and retain new information. This action of the gonadal hormones is probably one of the important mechanisms which explains the association between use of HRT and a lower incidence of Alzheimer's disease [11,12], a striking pathologic feature of which is a paucity of dendrites in hippocampal neurons. The magnitude of reduced risk is correlated with the duration of HRT use. Of interest is the fact that the Women's Health Initiative has added an Alzheimer's arm to its first-ever prospective randomized study of the impact of HRT on women.

The impact of estrogen on mood and memory is uncertain. Although HRT has been considered a remedy for the deterioration in recent memory that affects both sexes as they age, the positive impact of this intervention may be short-lived. While some investigators report improvement [13], others found that estrogen users had an amplified sense of well being and improvement in mood, but no impact on cognitive function or memory [14]. The much-touted depression of menopausal women seems to be more myth than fact; indeed, the Rancho Bernardo study documented an increased incidence of depression in treated compared with untreated women [15]. By the 6th and 7th age decade, however, treated patients had a lower incidence of depression.

Now legitimatized as a clinical disorder and not an hysterical symptom, premenstrual tension is probably best understood as an abnormal response to normal hormonal changes during the menstrual cycle. In their innovative protocol, Schmidt and colleagues [16] found that normal women had no perturbation of mood during treatment with leuprolide and subsequent replacement of either estradiol or progesterone, while over half the women with premenstrual tension were relieved by leuprolide and experienced a return of symptoms with the replacement of either one of the gonadal hormones. It may be worth asking a postmenopausal patient about the existence and severity of premenstrual syndrome before deciding to advise HRT. Postnatal depression, present in about 10% of women during the first few months after delivery [17], is probably the result of inadequate levels of gonadal hormones; it is significantly relieved by treatment with estrogen [18].

In contrast to the documented impact of gonadal steroids on the central nervous system, there are no data reporting any effect on nerve-skeletal muscle transmission or on skeletal muscle strength; indeed, users tend to fall more than past or never users, even after prolonged use [19].

Estrogen and the Cardiovascular System

Nowhere is the extensive and diverse impact of ovarian steroids better worked out than in the cardiovascular system. Estrogen has direct and multifaceted effects on the arterial wall, many of which have been worked out quite recently. Venkov et al. have recently demonstrated estrogen receptors in cultured endothelial cells, through which estrogen modulates the production of several endothelium derived vasodilators [5]. The latter include nitric oxide, which, by generating cyclic GMP inhibits calcium entry into and release within the vascular smooth muscle cell [20,21]. Neovascularization is also importantly enhanced through estrogen's effect on endothelial cells [22]. By lowering plasma endothelin levels, estrogen reduces smooth muscle proliferation in the arterial wall. It restores and modulates the response of vasculature to catecholamines, which stabilizes arterial flow and eliminates the vasomotor instability of women with hot flushes [23].

About 50% of the positive impact of estrogen on improving the resistance to and amelioration of coronary artery disease is due to its effects on serum lipids [24]. Unopposed estrogen raises high density lipoprotein (HDL) levels, particularly the HDL_2 subfraction; the latter is particularly important in the reverse cholesterol transport process, in which cholesterol is removed from cells and transported to the liver or transferred to other lipoproteins. Due to an impact on activity of hepatic apoprotein B and E receptors, estrogen promotes increased low density lipoprotein (LDL) uptake in the liver and levels fall although the particle formed is smaller and more dense than the original. Triglyceride levels rise with estrogen therapy, but the size of molecules is large and it is unlikely to be converted into the smaller particles which are atherogenic [25]. In general, progestins tend to reverse the impact of estrogen on serum lipids, particularly those which are more androgenic, i.e. the older 19 nortestosterone derivatives including norethindrone and dl-norgestryl. Newer 19-nortestosterone derivatives (such as desogestryl and norgestimate) have a less antagonistic effect on an optimal lipid profile [27].

These data help explain the association of estrogen use with a 50% lowered incidence of coronary artery disease. The benefit is most clear for women who suffer premature menopause (before the age of 40), whether naturally or surgically induced.

Estrogen and Bone

Through a combination of genomic and nongenomic mechanisms, estrogen retards bone resorbtion by osteoclasts [27]. It retards lysosomal enzyme production by mature cells and may also inhibit their differentiation [28]. Estrogen administration impacts osteoblasts as well, optimizing the pattern of bone deposition by these cells.

The impact of estrogen on osteoporosis is significant: treatment reduces the lifetime probability of hip fracture by 2/3 [29] and increases life expectancy by a year.

What are the Data about the Consequences of HRT?

Most of our information about the impact of estrogen/progesterone therapy in menopausal women comes from observational and/or retrospective studies. The results of such investigations are clouded by a variety of factors, including the unreliability of self-reporting, selection and surveillance bias, and uncontrolled confounding. Moreover, a significant amount of data comes from patients who used HRT before the late 1980s, when the effects of estrogen on the endometrium were not neutralized by progesterone administration: the long-term efficacy of combined therapy is still not available.

Two major prospective studies, the Postmenopausal Estrogen/Progestin Interventions (PEPI) and the Women's Health Initiative (WHI), are significant. The PEPI study concentrated on the impact of HRT on risk factors for CAD and did not focus on its effect on the incidence or on the mitigation of established CAD in the user [30]. The data showed that while progestins did not diminish the impact of estrogen on LDL, it did blunt the latter hormone's beneficial effects on HDL, no matter what preparation was used. Contrary to what had been held previously, estrogen did not increase blood pressure, and did decrease fibrinogen levels. The study was flawed by the fact that only 11% of subjects were minority women, of whom 4% were African-American.

The protocols of the Women's Health Initiative have been significantly expanded and amplified as the study goes on. It is the most ambitious prospective randomized study on the impact of hormonal replacement therapy on coronary artery disease, Alzheimer's disease, cancers of various types, and osteoporosis to date. Critics have called the study's protocols too ambitious and predicted that the study will be undoable as planned for a variety of reasons, including recruitment difficulties (plans call for 40% of patients to be from minority populations), lack of sufficient funding, and lack of compliance, given the difficult nature of stringent dietary restrictions and some of the side effects of HRT [31].

Deciding Who Should Have HRT

The multifactorial decision to advise HRT for the female patient should be made on an individual basis. Doctor and patient together must assess a whole variety of organ systems. A proper history must include an assessment of:

- menopausal symptoms,
- ability to think clearly, remember new information, and mood stability,

- risk factors for coronary artery disease,
- bone density, and
- sexual and urinary function.

If the aging patient's risks of severe morbidity or death from coronary artery disease, significant osteoporosis, or vulnerability to Alzheimer's disease are significant, even the specter of breast cancer or of other important but not lethal side effects of HRT should not prevent a recommendation to use hormones. If the patient is properly informed, and feels that there are compelling reasons for her to elect HRT, the physician will have no problem enlisting her collaboration. This is particularly true if her physician helps her understand the nature and magnitude of her vulnerability, and balance it with the consequences of taking the hormones. The patient will be further comforted by a promise to monitor her carefully, particularly in the first four months of therapy, and to re-assess her, whatever her decision is, after a two-year interval for whether the risk/benefit ratio of using HRT has changed since the initial discussions.

Suggestions for Future Research

It is fair to say that many questions about the value and risks of hormonal replacement therapy are still unanswered. When to begin HRT and how long to continue it remain important issues. Moreover, the new information about the ubiquity and power of the steroid hormones raise the issue of reopening the question of their usefulness in treating men, particularly men with coronary artery disease. It is likely that cognitive defects and osteoporosis men experience with aging might benefit from nonfeminizing analogues of estrogen. A study of the gender-specific aspects of the response to these hormones, particularly with regard to receptor biology would amplify our understanding of their precise mechanism of action.

In the best of all possible worlds, we will be able to improve the quality of life for our patients, postpone the time of death from preventable plagues like CAD, and avoid producing new categories and levels of disability by persuading them to accept hormonal replacement therapy.

References

1. Cauley JA, Cummings SR, Blad DM, et al. Prevalence and determinants of estrogen replacement therapy in elderly women. Am J Obstet Gynecol 1990;163:1438-44.
2. Berman RS, Epstein RS, Lydick E, et al. Risk factors associated with women's compliance with estrogen replacement therapy. J Women's Health 1996.5:213-220.
3. Landers SP, Spelsberg TC. New concepts in steroid hormone action: Transcription factors, proto-oncogenes, and the cascade model for steroid regulation of gene expression. Crit Rev Eukar Gene Exp 1992;2:19-63.
4. Komm BS, Terpenning CM, Benz DJ, et al. Estrogen binding, receptor mRNA and biologic response in osteoblast-like osteosarcoma cells. Science 1988;241:81-84.
5. Venkov CD, Rankin AB, Vaughan DE. Identification of authentic estrogen receptor in cultured endothelial cells. A potential mechanism for steroid hormone regulation of

endothelial function. Circulation 1996;94:727-33.
6. Guftafsson J, Kushner PJ, Scanlan TS. Differential ligand activator of estrogen receptors ER alpha and ER beta at AP1 sites. Science 1997;277:1508-10.
7. Meltzer PC, Anzick SL, Konenn J, Walker RL, Azorsa DO, Tanner MM. AIB, a steroid receptor coactivator amplified in breast and ovarian cancer. Science. 1997;277:965-68.
8. Bitan D, Lan NC, Gee KW. Anxiolytic effects of 3 sulpha hydroxy-5-alpha (beta)-pregnane-20-one. Endogenous metabolites of progesterone that are active at the $GABA_A$ receptor. Brain Res 1991;561:157-61.
9. McEwen BS, Woolley CS. Estradiol and progesterone regulate neuronal structure and synaptic connectivity in adult as well as developing brain. Exp Gerontol 1994;29:431-36.
10. Woolley CS, McEwen BS. Estradiol mediates fluctuation in hippocampal synapse density during the estrus cycle in the adult rat. J Neurosci 1992;12:2549-54.
11. Paganini-Hill A, Buckwalter JC, Logan CG, Henderson VW. Estrogen replacement and Alzheimer's disease in women. Soc Neurosci Abstr 1993;19:1046.
12. Tang M, Jacobs D, Sten Y, et al. Effects of oestrogen during menopause on risk and age at onset of Alzheimer's disease. Lancet 1996;348:429-32.
13. Robinson D, Friedman L, Marcus R, et al. Estrogen replacement therapy and memory in older women. J Am Geriatr Soc 1994:42:919-22.
14. Ditkoff EC, Crary WG, Crito M, et al. Estrogen improves psychological function in asymptomatic postmenopausal women. Obstet Gynecol 1991:78:991-95.
15. Palinkas LA, Barrett-Connor E. Estrogen use and depressive symptoms in postmenopausal women. Obstet Gynecol 1992:80:30-36.
16. Schmidt PJ, Nieman LK, Danaceau MA, Adams LF, Rubinow DR. Differential behavioral effects of gonadal steroids in women with and in those without premenstrual syndrome. N Engl J Med 1998:338:209-16.
17. O'Hara MW, Zekoski EM. Postpartum depression: A comprehensive review. In: Kuma R, Brockington IF, editors. Motherhood and mental illness. London: Wright, 1988:63.
18. Gregoire AJP, Kumar R, Everitt, B, Henderson AF, Studd JWW. Transdermal oestrogen for treatment of severe postnatal depression. Lancet 1996;347:930-33.
19. Seeley DG, Cauley JA, Grady D, et al. Is postmenopausal estrogen therapy associated with neuromuscular function or falling in elderly women? Arch Intern Med 1995;155:293-99.
20. Collins P, Rosano GM, Jian C, et al. Cardiovascular protection by oestrogen-a calcium antagonist effect. Lancet 1993;341:1264-65.
21. Gaulin-Glaser TL, Sessa W, Sarrel P, Bender J. The effect of 17 beta estradiol on human endothelial cell nitric oxide production. Circulation 1994;90(Suppl.I):I.30. Abst.
22. Morales DE, McGowan KA, Grant DS, et al. Estrogen promotes angiogenic activity in human umbilical vein endothelial cells *in vitro* and in a murine model. Circulation 1995; 91:755-63.
23. Kronenberg F, Cote L, Linkie DM, et al. Menopausal hot flashes: Thermoregulatory, cardiovascular and circulating catecholamine and LH changes. Maturitas 1984;6:31-43.
24. Bush TL. The epidemiology of cardiovascular disease in postmenopausal women. Ann NY Acad Sci 1990;592:263-71.
25. Walsh BW, Schiff I, Rosner B, Ravnikar V, Sacks FM. Effects of postmenopausal estrogen replacement on the concentrations and metabolism of plasma lipoproteins. N Engl J Med 1991;325:1196-1204.
26. Speroff L, DeCherney A, Burkman RT Jr, Carr BR, Comp PC, Crook D, et al. Evaluation of a new generation of oral contraceptives. Obstet Gynecol 1993;81:1034-47.

27. Oursler MJ, Landers JP, Riggs BL, Spelsperg TC. Oestrogen effects on osteoblasts and osteoclasts. Ann Med 1993;25:361-71.
28. Ransome LJ, Verma IM. Nuclear proto-oncogenes fos and jun. Annu Rev Cell Biol 1990;6: 539-44.
29. Grady D, Rubin SM, Petitti DB, et al. Hormone therapy to prevent disease and prolong life in postmenopausal women. Ann Int Med 1992;117:1016-37.
30. The Postmenopausal Estrogen/Progestin Interventions (PEP) Trial Investigators. Effects of estrogen or estrogen/progestin regimens onheart disease risk factors in postmenopausal women: The Postmenopausal Estrogen/Progestin Interventions (PEPI) Trial. JAMA 1996; 274:1676.
31. Thaul S, Hotra D, editors. An assessment of the NIH Women's Health Initiative. Committee to Review the NIH Women's Health Initiative. Food and Nutrition Board and Board on Health Sciences Policy of the Institute of Medicine. National Academy Press: Washington, D.C., 1993:142.

NONGENOMIC MECHANISMS OF SEX HORMONES

Jos H.H. Thijssen

Introduction, Genomic Mechanisms

Following the isolation and the identification of steroid hormones in the first half of this century, their mechanism(s) of action have been under investigation. The development of radiolabelled ligands allowed the identification of specific binding proteins, present intracellularly, that were translocated to the nucleus after binding with the steroid. The location of the steroid-protein complex suggested a link between this binding phenomenom and transcriptional activity as explanation of the physiological activity [1]. Further studies indeed demonstrated that steroid hormones exerted their biological activity after binding to these high-affinity binding proteins, named receptors, and that this binding induced an allosteric change in the receptor-protein that enabled the steroid-receptor complex to bind to high-affinity binding sites on the chromatin and to modulate transcriptional activity. It became established that steroid hormones act at the level of nuclear DNA and that they are involved in regulation of specific gene expression.

The steroid hormone receptors were identified and subsequently cloned, leading to a more profound understanding of this genomic mechanism of action. A family of steroid hormone receptors has been discovered, including mineralocorticosteroid-, glucocorticosteroid-, progesterone-, estrogen-, androgen- and 1α,25-vitamine-D3-receptors, as members of the superfamily of nuclear receptors [2] with unknown ligands for several members of this family [3,4], the so-called orphan receptors. In a recent series of articles [2,5-7] many aspects of this field have been reviewed and the rapidly expanding knowledge on the complicated regulation of transcriptional activity was illustrated.

The original, simplified model of steroid action

$$\mathbf{S + R \rightarrow SR \rightarrow SR^{*} \rightarrow SR^{*}DNA \rightarrow \uparrow mRNA \rightarrow \uparrow protein}$$

is being replaced by a much more complicated mechanism of regulation of the transcriptional activity. The steroid receptors are composed of a number of functional domains with different functions, the ligand binding domain possessing the essential property of hormone recognition that ensures specificity and selectivity of the responses. After binding of the steroid, the activated steroid-steroid receptor complexes do have to interact with co-regulatory proteins and thus are one of the factors in the regulation of transcription of

R. Paoletti et al. (eds.), Women's Health and Menopause, 47–51.

specific genes [8]. The increasing knowledge on the ligand-dependent but also on ligand-independent activations allows more insight in the observed discrepancies between agonistic and antagonistic effects of derivatives and analogues of steroid hormones, which could not easily be explained by the original model.

Criteria for Nongenomic Mechanisms

A large body of experimental data suggests the existence of steroid effects that cannot be explained by the genomic model of steroid-target cell interaction. These effects may rather be explained by the existence of signal-generating (steroid?) receptors on the cell surface, involved in these so-called nongenomic effects. In general, nongenomic effects have the following characteristics [9]:

1. are too rapid (from milliseconds to seconds to a few minutes) to be compatible with the involvement of changes in mRNA and protein synthesis, processes which are supposed to need several minutes and usually hours;
2. can be observed in highly specialized cells that do not accomplish mRNA and protein synthesis (e.g. spermatozoa) or by cells lacking the intracellular nuclear steroid receptors;
3. can be elicited even by steroids coupled with high-molecular weight substances that do not cross the plasma membrane and do not enter the cells;
4. are not blocked by inhibitors of mRNA or protein synthesis;
5. may not be blocked by antagonists of the genomic steroid receptors, unless these antagonists could react with structurally related steroid-binding domains that may be common to both the genomic and the nongenomic steroid receptors;
6. are highly specific, as steroids with very similar, but not identical, chemical structure may show various degrees of potency.

Nongenomic Mechanisms

In this contribution several aspects of the nongenomic mechanisms will be presented and discussed. Basically several options for effects of steroids on cellular functions are possible.

SPECIFIC CELL-SURFACE RECEPTORS (MEMBRANE-BOUND) FOR SEX HORMONES

On basis of a number of experimentally observed phenomena, steroid-second messenger-effector pathways have been postulated. The receptor(s?) involved must have their localization in the plasma membrane of the cells [10]. Membrane-binding sites for steroid hormones have been described for mineralocorticoids on kidney plasma membranes, for glucocorticoids on liver and pituitary membranes, for estrogens on endometrial membranes and a neuroblastoma cell line [11] and for progesterone on spermatozoa [9]. A convincing match between these binding sites and rapid steroid effects, supporting the causal relation, has rarely been described.

Cloning of the first membrane receptor for steroid hormones remains one of the major challenges. It would enable researchers to search for antagonists, to study proximal parts of the signaling and to validate many of the assumptions drawn on the ground of binding data and traditional cell physiology. Various attempts to purify these receptor proteins have already been described [10]. Using photoaffinity labeling and specificity studies of the proteins, determination of the molecular weight and partial characterization of the aldosterone membrane receptor and a partial sequence of a progesterone membrane receptor have been reported [12].

Apparent contradictions in studies using spermatozoa are interpreted as originating from the presence of different types of progesterone-receptors in the sperm plasma membrane [9]. Progesterone effects on spermatozoa appear to be mediated by multireceptor systems. When one member of this system is not operational, the biological response may fail even if the other types function normal. Obviously, future investigations on the rapid steroid actions will be needed in order to get answers to many of the existing questions.

INTERACTION WITH SPECIFIC RECEPTORS FOR NONSTEROIDAL LIGANDS

In particular studies on the effects of steroids on the central nervous system (CNS) have stimulated the search for interactions of steroid hormones with membrane-receptors for nonsteroidal ligands. Some effects of steroids can be attributed to the genomic mechanism of action but in the last decade, evidence has accumulated that other specific receptors, which are highly involved in the functioning of cells in the brain, are partly under the influence of steroid hormones (progesterone, estradiol) plus several of their metabolites [13]. The origin of the so-called neurosteroids, defined as steroids synthesized in the brain, will be discussed in the section on these components.

Neurosteroids and many of the neuroactive steroids bind preferentially to membrane bound components like neurotransmitter receptors, ion channels and ion pumps. Several studies have shown effects of steroids on different receptor-types but at this moment most evidence is pointing to interactions with the $GABA_A$-receptor and the glutamate-receptors.

$GABA_A$-receptor. The $GABA_A$/benzodiazepine-receptor Cl^- channel complex is a member of the ligand-gated ion channel superfamily of receptors. It contains several functional domains and it exists in a number of forms determined by its subunit composition. Certain steroids positively modulate GABA-induced Cl^- flux whereas [14] others do have a negative modulatory effect [15]. Progesterone plus several of its naturally occurring reduced metabolites, $3\alpha,5\alpha$- and $3\alpha,5\beta$-tetrahydroprogesterone (allopregnanolone) but not their 3β-isomers enhance GABA-induced currents at concentrations of 1 nM, whereas pregnenolone- and dehydroepiandrosterone-sulfate antagonize this effect [15]. Also some of the effects of estrogens on the functioning of neuronal cells have been attributed to their interaction with this receptor [16].

Glutamate-receptors. These receptors belong to one of the major excitatory transmitter systems, crucial for basic brain function. They can be classified into two broad categories [16], ionotropic and metabotropic receptors. The ionotropic ones have been subdivided into three subtypes, based upon pharmacological and electrophysiological data. In particular the NMDA (N-methyl-D-aspartate) receptor [17] has been implicated in the action of several steroid hormones and again also of their metabolites.

MECHANISMS BASED ON INFLUENCES ON THE COMPOSITION OF THE MEMBRANES (PROTEIN AND/OR LIPIDS)

Sometimes steroid actions directly on membrane composition have been described, related to the lipophylic character of the steroids. The anesthetic activities of steroids have been known for many years and have been attributed to these influences [14] because of the relatively high concentrations needed to achieve these actions. The partial steroid specificity may be due to variable lipophylicity and polarity of the compounds studied. These steroid-membrane interactions affect physicochemical properties such as fluidity and the microenvironment of proteins in the membrane. However, given the high specificity of many of the rapid steroid effects and their occurrence at low steroid concentrations, it is difficult to explain rapid effects on the basis of the rather nonspecific interactions.

METABOLITES OF STEROIDS (NEUROSTEROIDS)

Neurosteroids are steroids that can be produced locally in some specific areas of the brain. Their synthesis is independent from other steroidogenic organs like the gonads or the adrenal cortex and therefore the regulation of neurosteroids is different and not well understood yet [13,15]. Early studies have shown that the neurosteroids are persistent in the brain after removal of the peripheral steroid-producing organs [15]. The enzymatic machinery for the synthesis has been documented for many, but not all, of the necessary conversions of the supposed precursor cholesterol to the neurosteroids. Pregnenolone, dehydroepiandrosterone and several of their reduced metabolites are amongst the active components.

Effects of sex steroids on the release of gonadotropins have led to the discovery of an active metabolite of progesterone, 3α-hydroxy-4-pregnene-20-one or 3α-dihydro-progesterone, which has been described as the "missing link" in the progesterone biosynthetic/metabolic pathways [18].

Conclusions

We are in the midst of an explosion in our understanding of hormonal signaling that will give a dramatic increase in our knowledge. Evidence for nongenomic effects is now coming from all fields of steroid research and mechanisms of action are being studied with regard to the membrane receptors and second messengers. Potential new strategies for therapeutic procedures derived from these developments should emerge, such as antagonists that might block both the genomic and nongenomic aspects of steroid actions.

References

1. Jensen EV, Jacobson HI, Flesher JW, et al. Estrogen receptors in target tissues. In: Pincus G, Nakao T, Tait JF, editors. Steroid Dynamics. New York: Academic Press, 1966:133-156.
2. Mangelsdorf DJ, Thummel C, Beato M, et al. The nuclear receptor superfamily: The second decade. Cell 1995;83:835-39.
3. Truss M, Beato M. Steroid hormone receptors: Interaction with deoxyribonucleic acid and transcription factors. Endocr Rev 1993;14:459-79.
4. Ciocca DR, Roig LM. Estrogen receptors in human nontarget tissues: Biological and clinical implications. Endocr Rev 1995;16:35-62.
5. Mangelsdorf DJ, Evans RM. The RXR heterodimer and orphan receptors. Cell 1995;83:841-50.
6. Beato M, Herrlich P, Schütz G. Steroid hormone receptors: Many actors in search of a plot. Cell 1995;83:851-57.
7. Kastner P, Mark M, Chambon P. Nonsteroid nuclear receptors: What are genetic studies telling us about their role in real life. Cell 1995;83:858-69.
8. Shibata H, Spencer TE, Onate SA, et al. Role of co-activators and co-repressors in the mechanism of steroid/thyroid receptor action. Rec Progr Horm Res 1997;52:141-65.
9. Revelli A, Massobrio, Tesarik J. Nongenomic actions of steroid hormones in reproductive tissues. Endocr Rev 1998;19:3-17.
10. Wehling M. Specific, nongenomic actions of steroid hormones. Annu Rev Physiol 1997;59: 365-93.
11. Watters JJ, Campbell JS, Cunningham M, Krebs EG, Dorsa DM. Rapid membrane effects of steroids in neuroblastoma cells: Effects of estrogen on mitogen activated protein kinase signalling cascade and c-fos immediate early gene transcription. Endocrinology 1997;138: 4030-33.
12. Meyer C, Schmid R, Scriba PC, Wehling M. Purification and partial sequencing of a putative progesterone-receptor from porcine liver membranes. Eur J Biochem 1996;239:726-31.
13. Spindler KD. Interactions between steroid hormones and the nervous system. Neuro Toxicol 1997;18:745-54.
14. McEwen BS. Non-genomic and genomic effects of steroids on neural activity. Trends in Pharmacol Sci 1991;12:141-47.
15. Baulieu EE. Neurosteroids: Of the nervous system, by the nervous system, for the nervous system. Rec Progr Horm Res 1997;52:1-32.
16. Moss RL, Gu Q, Wong M. Estrogen: Nontranscriptional signalling pathway. Rec Progr Horm Res 1997;52:33-69.
17. Sucher NJ, Awobuluyi M, Choi Y-B, Lipton SA. NMDA-receptors: From genes to channels. Trends in Pharmacol Sci 1996;17:348-55.
18. Wiebe JP. Nongenomic actions of steroids on gonadotropin release. Rec Progr Horm Res 1997;52:71-101.

Action of Specific Estrogens on Vascular Cells

Callum S. Wingrove and John C. Stevenson

Introduction

Coronary heart disease (CHD) is a major cause of mortality in First World countries in both men and women. Men exhibit up to a tenfold greater risk of CHD than women of premenopausal age. In women undergoing natural [1] or surgical menopause [2], the risk of CHD increases and approaches that of men. Cessation of ovarian function and loss of endogenous sex hormone production has been suggested to be the primary cause of the increase in risk of CHD observed in postmenopausal women. Indeed, hormone replacement therapy (HRT) studies have indicated that hormone replacement is associated with a reduction in both morbidity and mortality of around 50% [3]. This discrepancy in risk of CHD between the genders has prompted intense research into the vascular effects of female sex hormones. Studies on coronary occlusion as assessed by selective coronary arteriography have revealed that HRT users had a relative risk of coronary artery disease of 0.37-0.44 compared to non-users [4]. The female sex hormones, in particular estrogen, are believed to cause their relative cardioprotection in part by alterations in plasma lipoprotein levels [5]. Estrogens are able to increase plasma high density lipoproteins (HDL) while lowering low density lipoproteins (LDL) [6]. Given the positive association of the latter with CHD risk and the association of the former with cardiovascular protection, it was at first believed that this was the primary mechanism by which female sex hormones mediated cardiovascular benefit. However, studies quantifying the relative contribution of these changes to cardioprotection, concluded that changes in lipid profiles at most accounted for only 25%-50% of the effect of female sex hormones [7].

Direct Actions of Estrogen on the Vessel Wall

The underlying mechanisms behind the action of estrogen on vascular disease risk are not fully understood. Direct effects on the blood vessel wall have been postulated to account for a significant proportion of the nonlipid-associated beneficial properties of estrogen on CHD risk. The vascular endothelium is responsible for the regulation of cell proliferation, inflammation, vascular tone, platelet aggregation, and coagulation. Specific receptors for both estrogen and progesterone have been demonstrated in a variety of vascular tissues including coronary vessels [8], aorta [9], and myocardium [10]. In this paper we will

R. Paoletti et al. (eds.), Women's Health and Menopause, 53–58.

discuss some of the key cellular actions of estrogen and how these effects vary dependent on the estrogen. In all our studies we have investigated the natural human estrogen, 17-β estradiol. We have also investigated a variety of other estrogens that represent important constituents of Premarin, the most widely used postmenopausal estrogen in the United States. Premarin represents a complex mixture of conjugated equine estrogens. The estrogenic components we have studied are equilin, 17-β dihydroequilin, estrone, Δ 8,9-dehydroestrone, 17-α dihydroequilenin, and 17-β Δ 8,9-dehydroestradiol. We will discuss the cellular actions of these estrogens on important vascular endothelial products. In addition we will briefly discuss a new finding relating to the action of 17-β estradiol on the vascular extracellular matrix.

Endothelial Nitric Oxide Synthase

There is substantial evidence that estrogen increases release of nitric oxide from endothelial cells. One study, aimed at elucidating the contribution of endothelial nitric oxide (NO) release to blood pressure regulation, demonstrated that in mice genetically altered to be deficient in the gene for endothelial nitric oxide synthase, mean arterial pressure was increased by 35%, showing the importance of NO regulation [11]. Studies on rabbit aorta have found greater production of NO in aortic rings from female rabbits than males [12]. Increases in NO production are in part related to increased expression of the endothelial nitric oxide synthase enzyme (eNOS) affected through the classical pathway of estrogen enhanced gene transcription. Studies of human aortic endothelial cell lines have shown increases in levels of nitric oxide synthase in response to physiological doses of 17-β estradiol [13]. We have shown a similar effect in primary cultures of human umbilical vein endothelial cells (HUVECs) and coronary artery endothelial cells (CAECs) [14]. Furthermore we found differential effects on eNOS depending upon the estrogenic compound tested. With the exception of estrone and 17-α dihydroequilenin, all the estrogens we tested significantly increased eNOS expression. Primate studies have suggested that the latter estrogen does not mediate proliferative or trophic effects on the uterus, mammary glands, or endometrium [15]. This, in turn, suggests that 17-α dihydroequilenin does not mediate its actions through the classical estrogen pathway. This hypothesis would be consistent with the lack of any discernible effect by this estrogen on eNOS expression.

Endothelin-1

The endothelium is also the source of production of the vasoconstrictor, endothelin. This highly conserved 21 amino acid peptide, is the most potent naturally occurring vasoconstrictor and is also a powerful mitogen for vascular smooth muscle cells [16]. Circulating plasma levels of ET-1 are very low (0.3-5 ρg/ml). This probably reflects secretion of ET-1 abluminally towards the underlying smooth muscle cells as well as a rapid clearance mechanism in the pulmonary circulation [17]. It also suggests that ET-1 is very much a localized effector of vasoconstriction rather than a systemic one. However, elevated circulating plasma ET-1 concentrations have been observed in a number of disease states including hypertension [18] and congestive heart failure [19]. Premenopausal women appear to have lower circulating levels of plasma ET-1 than men of comparable age [20]. Transdermal HRT results in a significant 23% reduction in circulating plasma ET-1 [21]. We have found that 17-β estradiol attenuates the stimulated release of ET-1 from both HUVECs and CAECs [14]. This effect has been reproduced in an animal cell line and shown to result from a negative transcriptional regulation of the ET-1 transcriptional program [22]. Our own studies have indicated that all estrogens thus far examined are capable of mediating this effect to equivalent degrees.

Angiotensin Converting Enzyme

Estrogens may affect arterial function through modulation of the renin-angiotensin system. An association between the angiotensin converting enzyme (ACE) gene deletion allele and increases in angiotensin-1 conversion has been observed [23]. Furthermore, an association between the ACE deletion allele and risk of CHD has been shown to occur and be especially prevalent in women [24]. Continuous combined estrogen and progestogen can significantly reduce the levels of circulating serum ACE in postmenopausal women thus potentially providing protection against CHD [25]. We have found that 17-β estradiol and estrone mediated significant increases in ACE associated with CAEC. Whether this represents an increase in ACE expression or a decrease in ACE release from the cell is at this time unclear. However, given that serum ACE levels are believed to originate by release or cleavage from the endothelium, and estrogen use *in vivo* has been associated with reductions in serum ACE, a reduction in ACE release from the endothelium would be consistent. It is of interest to note the effect of the equine estrogen, 17-β Δ 8,9-dehydroestradiol on cellular ACE. This estrogen caused a significant decrease in cellular ACE concentrations. Again this could have been due to an increased release of ACE from the cell or decreased ACE expression. We can speculate based on our data that estrone reduces release of ACE from the cell whilst 17-β Δ 8,9-dehydroestradiol reduces expression of ACE.

Matrix Metalloproteinases and Collagen Degradation

Despite the beneficial vascular effects of physiological estrogen concentrations it has long been recognized that excess estrogenic activity, whether through oral contraception or natural events such as pregnancy, is associated with harmful, even fatal, vascular events. Studies of the vascular pathology of oral contraceptive users, primarily those of Irey in the early 1970s, demonstrated lesions in the vasculature ranging from the pulmonary, systemic and portal circulations, and in veins and arteries [26]. The distinct vascular changes that were observed included endothelial proliferation and intimal thickening and these were associated with thrombotic occlusion. Crucially these changes preceded thrombus formation. Moreover, they were seen in the vasculature of pregnant and postpartum women [27]. In 1972, Fischer reported increases in collagen and elastin degradation in the aorta of estradiol treated ovariectomized rats [28]. The potential links between these disparate studies were not noted at the time. The enzymes mediating vascular collagen degradation are the matrix metalloproteinases (MMPs). Effects of the female sex hormones on vascular MMP have thus far remained unexplored. We have examined the effect of 17-β estradiol on MMP-2 expression in human coronary artery (CAVSMC) and umbilical artery vascular smooth muscle cells (UAVSMC) [29]. pro-MMP-2 was secreted by VSMCs and increasing levels of 17-β estradiol, from physiological through to supraphysiological, were associated with significant dose-dependent increases in MMP-2 levels in culture media. This effect was dependent on *de novo* protein synthesis and could be antagonized by the estrogen receptor antagonist, tamoxifen and the specific receptor antagonist ICI 182,780, indicating that this action is due to the classical estrogen receptor pathway of gene transcription. The dose dependency of this effect suggests that physiological estrogen levels may be associated with modest increases in MMP whereas supraphysiological levels would be expected to yield high expression of MMP. The latter may be responsible for the vascular lesion formation seen in oral contraceptive users whilst the former may serve to protect the vasculature through reductions in collagen deposition. This hypothesis is supported by a recent *in vivo* study of conjugated equine estrogens in atherosclerotic female monkeys which revealed an estrogen-associated decrease in aortic collagen content [30]. The question of whether this was due to decreased collagen synthesis or increased collagen degradation was not addressed. Increased activity of MMPs would be consistent with the observed decrease in collagen content. The authors concluded that estrogen potentially inhibits detrimental connective tissue alterations.

Conclusions

In summary, estrogens demonstrate a range of vasoactive properties on human vascular endothelial cells. Collectively, the effects demonstrated by estrogens on ET-1 release, eNOS expression and cell-associated ACE concentrations may make a significant contribution to the beneficial effects that estrogens demonstrate on the vasculature. *In vitro* effects of 17-β estradiol on MMP expression by vascular smooth muscle cells may be mirrored by conjugated equine estrogens *in vivo*, suggesting a novel means of atherosclerotic progression.

References

1. Colditz GA, Willett WC, Stampfer MJ, Rosner B, Speizer FE, Hennekens CH. Menopause and the risk of coronary heart disease in women. N Engl J Med 1987;316:1105-10.
2. Adams MR, Kaplan JR, Clarkson TB, Koritnik DR. Ovariectomy. Social status and atherosclerosis in cynomolgus monkeys. Arteriosclerosis 1985;5:192-200.
3. Grady D, Rubin SM, Petitti DB, et al. Hormone therapy to prevent disease and prolong life in postmenopausal women. Ann Intern Med 1992;117:1016-37.

4. Gruchow HW, Anderson AJ, Barboriak JJ, Sobocinski KA. Postmenopausal use of estrogen and occlusion of coronary arteries. Am Heart J 1988;115:954-63.
5. Stevenson JC. The metabolic and cardiovascular consequences of HRT. Br J Clin Pract 1995;49:87-90.
6. Sitruk-Ware R. Estrogen therapy during the menopause. Practical treatment recommendations. Drugs 1990;39:203-17.
7. Walsh BW, Schiff I, Rosner B, Greenberg L, Ravnikar V, Sacks FM. Effects of postmenopausal estrogen replacement on the concentration and metabolism of plasma lipoproteins. N Engl J Med 1991;325:1196-1204.
8. Horwitz KB, Horwitz LD. Canine vascular tissues are targets for androgens, estrogens, progestins, and glucocorticoids. J Clin Invest 1982;69:750-58.
9. Lin AL, McGill HC Jr, Shain SA. Hormone receptors of the baboon cardiovascular system: Biochemical characterization of aortic and myocardial cytoplasmic progesterone receptors. Circ Res 1982;50:610-16.
10. McGill H. Sex steroid hormone receptors in the cardiovascular system. Postgrad. Med April 1989;85(pt.2):64-68.
11. Huang PL, Huang Z, Mashimo H, et al. Hypertension in mice lacking the gene for endothelial nitric oxide synthase. Nature 1995;377:239-42.
12. Hayashi T, Fukuto JM, Ignarro LJ, Chaudhuri G. Basal release of nitric oxide from aortic rings is greater in female rabbits than in male rabbits: Implications for atherosclerosis. Proc Natl Acad Sci 1992;89:11259-63.
13. Hishikawa K, Nakaki T, Marumo T, Suzuki H, Kato R, Saruta T. Up-regulation of nitric oxide synthase by estradiol in human aortic endothelial cells. FEBS Lett 1995;360:291-93.
14. Wingrove CS, Stevenson JC. 17-β estradiol inhibits stimulated endothelin-1 release from human vascular endothelial cells. Eur J Endocrinol 1997;137:205-8.
15. Washburn SA, Honoré EK, Cline JM, et al. Effects of 17-α dihydroequilenin sulfate on atherosclerotic male and female rhesus monkeys. Am J Obstet Gynecol 1996;175:341-51.
16. Bobik A, Grooms A, Millar JA, Mitchell A, Grinpukel S. Growth factor activity of endothelin on vascular smooth muscle. Am J Physiol 1990;258:C408-C415.
17. de Nucci G, Thomas R, D'Orleans-Juste P, et al. Pressor effects of circulating endothelin are limited by its removal in the pulmonary circulation and by the release of prostacyclin and endothelium-derived relaxing factor. Proc Natl Acad Sci USA 1988;85:9797-9800.
18. Haak T, Jungmann E, Felber A, Hillmann U, Usadel KH. Increased plasma levels of endothelin in diabetic patients with hypertension. Am J Hypertens 1992;5:161-66.
19. Cody RJ, Haas GJ, Binkley PF, Capers Q, Kelley R. Plasma endothelin correlates with the extent of pulmonary hypertension in patients with chronic congestive heart failure. Circulation 1992;85:504-9.
20. Polderman KH, Stehouwer CD, van Kamp GJ, Dekker GA, Verheugt FW, Gooren LJ. Influence of sex hormones on plasma endothelin levels. Ann Intern Med 1993;118:429-32.
21. Ylikorkala O, Orpana A, Puolakka J, Pyorala T, Viinikka L. Postmenopausal hormonal replacement decreases plasma levels of endothelin-1. J Clin Endocrinol Metab 1995;80:3384-87.
22. Morey AK, Razandi M, Pedram A, Hu R-M, Prins BA, Levin ER. Oestrogen and progesterone inhibit the stimulated production of endothelin-1. Biochem J 1998;330:1097-1105.
23. Buikema H, Pinto YM, Rooks G, Grandjean JG, Schunkert H, van Gilst WH. The deletion

polymorphism of the angiotensin-converting enzyme gene is related to phenotypic differences in human arteries. Eur Heart J 1996;17:787-94.
24. Schuster H, Wienker TF, Stremmler U, Noll B, Steinmetz A, Luft FC. An angiotensin-converting enzyme gene variant is associated with acute myocardial infarction in women but not in men. Am J Cardiol 1995;76:601-3.
25. Proudler AJ, Ahmed AI, Crook D, Fogelman I, Rymer JM, Stevenson JC. Hormone replacement therapy and serum angiotensin-converting-enzyme activity in postmenopausal women. Lancet 1995;346:89-90.
26. Irey NS, Manion WC, Taylor HB. Vascular lesions in women taking oral contraceptives. Arch Path 1970;89:1-8.
27. Irey NS, Norris HJ. Intimal vascular lesions associated with female reproductive steroids. Arch Pathol 1973;96:227-34.
28. Fischer GM. *In vivo* effects of estradiol on collagen and elastin dynamics in rat aorta. Endocrinol 1972;91:1227-32.
29. Wingrove CS, Garr ED, Godsland IF, Stevenson JC. 17-β oestradiol enhances release of matrix metalloproteinase-2 from human vascular smooth muscle cells. Biochim Biophys Acta 1998;1406:169-74.
30. Register TC, Adams MR, Golden DL, Clarkson TB. Conjugated equine estrogens alone, but not in combination with MPA, inhibit aortic connective tissue remodeling after plasma lipid lowering in female monkeys. Arterioscler Thromb Vasc Biol 1998;18:1164-71.

Direct Actions of Estrogen on Vascular Cells Ameliorates Response to Injury

D.W. Losordo

Introduction

In spite of significant advances in technique and adjunctive medical treatment, restenosis after percutaneous coronary revascularization remains a major problem. Moreover, women have fared comparatively poorly after percutaneous transluminal coronary angioplasty (PTCA) with less long-term relief of symptoms and a trend towards reduced survival compared to men [1]. Interestingly, two recent retrospective studies have suggested that women who took estrogen replacement therapy (ERT) after their angioplasty had less restenosis [2] and better long-term outcome [3] than their nonestrogen taking counterparts.

If this beneficial effect is confirmed, this therapy could have a major positive impact on the treatment of cardiovascular disease in older women. It is the goal of this presentation to investigate the potential of ERT to prevent restenosis in older women. In addition, if the beneficial effect of estrogen after angioplasty can be confirmed, the continuing development of selective estrogen receptor modulators, capable of mimicking estrogen's actions in discrete tissue sites [4-9], could ultimately avail "estrogen-like" treatment to patients of both sexes.

Based on previous epidemiological and experimental data, our laboratory embarked on a series of experiments to determine the potential of ERT for prevention of restenosis and to identify potential mechanisms for this effect. In broad terms these efforts included attempts 1) to begin to establish the mechanisms of estrogen's direct effects on vascular cells and 2) to explore potential mechanisms responsible for an antirestenosis effect of estrogen. These studies are detailed in the following section.

Demonstration of Decreased Estrogen Receptor Expression in Atherosclerotic Coronary Arteries (see full manuscript [10])

A direct genomic effect of estrogen on arterial biology requires the expression of estrogen receptor (ER) by cells of the vessel wall. We performed an investigation to test the hypothesis that premature atherosclerosis in female patients may therefore be mediated by a failure of target cells in the vessel wall to adequately express ER, therefore abrogating the possibility of atheroprotection despite adequate levels of circulating estrogen. To test this hypothesis, we examined coronary artery specimens from female patients. Evaluation of

R. Paoletti et al. (eds.), Women's Health and Menopause, 59–66.

ER expression in these coronary artery specimens was performed using a monoclonal antibody to the estrogen receptor. Of 21 normal arteries studied, 15 of these showed evidence of ER expression, while in 19 atherosclerotic arteries, 13 showed no evidence of ER expression when assayed immunohistochemically. Contingency table analysis revealed that the differences between these groups were statistically significant with $p = .0117$.

This data provides further evidence suggesting a direct action of estrogen in protecting arteries from atherosclerosis, demonstrating a strong association between ER expression and the absence of coronary atherosclerosis in premenopausal women. The absence of this relationship in postmenopausal women could be explained by the estrogen-deficient state of the subjects, making the presence of the receptor alone insufficient to exert an atheroprotective effect. More recently Rubanyi and colleagues [11] have shown in an animal model that altered expression of ERα can have important functional consequences, thus corroborating our initial hypothesis and emphasizing the importance of direct actions of estrogen on the vessel wall.

In order to completely characterize the expression of ER in vascular smooth muscle cells (VSMC) we performed additional studies in cultured human cells to quantify receptor expression and binding affinity, and to establish the functional integrity of this receptor. These data are now published [10] and corroborated by others [12].

These data represented the first verification of high affinity estrogen receptors in human VSMC. We then proceeded to evaluate the functional integrity of this receptor by establishing its ability to act in a hormone responsive fashion, and bind to the estrogen responsive element which is found upstream of a number of estrogen responsive genes.

Steroid responsive elements are inducible regulatory DNA sequences which interact with their receptors and modulate transcription of target genes. The "estrogen responsive element" (ERE) has been shown to bind the ER only when the latter is induced by its ligand. The interaction of hormone bound ER with the ERE has been shown to alter the transcriptional efficiency of ERE containing genes. Efficient binding of ligand-activated ER to the ERE is therefore a useful measure of the functional integrity of putative receptors. We examined the ability of ERs from hVSMC to bind ERE using gel mobility assays. Specific and ligand-dependent binding of hVSMC ER to its cognate response element is demonstrated. Only the combination of extract from E_2 exposed cells and labeled ERE probe produces the "shifted" band corresponding to the DNA-receptor complex. Supershifting using antibody to ER confirmed the presence of receptor in the complexes.

Confirmation of the presence and functional integrity of ER in hVSMC suggests that a direct effect of E_2 on human arteries is a possibility.

In-Stent Restenosis Results from VSMC Hyperplasia (see full manuscript [13])

When restenosis occurs within stents, these lesions have been considered to result from intimal proliferation. To examine this phenomenon we studied a series of 10 specimens retrieved by directional atherectomy from patients with in-stent restenosis of noncoronary sites. [Clinical experience has suggested that directional atherectomy of stented sites in coronary arteries may result in target stent disruption (14,15).] Each of the 10 specimens contained extensive foci of hypercellularity composed predominantly of SMC. Evidence of ongoing proliferative activity was documented in all appropriately preserved specimens. Because one [16] of three previous studies of native vessel restenosis lesions [16-18] was interpreted to show a paucity of VSMC proliferation after immunostaining for proliferating cell nuclear antigen (PCNA), we used polyclonal antibodies to two additional cell cycle regulatory proteins (Cyclin-E and cdk-2) to confirm the extent of SMC proliferation observed for in-stent restenosis. Double immunostaining for all three cell cycle proteins and alpha-actin confirmed that SMCs accounted for most of the proliferative activity.

Estrogen Inhibits Proliferation of Human Vascular Smooth Muscle Cells [19]

Having established the expression of a functional estrogen receptor in human VSMC and documented significant degrees of human VSMC proliferation in restenosis lesions from stented arteries, our next goal was to provide evidence that the antiproliferative effect of estrogen on VSMCs observed in other species was also present in human cells. Accordingly, a series of experiments was performed to determine whether estrogen could inhibit human VSMC proliferation [19].

Human VSMC in their second passage were subcultured and grown in standard medium for 48 hours. After 48 hours, the medium was supplemented with serum containing varying concentrations of estradiol: 10% pooled serum from human male donors (estrogen concentration = 40 pg/ml $\approx 1.5 \times 10^{-10}$ M), 10% pooled serum from human female donors (estrogen concentration = 270 pg/ml $\approx 10^{-9}$ M, progesterone concentration 18.5 ng/ml $\approx 5 \times 10^{-8}$ M), 10% pooled serum from male donors with the addition of estradiol in concentrations ranging from 10^{-10} M to 10^{-7} M. The strategy for utilizing human serum in these studies was to evaluate the effect of estrogen upon hVSMC behavior in a growth medium which contains all of the elements of human serum. We felt that in a study evaluating the effect of estrogen on hVSMC, the possibility that co-factors, necessary for estrogen effect, may exist in human serum that do not exist in other nonhuman serum supplements for cell culture was a crucial consideration.

Male serum, with the lowest estrogen concentration, induced the greatest proliferative activity of cultured hVSMC. With the addition of E_2 the proliferative activity of the cells was significantly diminished in a dose-dependent fashion.

Further studies, using the pure antagonist ICI 182780 (generously supplied by Zeneca Pharmaceuticals), were performed to determine whether the antiproliferative effect of E_2 on hVSMC is mediated by ER.

In studies using human serum or dextran charcoal stripped FBS, the estrogen receptor antagonist (ERA) mitigated the antiproliferative effect of E_2 when it was added to the culture medium before the E_2 containing culture medium.

Thus far, our studies have revealed that: 1) human VSMC express a functional ER; 2) ER expression is decreased in atherosclerotic coronary arteries from premenopausal females; 3) VSMC proliferation is a prominent component of restenosis in stents; and 4) estrogen inhibits VSMC proliferation via the ER expressed by these cells.

We were next interested in evaluating the effect of estrogen on endothelial cells, since our laboratory and others [20,21] have identified ER expression in human endothelial cells.

Estradiol Inhibits Endothelial Cell Apoptosis (See full manuscript [22])

Endothelial injury is a postulated mechanism for the initiation of atherogenesis [23]. We therefore hypothesized that protection of endothelial cells (EC) from toxic insults could represent an alternative mechanism for the inhibition of atherogenesis. Accordingly, we investigated the potential of E_2 to act as a protective agent for EC, inhibiting programmed cell death (apoptosis). Recent data [24-31] suggests an important functional role for tumor necrosis factor alpha (TNF) in human atherosclerosis and restenosis and in experimental atherosclerosis. In addition, TNF is known to induce apoptosis (Ap) in EC. We therefore examined the potential for estrogen to inhibit Ap induced by TNF [22].

Apoptosis was induced in human umbilical vein EC (HUVEC) in culture by the addition of TNF(40 ng/ml) to the culture medium. Ap was then evaluated by several complimentary assays. Immunofluorescence microscopy was performed after labeling HUVEC using the TUNEL method to identify individual apoptotic cells. Cells were also simultaneously examined for immunologic evidence of interleukin-1β converting enzyme (ICE) protein expression. [The ICE gene has been shown to mediate Ap induced by TNF [32]. ICE is also known to be expressed in human atheroma [33]. TUNEL labeling identifies individual apoptotic EC after exposure to TNF. These cells also demonstrate morphologic features of Ap and expression of ICE.

Phase contrast microscopy was performed and, qualitatively, demonstrates inhibition of TNF induced Ap by E_2 treatment.

Fluorescence activated cell sorting (FACS) was also performed to detect Ap by assaying for the characteristic inter-nulceosomal DNA fragmentation which occurs as a result of Ap. Ap is clearly induced in EC exposed to TNF and markedly inhibited by simultaneous exposure of EC to E_2. Cell counting verifies these results.

Viability assays, which measure mitochondrial function, are capable of detecting cell death earlier than other techniques [34]. Using this technique TNF was shown to induce Ap in 35% of HUVEC after 24 hours. Once again, simultaneous treatment with E_2 resulted in a dose-dependent increase in EC survival. The reduction in cell death was most marked at an E_2 concentration of 10^{-9} M. The specific estrogen receptor antagonist ICI 182780 completely abrogated the protective effect of E_2.

These studies define a survival function for E_2 in endothelial cells, protecting them from programmed cell death. We next questioned whether this *in vitro* protective effect of estrogen also resulted in an *in vivo* change of endothelial cell biology. Specifically, we were curious to know if estrogen, like other endothelial cell mitogens [35] was capable of inducing more rapid healing of the endothelial "wound" induced by balloon angioplasty. In the light of the above findings, this question is of particular interest in the setting of balloon angioplasty since balloon injury of the arterial wall has been shown to increase the local expression of TNF [27,28,30].

Estrogen Replacement Therapy Accelerates Functional Endothelial Recovery After Angioplasty [36]

The ability of estrogen treatment to inhibit neointimal formation after injury has been demonstrated in a variety of animal models [37-41]. This effect has generally been ascribed to the inhibition of myointimal proliferation. However, estrogen has also been shown to exert direct angiogenic effects on EC [42] which may be mediated by the estrogen receptor expressed by EC [20,21]. Recent studies have suggested that recovery of injured endothelium may inhibit neointimal formation [35]. Accordingly, we performed a series of studies to examine the hypothesis that estrogen inhibits neointimal formation by accelerating endothelial recovery.

A one-centimeter segment of the left carotid artery of 60 Sprague-Dawley rats was denuded with a 2Fr Fogarty balloon catheter. The treatment group consisted of ovariectomized females receiving estrogen replacement in the form of a subcutaneously pellet designed to deliver 1.5 mg or 5.0 mg of E_2 over 30 days. The control group consisted of ovariectomized rats receiving a placebo pellet. Pellets were implanted one week prior to balloon injury. Animals were sacrificed one and two weeks post-injury. Re-endothelialization (ReEndo) was assessed by administering Evans blue dye 30 minutes prior to animal sacrifice. After perfusion fixation with methanol the injured arterial segments were opened longitudinally and the total surface area, non-ReEndo area (blue area) and ReEndo area (white area) were measured with a computerized planimetric analysis system. Similarly, the total neointimal area and medial area were measured on 5 serial elastic-trichrome stained longitudinal sections from each artery. At both the one- and two-week time points the E_2 treated arteries show a significantly greater area of ReEndo.

Neointimal formation was shown to be inhibited as in multiple previous studies, with the additional finding of a dose response effect. Similar findings have also been reported in a mouse carotid injury model [40].

Statistical analysis reveals a dose response effect on the absolute area re-endothelialized, the percentage of injured area re-endothelialized and neointimal thickening at both the one- and two-week time points.

Perhaps most importantly, these studies reveal that the recovery of endothelial function is also accelerated in arteries from animals treated with ERT compared to arteries from control animals. These results have been subsequently confirmed [43].

These findings are also consistent with the recent demonstration that agents, such as vascular endothelial growth factor(VEGF), which directly accelerate endothelial recovery are also capable of attenuating neointimal proliferation after balloon injury [35], and stenting [44].

Estradiol Increases Expression of VEGF by Vascular Smooth Muscle Cells

Interestingly, our data and others reveal that the expression of VEGF is increased in VSMC exposed to estrogen [36,45] suggesting an additional potential mechanism for the observed impact of estrogen treatment on endothelial recovery.

In summary, our preliminary *in vitro* and *in vivo* data reveal multiple estrogen receptor-mediated effects by which estrogen could favorably effect the response to balloon injury in arteries. Our studies reveal that estrogen is capable of inhibiting VSMC proliferation and inhibiting endothelial cell apoptosis. *In vivo*, these effects are associated with accelerated re-endothelialization of injured artery segments and attenuation of neointimal lesion formation.

References

1. Bell MR, Grill DE, Garratt KN, Berger PB, Gersh BJ, Holmes DR. Long-term outcome of women compared with men after successful coronary angioplasty. Circulation 1995;91: 2876-80.
2. O'Brien JE, Peterson ED, Keeler GP, et al. Relation between estrogen replacement therapy and restenosis after percutaneous coronary interventions. JACCb 1996;28:1111-18.
3. O'Keefe JH, Kim SC, Hall RR, Cochran VC, Lawhorn SL, McCallister BD. Estrogen replacement therapy after coronary angioplasty in women. J Am Coll Cardiol 1997;29:1-5.
4. Williams JK, Wagner JD, Li Z, Golden DL, Adams MR. Tamoxifen inhibits arterial accumulation of LDL degradation products and progression of coronary artery atherosclerosis in monkeys. Arterioscler Thromb Vasc Biol 1997;17:403-8.
5. Bjarnason NH, Haarbo J, Byjalsen I, Kauffman RF, Christiansen, C. Raloxifen inhibits aortic accumulation of cholesterol in ovariectomized, cholesterol-fed rabbits. Circulation 1997;96:1964-69.
6. Holm P, Shalmi M, Korsgaard N, Guldhammer B, Skouby SV, Stender S. A partial estrogen receptor agonist with strong antiatherogenic properties without noticeable effect on reproductive tissue in cholesterol-fed female and male rabbits. Arterioscler Thromb Vasc Biol 1997;17:2264-72.
7. Wantanabe T, Inoue S, Ogawa S, et al. Agonistic effect of tamoxifen is dependent on cell type, ERE-promoter context, and estrogen receptor subtype: Functional difference between estrogen receptors a and b. Biochem & Biophys Res Comm 1997;236:140-45.
8. Paech K, Webb P, Kuiper GGJM, et al. Differential ligand activation of estrogen receptors ERa and ERb at AP1 sites. Science 1997;277:1508-10.
9. Delmas PD, Bjarnason NH, Mitlak BH, et al. Effects of raloxifene on bone mineral density, serum cholesterol concentrations, and uterine endometrium in postmenopausal women. NEJM 1997;337:1641-47.
10. Losordo DW, Kearney M, Kim EA, Jekanowski J, Isner JM. Variable expression of the estrogen receptor in normal and atherosclerotic coronary arteries of premenopausal women. Circulation 1994;89:1501-10.
11. Rubayni GM, Freay AD, Kauser K, et al. Vascular estrogen receptors and endothelium-derived nitric oxide production in the mouse aorta. J Clin Invest 1997;99:2429-37.
12. Karas RH, Patterson BL, Mendelsohn ME. Human vascular smooth muscle cells contain

functional estrogen receptor. Circulation 1994;89:1943-50.
13. Kearney M, Pieczek A, Haley L, et al. Histopathology of in-stent restenosis in patients with peripheral vascular disease. Circulation 1997;95:1998-2002.
14. Mecander PJ, Roubin GS, Agrawal SK, Cannon AD, Dean LS, Baxley WA. Balloon angioplasty for treatment of in-stent restenosis: Feasibility, safety and efficacy. Cathet Cardiovasc Diagn 1994;32:125-31.
15. Bowerman RE, Pinkerton CA, Kirk B, Waller BF. Disruption of a coronary stent during atherectomy for restenosis. Cathet Cardiovasc Design 1993;71:364-66.
16. O'Brien ER, Alpers CE, Stewart DK, et al. Proliferation in primary and restenotic coronary atherectomy tissue: Implications for antiproliferative therapy. Circ Res 1993;3:223-31.
17. Pickering JG, Bacha P, Weir L, Jekanowski J, Nichols JC, Isner JM. Prevention of smooth muscle cell outgrowth from human atherosclerotic plaque by a recombinant fusion protein specific for epidermal growth factor receptor. J Clin Invest 1993;91:724-29.
18. Rekhter M, Ferguson SN, Gordon D. Cell proliferation in human arteriovenous fistulas used for hemodialysis. Arteriosclerosis and Thrombosis 1993;13:609-17.
19. Spyridopoulos I, Losordo DW. Cardiovascular estrogen receptors regulate vascular smooth muscle cell proliferation. In: Forte TM, editor. Hormonal, metabolic, and cellular influences on cardiovascular disease in women. Armonk, NY: Futura Publishing Company, Inc., 1997: 71-97.
20. Kim-Schulze S, McGowan KA, Hubchak SC, et al. Expression of an estrogen receptor by human coronary artery and umbilical vein endothelial cells. Circulation 1996;94:1402-7.
21. Venkov CD, Rankin AB, Vaughan DE. Identification of authentic estrogen receptor in cultured endothelial cells. Circulation 1996;94:727-33.
22. Spyridopoulos I, Sullivan A, Kearney M, Isner JM, Losordo DW. Estrogen receptor mediated inhibition of human endothelial cell apoptosis: Estradiol as a survival factor. Circulation 1997; 95:1505-14.
23. Ross R. The pathogenesis of atherosclerosis: A perspective for the 1990s. Nature 1993; 362:801-9.
24. Warner SJC, Libby P. Human vascular smooth muscle cells. Target for and source of tumor necrosis factor. J Immunol 1989;142:100-109.
25. Barath P, Fishbein CM, Cao J, Berenson J, Helfant RH, Forrester JS. Detection and localization of tumor necrosis factor in human atheroma. Am J Cardiol 1990a;65:297-302.
26. Barath P, Fishbein MC, Cao J, Berenson J, Helfant RH, Forrester JS. Tumor necrosis factor gene expression in human vascular intimal smooth muscle cells detected by in situ hybridization. Am J Pathol 1990b;137:503-9.
27. Clausell N, Correa de Lima V, Molossi S, et al. Expression of tumor necrosis factor and accumulation of fibronectin in coronary artery restenosis lesions retrieved by atherectomy. Br Heart J 1995;73:534-39.
28. Hancock WW, Adams DH, Wyner LR, Sayegh MH, Karnovsky MJ. (1994). CD4+ mononuclear cells induce cytokine expression, vascular smooth muscle cell proliferation, and arterial occlusion after endothelial injury. Am J Pathol 1994;145:1008-14.
29. Tanaka H, Sukhova G, Schwartz D, Libby P. Proliferating arterial smooth muscle cells after balloon injury express TNF-a but not interleukin-1 or basic fibroblast growth factor. Arterioscler Thromb Vasc Biol 1996;16:12-18.
30. Tanaka H, Swanson SJ, Sukhova G, Schoen FJ, Libby P. Smooth muscle cells of the coronary arterial tunica media express tumor necrosis factor-a and proliferate during acute

rejection of rabbit cardiac allografts. Am J Pathol 1995;147:617-26.

31. Clausell N, Molossi S, Sett S, Rabinovitch M. In vivo blockade of tumor necrosis factor-a in cholesterol-fed rabbits after cardiac transplant inhibits acute coronary artery neointimal formation. Circulation 1994;89:2768-79.
32. Duan H, Chinnaiyan AM, Hudson PL, Wing JP, He WW, Dixit VM. ICE-LAP3, a novel mammalian homologue of the caenorhabditis elegans cell death protein ced-3 is activated during fas- and tumor necrosis factor-induced apoptosis. J Biol Chem 1996;271:1621-25.
33. Geng YJ, Libby P. Evidence for apoptosis in advanced human atheroma. Am J Pathol 1995;147:251-66.
38. Chen SJ, Li H, Durand J, Oparil S, Chen YF. Estrogen reduces myointimal proliferation after balloon injury of rat carotid artery. Circulation 1996;93:577-84.
39. Hanke H, Hanke S, Finking, G, et al. Different effects of estrogen and progesterone on experimental atherosclerosis in female versus male rabbits. Circulation 1996;94:175-81.
34. Tepper CG, Jayadev S, Liu B, et al. Role of ceramide as an endogenous mediator of Fas-induced cytotoxicity. Proc Natl Acad Sci USA 1995;92:8443-47.
35. Asahara T, Bauters C, Pastore C, et al. Local delivery of vascular endothelial growth factor accelerates reendothelialization and attenuates intimal hyperplasia in balloon-injured rat carotid artery. Circulation 1995;91:2793-2801.
36. Krasinski K, Spyridopoulos I, Asahara T, van der Zee R, Isner JM, Losordo DW. Estradiol accelerates functional endothelial recovery after arterial injury. Circulation 1997;95:1768-72.
37. Wolinsky H. Effect of estrogen and progestogen treatment on the response of the aorta of male rats to hypertension. Circ Res 1972;30:341-49.
40. Sullivan TR, Karas RH, Aronovitz M, et al. Estrogen inhibits the response-to-injury in a mouse carotid artery model. J Clin Invest 1995;95:2482-88.
41. Foegh ML, Asotra S, Howell MH, Ramwell PW. Estradiol inhibition of arterial neointimal hyperplasia after balloon injury. J Vasc Surgery 1994;19:722-26.
42. Morales DE, McGowan KA, Grant DS, et al. Estrogen promotes angiogenic activity in human umbilical vein endothelial cells in vitro and in a murine model. Circulation 1995; 91:755-63.
43. White CR, Shelton J, Chen SJ, et al. Estrogen restores endothelial cell function in an experimental model of vascular injury. Circulation 1997;96:1624-30.
44. VanBelle E, Fermin OT, Chen D, Maillard L, Kearney M, Isner JM. Passivation of metallic stents after arterial gene transfer of phVEGF165 inhibits thrombus formation and intimal thickening. J Am Coll Cardiol 1997;29:1371-79.
45. Karas RH, Bieber HE, Baur WE, Mendelsohn ME. Estrogen enhances vascular endothelial growth factor (VEGF) gene expression in human vascular smooth muscle cells. Circulation 1996;94:I-595.(Abstract).

NEWER PROGESTOGENS

Göran Samsioe

Introduction

Sweats and hot flushes, the most abundant symptoms of the menopause, are effectively treated by estrogens, a fact that has been known for decades. However, estrogen treatment was fairly uncommon prior to 1960 when it started to increase, especially in the United States. Estrogens were almost always given as a monotherapy. When a sufficient number of women had used this estrogen monotherapy for several years, observational studies suggested in 1975 that there might be association between estrogen monotherapy and the occurrence of endometrial cancer. Several experimental and epidemiological studies have later confirmed this relationship. At this time it was also found that the estrogen-associated increase in endometrial cancer commonly occurred via endometrial hyperplasia. Several studies in the late seventies indicated that progestogen comedication would almost abolish the risk of hyperplasia of the endometrium when given in sufficient dosage over a period of at least ten days. All commercially available progestogens seem to confer endometrial protection against endometrial hyperplasia. This concept was also believed to protect from the estrogen-induced endometrial cancer, a fact that is confirmed by several studies today, although endometrial cancer may still occur also in women using combined hormone replacement therapy (HRT) as all cancers are not driven by the hormonal milieu. Thus doses of progestogens often used in sequential therapy such as medroxyprogesterone acetate 5-10 mg, norethisterone acetate 1 mg, levonorgestrel 75 μg, desogestrel 150 μg, dydrogesterone 10 mg, cyproterone acetate 1 mg, and natural progesterone 200-300 mg seem not to interfere with the estrogen-induced overall beneficial lipid changes. The corresponding doses in continuous combined therapy are medroxyprogesterone acetate 2.5 mg when combined with conjugated equine estrogens 0.625 mg and norethisterone acetate 1 mg when combined with estradiol 2 mg, possibly 0.25-0.5 mg, when combined with estradiol 1 mg. One large observational study provides evidence to include sequential levonogestrel 250 μg. One should keep in mind that most of the studies have been conducted with either conjugated equine estrogens 0.625 mg or estradiol (valerate) 2 mg or to a much lesser extent transdermal estradiol 50-100 μg. Other doses of these compounds might produce quite different effects.

Hence, the late seventies saw the introduction of progestogen comedication to minimize the risk of endometrial hyperplasia and subsequent development into endometrial

R. Paoletti et al. (eds.), Women's Health and Menopause, 67–74.

cancer.

The Continuous Combined Regimen

In 1982 a new concept was introduced in that a continuous medication with an estrogen and a progestogens seemed to be able to induce an atrophy of the endometrium resulting clinically in a bleed-free regimen [1]. It is now apparent that women prefer bleed-free regimens especially with advancing age. The progestogens available for comedication with estrogens in the mid-seventies were levonorgestrel, norethisterone, and medroxy-progesterone and, to some extent, the retrosteroid dydrogesterone. These progestogens were developed mainly for contraception and adjuvant treatment of gynecological disorders rather than to be the comedication with estrogens in the treatment of climacteric women. However, much effort has been spent to find new delivery systems, such as the progestogen-releasing IUD or implants, vaginal creams, patches, etc. in order to use these progestogens. Until very recently progestogens had not been primarily developed as a comedication with estrogens. Most progestogens are developed for other general purposes, mainly contraception but also for treatment of endometriosis, etc.

Metabolic Effects of Progestogens

Apart from the important effect on the endometrium, it also became apparent that progestogens have metabolic influences. This is of importance as long-term estrogen therapy seems to prevent cardiovascular disease, especially by reducing atherosclerosis, and to reduce osteoporotic fractures. In both of these metabolic conditions the progestogen comedication may well play a role.

Some Pharmacological Aspects of Progestogens

Natural progesterone has a poor bioavailability as an oral adjunct to estrogen replacement; Hence, we have few data on natural progesterone [2]. Another crucial point is how to compare different progestogens (see Table 1). Using biological tests it is quite simple to establish a potency ratio on the transformation of the endometrium for example. However, it has become increasingly clear that this potency ratio for one clinical important endpoint may not hold true for other significant metabolic effects. Furthermore, potency ratios may vary from one patient to another, a fact which may explain why some women can do well on norethisterone acetate and not on medroxyprogesterone acetate and vice versa. The pharmacokinetics and to some extent also pharmacodynamics of progestogens may vary. Some, like lynestrenol and desogestrel, are prodrugs and the main progestogenic activity is exerted by their metabolites norethisterone and 3-ketodesogestrel. Most progestogens are metabolized in steps via several intermediate compounds which in turn may show various degrees of progestogenic activity.

Another important aspect is the prevailing estrogen influence. There is no natural situation in which progestogens act on their own without estrogens. Progestogens are regarded to be antiestrogenic. This is due to the fact that progestogens downregulate estrogen receptors and much of the effects induced by progestogens could well be an estrogen antagonistic mode of action. In addition, by decreasing estrogen activity the major carrier protein sex hormone binding globulin levels (SHBG) could be reduced, which in turn increases androgen activity as free androgens are now increased and several progestogenic molecules compete for binding sites on the SHBG-molecules with the natural androgens. Hence, several progestogens are believed to have an androgenic effect, but in most cases this is a secondary effect mediated via SHBG. It is very difficult to explore a pure progestogenic effect as this is mostly a combination of antiestrogenic and androgenic actions. The metabolism of progestogens is to some extent different from natural estrogens. There is a metabolic capacity to cope with high doses of natural estradiol and progesterone, but several of the synthetic progestogens do not induce hepatic catabolism and hence may accumulate if the doses are too high.

Several papers indicated in the early 1980s that progestogens mainly are antiestrogenic. Progestogens may also display some androgenic properties on important metabolic parameters. The addition of certain progestogens to an estrogen monotherapy result in lipid metabolic effects such as a reduction of estrogen-induced increase of high density lipoprotein (HDL) cholesterol and triglycerides. However, low density lipoprotein (LDL) cholesterol is further reduced by the addition of most progestogens that are commonly used today. Especially in the 1980s HDL cholesterol was given a pivotal role in the assessment of the risk of cardiovascular disease. Much effort was made in trying to develop progestogens with minimal or no detrimental effect on HDL. The third generation progestogens in oral contraceptives were part of these concepts, such as desogestrel, gestodene, and norgestimate. The sequential administration of norethisterone acetate or medroxyprogesterone acetate seems to diminish the effect of estrogens on the vascular resistance of uterine arteries [3]. However, vaginal natural progesterone or nomegestrol apparently do not attenuate the estrogen action [4].

Another important aspect is the indication used for HRT. In the past climacteric symptoms was almost the sole indication. Such treatment may last for 1-2 years and often substantially less. Today osteoporosis prevention and treatment is becoming more a prevalent indication. This calls for an increasing awareness of long-term effects. Even if such data are available for progestogens used in oral contraceptives, little long-term data are available in women beyond 50 inclusive of efficacy, metabolic effects, and safety [5]. As far as bone metabolism is concerned, most progestogens seem to be neutral although some data on norethisterone suggest an additive effect [6], although it is unclear whether this possible effect is due to the androgenic properties of norethisterone or result from additional estrogenic activity from aromatization of norethisterone to ethinyl estradiol.

Another concept was the use of substances with clearly antiandrogenic properties such as cyproterone acetate. This drug had its first indication as an antiandrogenic drug in hypersexual males. However, it soon became evident that it also had progestogenic properties and an oral contraceptive and later an HRT combination with cyproterone acetate was developed. Mainly for historical reasons the introduction of newer progestogens has been slow, especially in the United States.

Table 1. Classification of Progestogens

A	Natural progesterone		
B	Synthetic derivatives of natural progesterone		
Structurally related to progesterone		Structurally related to testosterone	
Pregnane derivatives		Ethinylated	
Acetylated	Nonacetylated	Estrane	Gonane derivatives
Medroxyprogesterone acetate	Dydrogesterone	Norethisterone (acetate)	Levonorgestrel
Megestrol acetate		Norethynodrel	Desogestrel
Cyproterone acetate		Lynestrenol	Gestodene
Chlormadinone acetate		Ethynodiol diacetate	Norgestimate
Medrogestone		Nonethinylated	
		Dienogest	
Norpregnane derivatives			
Demegestone			
Promegestone			
Trimegestone			
Nomegestrol acetate			

Adapted from [11].

PMS-like symptoms are not uncommon and some progestogens may worsen or even induce premenstrual tension. From what has been discussed above it was quite clear that a rationale existed for an improvement of progestational agents. Mainly development has gone along two lines. Firstly, to introduce new delivery systems, especially the progestogen-releasing IUD by which progestogens exert their activity only close to the only organ where it has shown to be of clear-cut benefit, i.e. the uterine endometrium. Secondly, the development of new substances with other qualities more suitable for progestogen comedication, especially in the continuous combined regimens. The third line of development was to create a steroidal molecule combining estrogen and progestogen qualities. Indeed, such a molecule, tibolone, has been introduced onto the market.

Tibolone is a synthetic steroid with a structural similarity to norethisterone, danazol, and stanozolol which combines effects of a weak androgen, weak progestogen, and weak estrogen as shown in animal studies compared with ethinylestradiol and norethisterone acetate. As it does not appear to stimulate endometrial growth it seems to be an alternative in postmenopausal women with an intact uterus. It slightly increases LDL cholesterol in the presence of HDL cholesterol reduction [7] predominantly due to its impact on HDL_3. The effect on triglyceride seems to be dependent on the length of administration. In long-term studies triglycerides were unchanged [8]; in short-term studies, the levels decreased [9]. As with oral estradiol therapy tibolone decreases Lp(a) substantially. It is difficult to draw comparisons and firm conclusions as epidemiological data and larger comparative studies with this compound are lacking.

An additional concept is represented by the so-called selective estrogen receptor modulators (SERM) whereby undesirable effects on the endometrium can be avoided while maintaining the positive effects of estrogens such as fracture prevention and reduction of total and LDL cholesterol. One example is raloxifene which was recently introduced as a commercial product However, this concept will not be discussed further in this paper.

New Delivery Systems

One of the problems with natural progesterone is its poor oral bioavailability. This forms the basis for the manufacturing of 17-alpha hydroxy progesterones which display a much improved bioavailability when used as tablets. Modern microcystallization techniques have improved oral bioavalability of progesterone and progesterone tablets are available in some countries. However oral daily doses of about 100-300 mg must be used to induce secretory transformation of the endometrium which makes oral progesterone fairly expensive when considering its oral progestogenic effect [2]. For this reason progesterone delivery exists as a (vaginal) cream and as a progesterone-containing IUD. The constant release of progesterone from the IUD is an additional advantage as variations both inter- and intraindividual need not to be considered. The levonorgestrel-containing IUD can also be used. However the amount of levonorgestrel released from the commercially available is more suitable for contraception and does have metabolic implications exemplified here by a decrease in HDL cholesterol [10].

Few progestogens penetrate the skin. Apart from progesterone only norethisterone is as yet available in patches, soon to be followed by levonorgestrel. Furthermore, some metabolites of norgestimate with progestogenic activity seem also to penetrate the skin in sufficient amounts to be considered in HRT.

Newer Progestogenic Compounds in HRT

Norgestimate itself could also be used and data have been published on an oral bleed free combination of 2 mg estradiol + 90 μg norgestimate. Apart from a new progestogen this product also introduces a new concept in minimizing progestogenic effects in that norgestimate is given in a cyclophasic manner, i.e added to the estradiol for 3 days, then discontinued for 3 days, and then added in for 3 days, etc. [11,12].

Several new compounds are under development. Most of these have not been used much for HRT as yet some have only undergone minor clinical testing. However, some compounds may well be interesting also for HRT. Some data exist on medrogestone-containing HRT regimens. One report [13] looked at effects on bone mass and concludes that the progestogen does not influence the bone sparing effects of estrogen when bone mass was evaluated by ultrasound. Another paper [14] describes effects on cardiovascular risk markers and concludes that medrogestone is fairly neutral in this respect.

Nomegestrol is used mainly for contraception . Of particular interest in this context is a nomegestrol-releasing implant which could be used also in HRT. One study [15] reported favorable results and user satisfaction with this product. Nomegestrol has also been evaluated in clinical use in menopausal women. The study by Reginster et al. [16] reported on the improvement of quality of life in menopausal women using a nomegestrol-containing HRT regimen. An overall improvement of quality of life was found and little or no negative effects by this new progestogen.

Dienogest is an interesting substance which may be of value also in HRT. One particular feature is that it seems to inhibit hormone-dependant cancer cell lines in a different manner compared to medroxyprogesterone [17]. However this observation needs further confirmation. In clinical terms the bulk of data on dienogest is its function as a contraceptive steroid. A combined oral contraceptive (OC) with 30 μg ethinylestradiol and 2.0 mg dienogest was reported to be highly efficacious with few side effects [18]. In a double-blind, placebo-controlled study on hemostatic factors with this OC few differences compared to placebo were found and the clinical significance of these changes less clear [19].

Trimegestone has been under development for some time. Preclinical data show that it is a molecule with clear progestogenic properties. In clinical terms only a dose finding study on endometrial effects has been published [20]. The study concludes that 0.5 mg seems to be the most appropriate oral dose when used in combination with oral estradiol in menopausal women.

References

1. Mattsson L-Å, Cullberg G, Samsioe G. Evaluation of a continuous oestrogen-progestogen regimen for climacteric complaints. Maturitas 1982;4:95-102.
2. The Writing Group for the PEPI trial. Effects of estrogen or estrogen/progestin regimens on heart disease risk factors in postmenopausal women: The Postmenopausal Estrogen/Progestin Interventions (PEPI) trial. JAMA 1995;273:199-208.
3. Hillard TC, Bourne TH, Whitehead MI, Crayford TB, Collins WP, Campbell SC. Differential effects of transdermal estradiol and sequential progestogens on impedance to flow within the uterine arteries of postmenopausal women. Fertil Steril 1992;58:959-63.
4. de Ziegler D, Zartarian M, Micheletti MC, Linh CG, Schaerer E. Cyclical administration of nomegestrol acetate does not alter the vasodilatation effects of estradiol on the uterine

artery. Contracept Fertil Sex 1994;22:767-70.

5. Christiansen C, Riis J. Five years with continuous combined estrogen/progestogen therapy. Effects on calcium metabolism, lipids and bleeding pattern. Br J Obstet Gynaecol 1990;97: 1087-92.
6. Marslew U, Overgaard K, Riis BJ, Christiansen C. Two new combinations of estrogen and progestogen for prevention of postmenopausal bone loss. Long-term effects on bone, calcium and lipid metabolism, climacteric symptoms, and bleeding. Obstet Gynecol 1992; 79:202-10.
7. Crona N, Silfverstolpe G, Samsioe G. A double-blind cross-over study on the effects of Org OD 14 compared to estradiol valerate and placebo on lipid and carbohydrate metabolism in oophorectomized women. Acta Endocrinol Copenh 1983;102:451-55.
8. Hänggi W, Lippuner K, Riesen W, Jaeger P, Birkhäuser M. Long-term influence of different postmenopausal hormone replacement regimen on serum lipids and lipoprotein(a). A randomised study. Br J Obstet Gynaecol 1997;104:708-17.
10. Andersson K, Stadberg E, Mattson LÅ, Rybo G, Samsioe G. Intrauterine or oral administration of levonorgestrel in combination with estradiol to perimenopausal women - effects on lipid metabolism during 12 months of treatment. Int J Fertil 1996;41:476-83.
9. Netelenbos JC, Siregar-Emck MT, Schot LP, van Ginkel FC, Lips P, Leeuwenkamp OR. Short-term effects of Org OD 14 and 17β oestradiol on bone and lipid metabolism in early post-menopausal women. Maturitas 1991;13:137-42.
11. Dören M, Samsioe G. Metabolic effects of progestins in the replacement therapy of postmenopausal women. Israel J Obstet Gynecol 1998;9:36-46.
12. Casper RF, Chapdelaine A. Estrogen and interrupted progestin. A new concept for menopausal hormone replacement therapy. Am J Obstet Gynecol 1993;168:1188-94.
13. Cameron ST, Critchley HO, Glaiser AF, Williams AR, Baird DT. Continuous transdermal estrogen and interrupted progestogen as a novel bleed-free regimen of hormone replacement therapy for postmenopausal women. Br J obstet Gynaecol 1997;104:1184-90.
14. De Aloysio D, Rovati LC, Cadossi R, et al. Bone effects of transdermal hormone replacement therapy in postmenopausal women as evaluated by means of ultrasound: An open one-year prospective study. Maturitas 1997;27:61-68.
15. Van der Mooren MJ, Demacker PN, Blom HJ, et al. The effect of sequential three monthly hormone replacement therapy on several cardiovascular risk estimators in postmenopausal women. Fertil Steril 1997;67:67-73.
16. Coutinho EM, Athayde C, Barbosa I, et al. Results of a user satisfaction study carried out in women using Uniplant contraceptive implant. Contraception 1996;54:313-17.
17. Reginster JY, Zartarian M, Colau JC. Influence of nomegestrol acetate on the improvement of quality of life induced by estrogen therapy in menopausal women. Contracept Fertil Sex 1996;24:847-51.
18. Katsuki Y, Shibutani Y, Aoki D, Nozawa S. Dienogest, a novel steroid, overcomes hormone-dependent cancer in a different manner than progestins. Cancer 1997;79:169-76.
19. Spona J, Feichtinger W, Kindermann C, et al. Modulation of ovarian function by an oral contraceptive containing 30 microgram ethinyl estradiol in combination with 2.00 mg dienogest. Contraception 1997;56:185-91.
20. Spona J, Feichtinger W, Kindermann C, et al. Double-blind, randomized, placebo controlled study on the effects of the monophasic oral contraceptive containing 30 micrograms ethinylestradiol and 2.00 mg dienogest on the hemostatic system.

Contraception 1997;56:67-75.
21. Ross D, Gofree V, Cooper A, et al. Endometrial effects of three doses of trimegestone, a new orally active progestogen, on the postmenopausal endometrium. Maturitas 1997;28: 83-88.

Cardiovascular Disease: Risk Factors Related to Thrombosis

Giovanni de Gaetano, Maria Benedetta Donati, and Licia Iacoviello

Introduction

The risk of coronary heart disease (CHD) is consistently higher among men than women. Mortality rates among premenopausal women in developed countries are about 20-30% lower than rates among men of similar age. Hormonal factors play a critical role in protecting women from the risk associated with other genetic and/or environmental factors. Gender differences in mortality, indeed, diminish following natural or surgical menopause. Among others, the systems of coagulation and fibrinolysis are related to CHD development. Increased levels of fibrinogen and coagulation factor VII (F VII) have been all consistently and independently related to the risk of myocardial infarction (MI). Moreover, the plasma levels of these proteins can be modulated by sex hormones. In particular the levels of F VII are lower in premenopausal women than in males. They increase after menopause or under contraceptives, while conflicting effects have been reported after hormone replacement therapy (HRT).

Recently, genetic modulation of both fibrinogen and F VII has been documented. Polymorphisms in the genes codifying F VII or fibrinogen contribute to their variability in blood within their normal range of distribution. They are also related to the risk of CHD. Two polymorphic alleles of the F VII gene have been recently associated with a protective effect on the risk of MI. These polymorphisms are major determinants of F VII variability in humans. In particular, the alleles Q and H7 were both associated with low levels of F VII in healthy and diseased populations. A gender-dependent genetic regulation of F VII, probably related to sex hormones, has also been described. Similarly, polymorphisms in the fibrinogen gene have been related both to the risk of CHD and to plasma levels of fibrinogen. A gender-dependent association between polymorphisms and fibrinogen levels has also been shown. Gender-specific hormones, indeed, can upregulate the effect of genotype on fibrinogen levels.

Coronary Heart Disease and Gender

In Italy, the overall age-adjusted CHD incidence rates for women are about half those for men [1]. The incidence of the disease is age-dependent in both sexes; however, the effect of age in women is greater. The age distribution of in-hospital acute myocardial infarction

R. Paoletti et al. (eds.), Women's Health and Menopause, 75–82.

(AMI) from the GISSI-3 study shows that in patients over the age of 65 years the rates for women increase progressively over the rates for men of similar age [2]. Menopause is critical in determining this difference in the distribution of CHD in men and women, over the age of 65. The role of hormonal changes related to menopause is underlined, in addition to the reduction in gender mortality differences after menopause, by the dramatic increase in CHD risk after oophorectomy [3] and by the reduction in CHD risk in postmenopausal women treated with HRT [4].

Several factors (such as environmental or genetic) may affect the development of CHD in both sexes (Table 1). Glucose intolerance, cigarette smoking, sedentary life-style, elevated blood pressure, increased cholesterol, low-density lipoprotein or triglyceride levels, decreased high-density lipoprotein levels, increase in coagulation factors, and old age have been identified as relevant risk factors for CHD in men and women. On the other hand, in women estrogen deprivation after natural or surgical menopause is an additional severe risk factor [5]. Certain risk factors may also affect women differently from men; diabetes, hypertriglyceridemia, high levels of fibrinogen, and low levels of HDL appear to be more severe risk factors for women than for men [5], probably because they interact with hormonal factors in determining such a risk.

Table 1. Risk factors of CHD in men and women.

	Men	Women
Glucose intolerance	+	++
Cigarette smoking	+	+
Increased blood pressure	+	+
Increased cholesterol	++	++
Increased LDL	+	+
Low HDL	+	++
Increased triglycerides	+	++
Sedentary lifestyle	+	+
Age	+	++
Estrogen deprivation	–	+++

Coagulation and Risk of CHD

Among others, blood coagulation factors can be relevant in mediating the effect of sex hormones on CHD development. Indeed, their levels have been consistently related to the risk of CHD and can be modulated by natural sex hormones.

The coagulation system plays a pivotal role in thrombus formation, being a "cascade" of events that leads, through amplification of enzymatic reactions, to the formation of thrombin. The "cascade" is started from the prompt binding of circulating F VII to tissue factor locally made available by damaged endothelial cells and/or infiltrating macrophages, i.e. at the level of an atheromatous plaque. The complex F VII-tissue factor activates factor X to factor Xa, that continues the cascade in the presence of factor V by activating prothrombin into thrombin, the specific enzyme that converts soluble fibrinogen into insoluble fibrin. An increased expression of tissue factor within atheromatous plaques was reported in patients with coronary artery disease [6] and one molecule of the F VII-tissue factor complex promotes the formation of several billions of fibrin molecules. The yield of the coagulation cascade is exceptionally high and could justify how slight changes in its components may result in significant influence on thrombus formation. In the last decade, several epidemiological studies have shown that enhanced F VII activity represents a risk factor for CHD [7-9].

Coagulation F VII and the Risk of CHD in Women

F VII levels, measured as antigen, clotting activity, or activated F VII, are influenced by gender [10-12]. In particular, they are lower in premenopausal women than in males, commensurate with the higher risk of CHD in the latter. F VII levels increase with age in both sexes, although the increase is greater in women, so that after 45 years of age, women show the highest levels. These changes are dependent on the hormonal status of women; premenopausal women who did not receive HRT showed F VII levels higher than premenopausal women. The effect of HRT on F VII seems to be different from that of natural estrogens. Although Scarabin et al. reported a decrease in F VII levels in postmenopausal women taking percutaneous estrogens in respect to those not receiving this treatment [12], several studies showed an increase [13]. The effect, in any case, seems to be dependent on estrogen concentrations, the route of administration, and the content of progestogens.

Recently, it has been shown that the levels of F VII, as those of many other factors related to cardiovascular disease, can be genetically determined. In particular, common variations (called polymorphisms) in the genes codifying such factors can contribute to their variability in blood within their normal range of distribution. In other words, to belong to the higher, intermediate, or lower part of the normal distribution of F VII levels, depends, at least in part, on the genotype for a specific locus at the F VII gene. Three polymorphisms have been described in different portions of the FVII gene, all associated with the levels of FVII [14-15]: a common polymorphism in the exon 8 of FVII gene R353Q, leading to a substitution of the arginine residue at position 353 by a glutamine, a decanucleotide insertion at position -323 (-323 0/10 bp) in the promoter and a tandem repeat unit polymorphism in the hypervariable region 4 (HVR 4) of the intron 7 of FVII gene.

The impact of F VII polymorphisms on F VII levels (both activity and antigen) is different in male and female subjects [16, Figure 1].

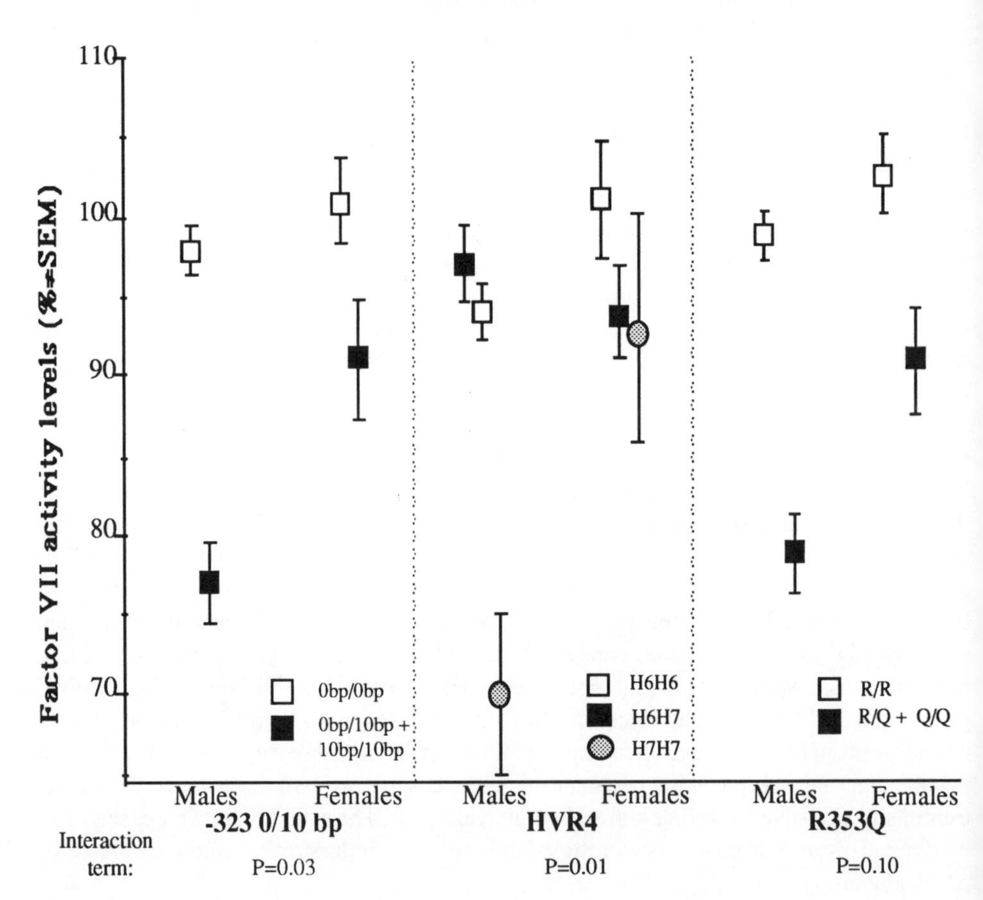

Figure 1. Effect of F VII gene polymorphisms on F VII clotting activity according to gender.

In males, all the three polymorphisms were associated with F VII levels. On the contrary, in females only the R353Q polymorphism was strongly associated with F VII levels, while the HVR4 polymorphism was not and the effect of the -323 0/10bp polymorphism was weaker. These findings suggest that hormones or other gender-specific factors could be important in the phenotypic expression of these genetic variants, by interacting with regulatory elements of the gene. F VII promoter contains hormone responsive elements that can upregulate the synthesis of F VII in females. These elements are close to two polymorphisms recently described [17], found to be in complete linkage disequilibrium between them and with -323 0/10 bp and significantly associated with FVII levels. Experimental studies will be required to clarify their relevance in the regulation of F VII levels by different genotypes in males and females.

The association between F VII levels and menopause is also dependent on F VII genotypes. Meilahn et al. showed that the increase in F VII clotting activity after menopause or HRT was present only in women with the RR F VII genotype, but not in those carrying the Q allele [18].

It has been recently shown that the alleles Q and H7 of R353Q and HVR4 polymorphisms of F VII gene had a protective effect on the risk of myocardial infarction and were associated with relatively low levels of F VII [19]. The presence of one of these alleles reduced about 50 % the risk of myocardial infarction; moreover, the risk was reduced 4 and 11 times, respectively, in subjects homozygous for the H7 and the Q allele, and more than 16 times in double homozygotes.

The Thrombosis Prevention Trial, conducted in UK, by the Medical Research Council's General Practice Research Framework, has lately reported that low-intensity oral anticoagulant therapy (average 4.1 mg warfarin daily) confers protection against ischemic heart disease (IHD), reducing mortality in high-risk men [20]. Low-dose aspirin was also effective, but only for nonfatal events, suggesting that the two treatments protected different high-risk subjects. The plasma levels of F VII obtained with low-dose warfarin were in the same low-normal range (60-80%) as those associated with the "protective" genotypes.

Few studies have investigated the effect of antithrombotic therapies in women. There is some evidence that aspirin may protect women from MI as well as men. The identification of subjects carrying the "protective" F VII genotypes could help define subgroups of individuals who might differentially benefit not only from different antithrombotic treatments, but also from HRT.

Fibrinogen and the Risk of CHD in Women

Epidemiological studies, reviewed in a meta-analysis [21], have shown that high plasma fibrinogen levels are a predictor of primary and secondary ischemic vascular events.

There are many pieces of evidence that make plausible a causal contribution of fibrinogen to cardiovascular disease. Indeed, fibrinogen plays a central role in determining the process of atherothrombosis, by promoting thrombosis in several ways [22]. The possible mechanisms include: 1) a relative increase in high molecular weight fibrinogen content; 2) increase in fibrin(ogen) deposition in atherosclerotic plaques, 3) increase in plasma viscosity; 4) increase in platelet aggregation; and 5) decrease in clot lysability.

Moreover, fibrinogen can be considered as the common pathway through which many of the established risk factors may act in promoting arterial disease [22]. Cigarette smoking and infections are the most important lifestyle correlates of fibrinogen. Moreover, fibrinogen levels increase with age, hypertension, diabetes, obesity, and menopause.

Similar to F VII, fibrinogen levels vary according to sex and menopausal status [23-

24]. They are lower in premenopausal women, as compared to men of the same age, and increase after menopause. HRT, however, may reverse the effect of menopause on fibrinogen levels.

Genetic factors seem to be equally important as environmental factors in determining fibrinogen levels in blood.

DNA variations of the beta fibrinogen gene have been reported to influence the levels of fibrinogen in several studies. Subjects with the rarest genotype had fibrinogen levels 15-20% higher than those with more common genotypes, while heterozygotes showed intermediate levels. However, the effect of the rare allele was different in women and men [24-25]. The overall effect of the A allele of the -455 G/A polymorphism of b-fibrinogen gene was additive in men, while it appeared to be dominant in women, with no additional effect on fibrinogen levels in the homozygotes. After stratification for menopausal status and HRT, the dominant effect of the A-allele in women was evident only in postmenopausal women not taking hormones. In both premenopausal and postmenopausal women treated with HRT, the effect of the A-allele was additive as observed in men [25]. These observations suggest that hormones or other gender-specific factors modulate the increase in fibrinogen associated with fibrinogen genotypes, by either downregulating hepatic fibrinogen synthesis or upregulating fibrinogen clearance.

The magnitude of the difference between genotypes was comparable to that found in epidemiological studies between subjects at high and low risk and makes conceivable a role of such polymorphisms in causing cardiovascular disease.

Some studies have associated fibrinogen polymorphisms with the occurrence of peripheral artery disease, the severity of coronary artery disease, and the occurrence of familial myocardial infarction [25-27]. However, whether gender differences are important in determining the effect of fibrinogen polymorphisms on the risk of CHD remains to be established.

Acknowledgements

This work has been partially supported by the Italian National Research Council (CNR, Rome, Italy) Progetto Strategico "Infarto del Miocardio" CTR n.970054674.

References

1. La Vecchia C, Levi F, Lucchini F, Negri E. Trends in mortality from major diseases in Europe, 1980-1992. Eur J Epidemiol 1998;14:1-8.
2. Bobbio M, Bergerone S, Maggioni AP, et al. Administration of thrombolytic therapy to 17,944 patients with myocardial infarction: The GISSI-3 database. Am Heart J 1998;135:443-48.
3. Godsland IF, Wynn V, Crook D, Miller NE. Sex, plasma lipoproteins and atherosclerosis: Prevailing assumptions and outstanding questions. Am Heart J 1987;114:1467-503.
4. Chae CU, Ridker PM, Manson JE. Postmenopausal replacement therapy and cardiovascular disease. Thromb Haemost 1997;78:770-80.

5. Eaker ED, Castelli W. Coronary heart disease and its risk factors among women in the Framingham Study. In: Eaker ED, Packard B, Wenger NK, Clarkson TB, Tyroler HA, editors. Coronary heart disease in women. Bethesda, Maryland: National Heart, Lung and Blood Institute, National Institutes of Health, 1987.
6. Ardissino D, Merlini PA, Ariens R, Coppola R, Bramucci E, Mannucci PM. Tissue-factor antigen and activity in human coronary atherosclerotic plaques. Lancet 1997; 349:769-71.
7. Meade TW, Brozovic M, Chakrabarti RR, et al. Haemostatic function and ischemic heart disease: Principal results of the Northwick Park Heart Study. Lancet 1986;2:533-37.
8. Heinrich J, Balleisen L, Schulte H, Assmann G, van de Loo J. Fibrinogen and factor VII in the prediction of coronary risk. Results from the PROCAM study in healthy men. Arterioscler Thromb 1994;14:54-59.
9. Junker R, Heinrich J, Schulte H, van de Loo J, Assmann G. Coagulation factor VII and the risk of coronary heart disease in healthy men. Arterioscler Thromb Vasc Biol 1997;17:1539-44.
10. Scarabin PY, Bonithon-Kopp C, Bara L, et al. Factor VII activation and menopausal status. Thromb Res 1990;57:227-34.
11. Meade TW, Imeson JD, Haines AP, Stirling Y, Thompson SG. Menopausal status and haemostatic variables. Lancet 1986;1:22-4.
12. Scarabin PY, Vissac AM, Kirzin JM, et al. Population correlates of coagulation factor VII. Importance of age, sex, and menopausal status as determinants of activated factor VII. Arterioscler Thromb Vasc Biol 1996;16:1170-76.
13. Kluft C, Lansink M. Effect of oral contraceptives on haemostasis variables. Thromb Haemost 1997;78:315-26.
14. Green F, Kelleher C, Wilkes H, Temple A, Meade T, Humphries S. A common genetic polymorphism associated with lower coagulation factor VII levels in healthy individuals. Arterioscler Thromb 1991;11:540-46.
15. Bernardi F, Arcieri P, Chiarotti F, et al. Contribution of factor VII genotype to activated FVII levels. Differences in genotype frequencies between Northern and Southern European populations. Arteriosc1 Thromb Vasc Biol 1997;17:2548-53.
16. Di Castelnuovo A, D'Orazio A, Amore C, et al. Genetic modulation of coagulation factor VII plasma levels: contribution of different polymorphisms and gender-related effects. Thromb Haemost 1998; in press.
17. Dell'Acqua G, Iacoviello L, D'Orazio A, Di Castelnuovo A, Donati MB. A polymorphic cluster in the 5' region of the human coagulation factor VII gene: Detection, frequency and linkage analysis. Thromb Res 1997;88 445-48.
18. Meilahn E, Ferrell R, Kiss J, et al. Genetic determination of coagulation factor VIIc levels among healthy middle-aged women. Thromb Haemost 1995;73:623-25.
19. Iacoviello L, Di Castelnuovo A, De Knijff P, et al. Polymorphisms in the coagulation factor VII gene and risk of myocardial infarction. New Engl J Med 1998;338:79-85.
20. The Medical Research Council's General Practice Research Framework. Thrombosis prevention trial: Randomised trial of low-intensity oral anticoagulation with warfarin and low-dose aspirin in the primary prevention of ischemic heart disease in men at increased risk. Lancet 1998;351:233-41.
21. Ernst E, Resch K. Fibrinogen as a cardiovascular risk factor: A meta-analysis and review of literature. Ann Intern Med 1993;118:956-63.
22. Heinrich J, Assmann G. Fibrinogen and cardiovascular risk. J Cardiovasc Risk

1995;2:197-205.
23. Scarabin PY, Oplu-Bureau G, Bara L, et al. Haemostatic variables and menopausal status; influence of hormone replacement therapy. Thromb Haemost 1993;70:584-87.
24. Tybjaerg-Hansen A, Agerholm-Larsen B, Humphries SE, Abildgaard A, Schnohr P, Nordestgaard BG. A common mutation in the b-fibrinogen promoter is an independent predictor of plasma fibrinogen, but not of ischemic heart disease. A study of 9,127 individuals based on the Copenhagen City Heart Study. J Clin Invest 1997;99:3034-39.
25. Zito F, Di Castelnuovo A, Amore C, D'Orazio A, Donati MB, Iacoviello L. BclI polymorphism in the gene of the fibrinogen b-chain is associated with the risk of familial myocardial infarction by increasing plasma fibrinogen levels: A case-control study in a sample of GISSI-2 patients. Arterioscler Thromb Vasc Biol 1997;17:3489-94.
26. Benhague I, Poirier O, Nicaud V al. b-fibrinogen gene polymorphisms are associated with plasma fibrinogen and coronary artery disease in patients with myocardial infarction. The ECTIM Study. Circulation 1996;93:440-49.
27. Fowkes FGR, Connor JM, Smith FB, Wood J, Donnan PT, Lowe GDO. Fibrinogen genotype and risk of peripheral atherosclerosis. Lancet 1992;339:693-96.

ESTROGEN EFFECT UPON CORONARY VASCULATURE

Giuseppe M.C. Rosano, Filippo Leonardo, and Gaia Panina

Introduction

Death from cardiovascular disease is rare in women before menopause while it represents the leading cause of death after either natural or surgically induced menopause [1]. The fact that the increased incidence of cardiovascular disease after menopause is not simply due to aging, but is strictly dependent upon estrogen deficiency and that, therefore, menopausal status influences the development of coronary artery disease independently from age, is demonstrated by the fact that, at any age, menopausal women have a significantly higher incidence of cardiovascular disease than premenopausal ones. Several large-scale trials have shown that estrogen replacement therapy reduces the occurrence of coronary disease and perhaps of cerebrovascular disease by nearly 50% in treated women compared to nonusers. These findings are supported by the evidence that estrogens have a beneficial effect on cholesterol metabolism and deposition, contributing to the inhibition of atherosclerotic plaque formation in arterial walls [2-3]. Early reports suggested that up to 60% of the protective effect of estrogens on coronary artery disease was attributable to favorable changes in plasma lipids. Reanalysis of the data indicated that the lipid changes probably account for approximately 20% of the cardioprotective effect of estrogen, and that other effects are therefore likely to be important [4]. Therefore, the favorable effect of estrogen replacement therapy on plasma lipids cannot fully explain the beneficial effects of estrogens upon cardiovascular disease. The influence of estrogens upon carbohydrate metabolism, atheroma formation, and cardiovascular hemodynamics may also play an integral role in the overall beneficial effect of the hormones. Estrogens are now regarded as cardioprotective agents as important as were aspirin and antihypertensive drugs in the past. Short-term effects of estrogens upon the coronary vascular reactivity and peripheral blood flow may warrant their consideration as cardioactive drugs with a possible beneficial effect in myocardial ischemia. *In vitro* studies have shown an endothelium-independent mechanism of action of the hormones while *in vivo* experiments have shown an endothelium-dependent mechanism of action upon coronary and peripheral vasculature. Furthermore estrogens have been shown to have antiadrenergic and calcium antagonistic effects which may be of importance in their beneficial effect upon the cardiovascular system.

R. Paoletti et al. (eds.), Women's Health and Menopause, 83–88.

Estrogens as Vasoactive Hormones

In addition to the development of atheroma and thrombosis, vessel spasm and changes in arterial tone are of importance in the pathogenesis of ischemic heart disease and myocardial infarction. Recent evidence suggests that estrogens are vasoactive substances that influence vascular tone in the reproductive and nonreproductive circulations. Animal studies have shown that the administration of estradiol 17β causes an increase in cardiac output and arterial flow velocity, and a decrease in systolic and diastolic blood pressure [5]. Human studies have reported that estrogens reduce peripheral vascular resistance and increase blood flow in several vascular districts such as the vagina, vulva, uterine, and carotid vessels, as well as in the coronary arteries [6-7]. Williams et al. have shown a reversal of the acetylcholine-induced vasoconstriction by estradiol 17β implants in ovariectomized monkeys fed with an atherogenic diet for 30 months [8]. Using quantitative coronary angiography, vasomotor responses to intracoronary acetylcholine were measured and compared in treated and untreated animals. Acetylcholine caused constriction of the coronary arteries in the estrogen-deficient monkeys and dilatation in those receiving estradiol 17β implants therefore suggesting that estrogen modulates vasomotor responses of atherosclerotic arteries by an endothelium-dependent mechanism. The authors speculated that a similar effect may be present in postmenopausal women taking estrogen replacement. Indeed, similar effects have been reported in humans [9-10]. Collins et al. have also shown that the reversal of acetylcholine-induced vasoconstriction is gender dependent. They have evaluated coronary artery diameters and blood flow after intracoronary infusion of acetylcholine, before and after the intracoronary infusion of estradiol 17β in 9 postmenopausal women and 7 men with proven coronary artery disease. Female patients showed coronary vasoconstriction after baseline infusion of acetylcholine which was converted in vasodilation after acute administration of estradiol 17β, blood flow was significantly increased after estradiol 17β exposure. Conversely to what was observed in women, estradiol 17β did not have any effect upon either acetylcholine-induced coronary artery constriction or blood flow in male patients [11]. The vasoactive effect of estrogens upon the peripheral vascular system in humans have been demonstrated by Volterrani et al. who showed that acute administration of sublingual estradiol 17β affects the peripheral vasculature in humans increasing forearm blood flow and reducing forearm vascular resistance compared to placebo [7]. The results of these studies suggest that the effect of estrogens upon the peripheral vascular system may involve one or a number of mechanisms.

Endothelium-Independent Effect of Estrogens

Calcium antagonistic properties of estrogen have been demonstrated in uterine arteries, cardiac myocytes, and vascular smooth muscle cells [12-13]. Since it has been proposed that calcium channel blockers such as nifedipine may reduce the progression of atherosclerosis, it has been suggested that the calcium antagonistic properties of estrogens may be responsible not only for the direct effect of the hormones upon vascular tone but may also influence the progression of coronary artery disease by influencing all stages of the progression of atheromatous plaque [14-15]. Jiang et al. have shown that estradiol 17β induces relaxation of coronary arterial rings contracted by both high extracellular potassium and Bay K 8644, a dihydropyridine derivative which opens potential-sensitive calcium channels [16-17]. This suggests that estradiol 17β may inhibit activation of potential-sensitive calcium channels. Estradiol 17β also induces relaxation of coronary arteries precontracted with PGF2α, suggesting that the relaxing effects of estradiol 17β may be dependent upon both activation of receptor-operated and potential-operated calcium channels [17]. Confirmation of a calcium antagonist property of estrogen was provided by experiments on isolated guinea pig cardiac myocytes [18]. Estradiol 17β has at pharmacological doses a negative inotropic effect on single ventricular myocytes by inhibiting inward calcium current and consequently reducing intracellular free calcium. Recent studies have confirmed that estrogens inhibit the contraction of epicardial coronary arteries by blocking calcium influx without changing calcium sensitivity of contractile elements [19]. In addition to their effects on calcium channels, estrogens have also been shown to affect large conductance chloride channels by a direct membrane effect in cultured fibroblasts [20] suggesting another regulatory role for estrogens. The hormones affect plasma ion channels via membrane binding sites distinct from the classical genomic pathway. A proof that ovarian hormones have an *in vivo* endothelium-independent effect and that this is likely related to the calcium-antagonistic effect of the hormones, has been recently provided by our group in humans. In fact, Rosano et al. have shown that acute intracoronary administration of estradiol 17β reduces the degree of methylergometrine-induced coronary artery constriction in menopausal women with coronary artery disease [21]. This effect of estradiol 17β upon coronary artery reactivity seems to be attributable more to the calcium antagonistic effect of the hormone than to its effect upon the endothelial function. The relief of ergonovine-induced vasoconstriction is an immediate effect, thus supporting the hypothesis of an acute modulation of coronary vasomotion due to a calcium antagonistic effect.

Endothelium-Dependent Relaxation

The endothelium plays an important role in the modulation of coronary artery tone by formation and secretion of vasoactive substances. Endothelium-derived relaxing factor (EDRF), known to be nitric oxide (NO) is one of this substances. NO causes vasorelaxation in coronary arteries with intact endothelium and is a product of the conversion of L-arginine by nitric oxide syntheses (NOS) to nitric oxide. NO has also been observed to slow the development of atheroma by inhibiting smooth muscle cell proliferation while stimulating proliferation of endothelial cells [22]. NO is released in response to many factors including acetylcholine and substance P, and causes relaxation of the vessel. Acetylcholine causes constriction in atherosclerotic coronary arteries and dilatation in normal coronary arteries suggesting that atheroma impairs endothelium-mediated dilatation of coronary arteries. The endothelium has been shown to mediate the relaxing effect to estradiol 17β in isolated vessels such as rabbit femoral artery [23] and rat aorta [24]. This evidence suggests that steroid hormones may modulate, *in vitro*, the vascular endothelium dependent responses. Recent studies have shown that estrogen induces a sixfold increase of basal NOS in human umbilical vein endothelial cultured monolayers within 30 minutes [25]. This may be due to an acute effect on the constitutive enzyme or on the inducible enzyme. Acetylcholine (Ach) induces endothelium-dependent vascular relaxation mediated by the release of EDRF [26]. As mentioned treatment with estradiol 17β modulates responses to Ach in coronary arteries in cynomolgus monkeys [27] and humans [11-14] suggesting an endothelium-dependent mechanism *in vivo*. These studies indicate that estrogen influences vascular tone also by an EDRF-dependent mechanism *in vivo*.

Rosano et al. have shown a beneficial effect of sublingual estrogen on myocardial ischemia in women with established coronary artery disease [28]. The mechanism of this effect is unknown, but an improvement in blood supply to the myocardium is likely, since the beneficial effect of estrogen was associated with an increased heart rate at the onset of myocardial ischemia (1 mm ST-segment depression) compared with placebo. The increased myocardial perfusion induced by estrogens during stress may be related to both an endothelium-dependent increase in coronary blood flow due, in part, to a facilitation of endothelium-dependent relaxation, which may be gender specific and to an endothelium-independent mechanism. Such effect may be of clinical benefit in the short- and long-term treatment of atherosclerotic coronary heart disease in postmenopausal women.

Conclusion

In conclusion, several mechanisms contribute to the cardioprotective effect of estrogens. The importance of changes in lipid profiles has been overestimated in the past. It is now clear that estrogens have vasoactive properties which are dependent upon their direct and indirect effect upon the different vascular layers. The importance of the endothelium-dependent effect of the ovarian hormones may have been overestimated in the past and it is possible that endothelium-independent effect may be of importance in the modulation of coronary artery tone and limiting the progression of atherosclerosis.

References

1. Wuest JH, Dry TJ, Edwards JE. The degree of coronary atherosclerosis in bilaterally oophorectomized women. Circulation 1953;7:801-8.
2. Sarrel PM. Ovarian hormones and the circulation. Maturitas 1990;12:287-98.
3. Bush TL. Noncontraceptive estrogen use and cardiovascular disease. Epidemiol Rev 1985;7:80-104.
4. Bush TL, Barrett-Connor E, Cowan LD. Cardiovascular mortality and noncontraceptive use of estrogen in women: Results from the Lipid Research Clinics Program Follow up Study. Circulation 1987;75:1102-9.

5. Magness RR, Rosenfeld CR. Local and systemic estradiol 17 beta: Effects on uterine and systemic vasodilation. Am J Physiol 1989;256:E536-E542.
6. Killam AP, Rosenfeld CR, Battaglia FC, Makowski EL, Meschia G. Effect of oestrogens on the uterine blood flow of oophorectomized ewes. Am J Obstet Gynecol 1973;115:1045-52.
7. Volterrani M, Rosano GMC, Coats A, Beale C, Collins P. Estrogen acutely increases peripheral blood flow in post-menopausal women. Am J Med 1995;99:119-22.
8. Williams JK, Adams MR. Oestrogen modulates responses of atherosclerotic coronary arteries. Circulation 1990;81:1680-87.
9. Reis SE, Gloth ST, Blumenthal RS. Ethinyl estradiol acutely attenuates abnormal coronary vasomotor responses to acetylcholine in postmenopausal women. Circulation 1994;89:52-60.
10. Gilligan DM, Quyumi AA, Cannon RO III. Effects of physiological levels of oestrogen on coronary vasomotor function in post-menopausal women. Circulation 1994;89:2545-51.
11. Collins P, Rosano GMC, Sarrel PM, et al. Oestradiol 17 beta attenuates acetylcholine induced coronary arterial constriction in women but not in men with coronary artery disease. Circulation 1995;92:24-30.
12. Stice Sl, Ford SP, Rosazza JP, Van Orden DE. Role of 4-hydroxylated oestradiol in reducing Ca^{2+} uptake of uterine arterial smooth muscle cells through potential-sensitive channels. Biol Reprod 1987;36:361-68.
13. Stice Sl, Ford SP, Rosazza JP, Van-Orden DE. Interaction of 4-hydroxylated oestradiol and potential-sensitive Ca2+ channels in altering uterine blood flow during the oestrous cycle and early pregnancy in gilts. Biol Reprod 1987;36:369-75.
14. Henry PD, Bentley KI. Suppression of atherogenesis in cholesterol fed rabbits treated with nifedipine. J Clin Invest 1981;68:1366-69.
15. Lichtlen PR, Hugenholtz PG, Hecker H, Jost S, Deckers JW. Retardation of angiographic progression of coronary artery disease by nifedipine. Lancet 1990;335:1109-13.
16. Brown AM, Kunze DL, Yatani A. The agonist effect of dihydropyridines on calcium channels. Nature 1984;311:570-72.
17. Jiang C, Sarrel PM, Lindsay DC, Poole-Wilson PA, Collins P. Endothelium-independent relaxation of rabbit coronary artery by 17 beta estradiol in vitro. Br J Pharmacol 1991;104: 1033-37.
18. Jiang C, Poole-Wilson PA, Sarrel PM, Collins P. Effect of estradiol 17 beta on contraction, calcium current and intracellular free calcium in guinea-pig isolated cardiac myocytes. Br J Pharmacol 1992;106:739-45.
19. Han SZ, Karaki H, Ouchi Y, Orimo H. 17 beta estradiol inhibits calcium influx and calcium release induced by thromboxane A2 in porcine coronary artery. Circulation 1995;91:2619-26.
20. Hardy SP, Valverde MA. Novel plasma membrane action of estrogen and antiestrogens revealed by their regulation of a large conductance chloride channel. FASB J 1994;8:760-65.
21. Rosano GMC, Lopez Hidalgo M, Arie S, et al. Acute effect of estradiol 17β upon coronary artery reactivity in menopausal women with coronary artery disease. Eur Heart J 1996;17 (Suppl.):3112 Abst.
22. Dubey RK, Overbeck HW. Culture of rat mesenteric arteriolar smooth muscle cells:

Effects of platelet derived growth factor, angiotensin, and nitric oxide on growth. Cell Tissue Res 1994;275:133-41.

23. Gisclard V, Miller VM. Effect of estradiol 17 beta on endothelium-dependent responses in the rabbit. J Pharmacol Exp Ther 1988;244:19-22.
24. Williams SP, Shackelford DP, Iams SG, Mustafa SJ. Endothelium-dependent relaxation in estrogen treated spontaneously hypertensive rats. Eur J Pharmacol 1988;145:205-7.
25. Caulin-Glaser TL, Sessa W, Sarrel PM, Bender J. The effect of 17 beta estradiol on human endothelial cell nitric oxide production. Circulation 1994;90:1-30.
26. Furchgott RF, Zawadzki JV. The obligatory role of endothelial cells in the relaxation of arterial smooth muscle by acetylcholine. Nature 1980;288:373-76.
27. Williams JK, Adams MR, Klopfenstein HS. Oestrogen modulates responses of atherosclerotic coronary arteries. Circulation 1990;81:1680-87.
28. Rosano GMC, Sarrel PM, Poole-Wilson PA, Collins P. Beneficial effect of oestrogen on exercise-induced myocardial ischemia in women with coronary artery disease. Lancet 1993; 342:133-36.

Estrogen and Endothelial Function

Agostino Virdis, Lorenzo Ghiadoni, Isabella Sudano, Stefania Pinto, Stefano Taddei, and Antonio Salvetti

Summary

Although observational studies have shown that estrogen can protect women against cardiovascular diseases (CVD), the underlying mechanisms are not fully clarified. Endothelium plays a pivotal role in the control of vascular function and structure; endothelial dysfunction, characterized by reduced or absent nitric oxide (NO) bioavailability linked and/or due to the production of cyclooxygenase-dependent endothelial-derived contracting factors (EDCFs), mainly represented by oxygen free radicals, is now viewed as an important mechanism causing atherothrombosis. Data in humans indicate that endogenous estrogen plays a protective role on endothelial function. Thus premenopausal women are protected against the adverse effect of hypercholesterolemia, with age-related impairment of endothelial function occurring only after the menopause in normotensive women and mainly after the menopause in hypertensive women. Moreover endogenous estrogen deprivation selectively impairs endothelial function in normotensive women. There is also evidence that exogenous estrogen can improve endothelial function in postmenopausal women. Thus acute intra-arterial or intravenous infusion of estrogen potentiates endothelium-dependent vasodilation both in coronary circulation and the forearm. Estrogen replacement therapy (ERT) seems to improve endothelial function both in coronary arteries of postmenopausal women with coronary atherosclerosis and in brachial arteries of postmenopausal women with mild hypercholesterolemia. Although short-term ERT did not improve endothelial function in the forearm of perimenopausal women and in postmenopausal women with various cardiovascular risk factors, 3-month ERT restored endothelial function in ovariectomized women. It is still controversial whether concomitant administration of different progestins can impair the effect of estrogen on endothelial function. Finally, the effect of estrogen on endothelial function seems to be mediated both by activation of the L-arginine-nitric oxide (NO) pathway and by the inhibition of EDCF production, probably through its antioxidant action. In conclusion, available data in human indicate that estrogen preserves endothelial function, an effect which may be at least partially responsible for the protective role of estrogen against the development of CVD in women.

R. Paoletti et al. (eds.), Women's Health and Menopause, 89–98.

Introduction

Observational studies have shown that estrogen can exert a protective action against CVD in women [1,2]. Although estrogen can favorably influence the lipid profile by reducing low density lipoprotein (LDL) cholesterol and increasing high density lipoprotein (HDL) cholesterol, by interfering with cholesterol deposition in the arterial wall and preserving LDL from oxidation [3-5], these effects on lipids have been calculated to be insufficient to fully explain the benefit of estrogen [1]. Thus additional mechanisms of estrogen acting on the arterial wall must be considered, among which the effect on endothelial function is a strong candidate for the following reasons.

It is now well documented that endothelium is an autocrine-paracrine organ, which plays a primary role in the modulation both of vascular tone and structure [6]. In physiological conditions endothelium produces various vasodilating substances, of which the most important is nitric oxide (NO) [7], a labile substance released under the stimulus of various substances, such as acetylcholine, bradykinin, thrombin, serotonin, etc., which interact with specific receptors, and by mechanical forces, such as shear stress [8-11]. In pathological conditions the same stimuli can also produce cyclooxygenase-dependent endothelium-derived contracting factors (EDCFs) [12,13], identified as prostanoids, thromboxane A_2 and prostaglandin H_2 [14,15], and free radicals, such as superoxide anions which are known to destroy NO, thus reducing its bioavailability [16]. NO and EDCFs not only exert an opposite effect on vascular tone, but also inhibit and stimulate respectively those mechanisms, such as platelet aggregation, vascular smooth muscle cell migration and proliferation, monocyte adhesion and molecule adhesion expression, which are known to play a primary role in the genesis of atherothrombosis [17]. Thus endothelial dysfunction, characterized by reduced and/or absent NO bioavailability linked to and/or caused by EDCFs production, is now viewed as an important mechanism causing atherosclerosis [17]. Therefore in the present paper we will examine the effects of estrogen on endothelial function following these lines of reasoning.

Effects of Endogenous Estrogen

The possibility that estrogen can modulate endothelial function during the menstrual cycle has been evaluated using two different approaches. First, Rosselli et al. found that serum nitrite/nitrate ($NO_2 + NO_3$) levels, a crude index of NO production, rise from the early to the mature phase of follicular development, showing a significant direct correlation with 17-β estradiol values but not with those of progesterone [18]. Second, it was found that in women flow mediated dilation (FMD) of the brachial artery, which is an endothelial-dependent phenomenon, was higher than that of men during the follicular and luteal phases, but not during the menstrual phase [19]. However in this study the response to nitroglycerin (NTG), an endothelial-independent vasodilator, showed a similar pattern [19]. Taken together these data suggest, but do not prove, that estrogen can stimulate NO production and modulate vascular tone during the menstrual cycle. Endothelial function is altered in the presence of several cardiovascular risk factors, such as hypercholesterolemia [20] and aging in normotensive subjects [21,22] and hypertensive patients [21,22]. However, when compared to hypercholesterolemic men, hypercholesterolemic premenopausal women did not show an impairment of endothelial function, as evaluated by forearm blood flow (FBF) response to intrabrachial infusion of acetylcholine [23], an endothelium-dependent vasodilator [24]. In normotensive women, age-related impairment of endothelial function both in the brachial artery [25] and the forearm [26] occurred only after menopause, while in hypertensive women aging only slightly reduced endothelial function before menopause but sharply impaired it after menopause [26]. Taken together these data indicate that endogenous estrogen can protect endothelial function against the adverse effect of hypercholesterolemia and prevents, completely in normotensive women and partially in hypertensive women, the negative effect of aging on endothelial function.

Finally, we found that in normotensive women 1 month after bilateral ovariectomy the forearm blood flow response to acetylcholine was significantly reduced as compared to preintervention evaluation, while the response to sodium nitroprusside, an endothelium-independent vasodilator, was unchanged, a finding which indicates that endogenous estrogen deprivation leads to selective impairment of endothelial function [27]. These data are in agreement with the finding that FMD is reduced in postmenopausal as compared to premenopausal women [28,29].

Effects Of Exogenous Estrogen

Intra-arterial infusion of physiological doses of 17-β estradiol potentiates endothelium-dependent vasodilation both in large epicardial arteries and coronary microcirculation [30], as well as in the forearm [31], of postmenopausal women. However, while in the coronary circulation estrogen selectively improved endothelial function both in healthy women and in those with coronary atherosclerosis [30], in the forearm the selective effect of estrogen on endothelial function was detectable only in healthy women while in those with various cardiovascular risk factors estrogen potentiated the response both to acetylcholine and sodium nitroprusside [31]. Thus it is unclear whether acute administration of physiological doses of estrogen can selectively improve endothelial function in the forearm of postmenopausal women with various cardiovascular risk factors, which can *per se* impair endothelial function. Moreover it has been reported that intravenous infusion of supraphysiological doses of conjugated estrogen, which increased estrogen plasma levels to values similar to those of pregnant women, selectively improved endothelial function, i.e. the vasodilating response to acetylcholine but not to sodium nitroprusside in the forearm of postmenopausal women with effort angina [32]. Interestingly, in this study the potentiating effect of estrogen on acetylcholine-induced vasodilation was abolished by intrabrachial infusion of L-NMMA, a competitive inhibitor of nitric oxide synthase (NOS). Taken together, these data suggest that higher doses of estrogen might be needed in order to improve endothelial function in postmenopausal women, whose endothelial function is further reduced by the presence of other cardiovascular risk factors. Although the acute effect of estrogen is interesting since it can indicate a direct and/or nongenomic endothelial effect, the action of prolonged estrogen administration, i.e ERT, seems to be more relevant from a clinical point of view. In a study performed on a small number of postmenopausal women with early coronary atherosclerosis, current ERT was associated with an attenuation or reversal of the vasoconstrictor response to intracoronary infusion of acetylcholine observed in the control group [33]. In a randomized, double-blind crossover study, in which postmenopausal women with mild hypercholesterolemia received placebo, or oral 17-β estradiol 1 and 2 mg o.d., for 9 weeks, FMD in the brachial artery was significantly increased by estrogen [34]. These data indicate that ERT can improve endothelial function both in large coronary arteries of postmenopausal women with coronary atherosclerosis and in brachial arteries of postmenopausal women with mild hypercholesterolemia.

However the data concerning the effect of ERT on endothelium-dependent vasodilation in the forearm are quite conflicting. In a small number of healthy perimenopausal women randomized to receive placebo (n=5) or oral estradiol 2 mg o.d. (n=6) for 8 weeks, estrogen did not improve the vasodilating response to intrabrachial infusion of acetylcholine [35]. Another intriguing finding of this study was that in the estrogen group blood pressure was significantly reduced and therefore, as expected [36], basal release of NO, as evaluated through the vasoconstrictor response to intrabrachial infusion of L-NMMA, was increased [35].

In a second study conducted in 33 postmenopausal women, either normal (n= 15) or with other cardiovascular risk factors (n=18), 3-week treatment with transdermal estradiol administration did not increase the vasodilating response to intrabrachial infusion of acetylcholine [37]. However in 31 out of these 33 women, intra-arterial infusion of estradiol, which achieved plasma levels of estrogen higher than those under transdermal estrogen, improved endothelial function. Finally, repeat intra-arterial infusion of estradiol in 8 women receiving transdermal estradiol potentiated the vasodilating response to acetylcholine [37]. In contrast, we found that 3-month treatment with transdermal estrogen restored the vasodilating response to intrabrachial infusion of acetylcholine in 10 ovariectomized normal women [27], a finding which indicates that ERT normalized endothelial function in these postmenopausal women. These data from these three studies, as reported in Table 1, can offer the following possible interpretations concerning the lack of effect of ERT on endothelial function in the forearm microcirculation. First, since there were no data concerning the presence of endothelial dysfunction in perimenopausal women [35], who showed plasma estrogen levels similar to those achieved after ERT in ovariectomized women, it is tempting to speculate that ERT cannot improve an already normal endothelial function. Second, the data from the second study [37] suggest that a more prolonged period of ERT or higher doses of estrogen may be needed in order to improve endothelial function in older postmenopausal women and in postmenopausal women with additional cardiovascular risk factors.

Another important question is whether concomitant administration of progestins can influence the effect of estrogen on endothelial function. A study in 26 postmenopausal women randomized to receive either hormone replacement therapy (HRT), transdermal 17-

β estradiol 50 mg every 3-4 days together with oral norethisterone acetate (NETA) 1 mg/day from day 1 through 12 of each month, or placebo for 2 years, showed that serum $NO_2 + NO_3$ levels were significantly higher in the HRT group as compared to the placebo group when blood samples were collected in estradiol-treated subjects not taking NETA [38]. However serum $NO_2 + NO_3$ values in estradiol-treated women taking NETA did not differ from the placebo group. These data suggest that HRT can increase NO levels in postmenopausal women and that concomitant administration of this type of progestin may attenuate estrogen-induced NO production.

Table 1. Effect of ERT on Acetylcholine-Induced Endothelial Function in Forearm Microcirculation

Reference	Subjects	Age (years)	ERT	Plasma Estradiol (pg/ml)
Negative Studies				
Sudhir K et al. *Hypertension* 1996	6 perimenopausal women	47 ± 1.9	8 weeks, oral	B = 37.3 ± 15.5 A = 405.8 ± 55.5
Gilligan DM *Am J Cardiol* 1995	15 normal 18 with ≥ 1 CV risk factor (HT, H-chol, DM)	57 ± 7 60 ± 7	3 weeks, transdermal	B = 16 ± 11 A = 120 ± 57
Positive Studies				
Pinto S et al. *Hypertension* 1997	10 normal ovariectomy	47.4 ± 1.6	12 weeks, transdermal	B = undetectable A = 48.6 ± 13.9

HT: hypertension; H-chol: hypercholesterolemia; DM: diabetes mellitus; B: before; A: after

In agreement with this last hypothesis, a recent randomized prospective study, in which normotensive postmenopausal women were randomized to receive HRT consisting of 2 mg/d estradiol plus 1mg/d norethisterone for 10 days, or no therapy for a mean follow-up of 3 years, showed that FMD, which was reduced in postmenopausal women as compared to a control premenopausal group, did not differ between treated and nontreated groups [28]. However another study, in which healthy postmenopausal women receiving HRT consisting of estrogen alone or combined with a progestin, mainly medroxyprogesterone, for around 5.5 years, were compared with similar women not taking ERT and with young premenopausal women, FMD, which was reduced in postmenopausal women, was improved by ERT both in the unopposed estrogen treatment and combined hormone replacement groups [29]. Thus the above data suggest that the kind of progestin added to ERT can differently influence the effect of estrogen on endothelial function in postmenopausal women, a hypothesis which is still to be validated.

Mechanism(s) Through Which Estrogen Can Influence Endothelial Function

There is evidence that acute estrogen administration improves endothelial function in the coronary circulation [39] and in the brachial artery [40] in postmenopausal women but not in men, a finding which indicates a gender differences in the effect of estrogen on endothelial function. However recent studies [41,42] have shown that high-dose long-term estrogen therapy in male to female transsexuals with concomitant orchidectomy or antiandrogen therapy increased both endothelium-dependent (FMD) and endothelium-independent (response to GTN) vasodilation in the brachial artery, a finding which suggests an aspecific effect of estrogen or of androgen depletion on vascular function in males.

As already pointed out, ERT increased serum levels of NO_2 + NO_3 [38] and the potentiating effect on endothelial function of acute estrogen administration was inhibited by L-NMMA [32]. These findings suggest that estrogen can stimulate NO production and/or NO bioavailability and, in agreement with experimental data, these effects can be exerted by stimulating NOS activity [43] and/or by inhibiting NO breakdown through their antioxidant effect [44]. To test this hypothesis we performed the following studies in ovariectomized women.

First, we evaluated activation of the L-arginine NO pathway through intrabrachial infusion of L-arginine and NO bioavailability through local infusion of L-NMMA during acetylcholine and sodium nitroprusside intrabrachial administration, both before and after ovariectomy and after ERT. Second, using the same time sequence and protocol we evaluated both the production of EDCFs, through intrabrachial infusion of indomethacin, an inhibitor of cyclooxygenase, and the role of oxidative stress, through local infusion of vitamin C, a scavenger of oxygen free radicals. Preliminary data obtained so far [45] indicate that estrogen deprivation abolished the potentiating effect of L-arginine and the inhibiting effect of L-NMMA on acetylcholine-induced vasodilation, without affecting the response to sodium nitroprusside, and that these effects were restored by ERT. Moreover indomethacin and vitamin C similarly potentiated the vasodilating response only to acetylcholine after ovariectomy, while they exerted no effect before ovariectomy and after ERT. Taken together these data indicate (Figure 1) that estrogen can improve and/or preserve endothelial function by activating the L-Arginine-NO pathway and preserving NO bioavailability through inhibition of production of EDCFs, which are mainly represented by oxygen free radicals [46].

Conclusions

Available data indicate that endogenous estrogen preserves endothelial function in premenopausal women and that estrogen administration can restore and/or improve endothelial function in postmenopausal women. Thus it is tempting to speculate that these effects might be at least partially responsible for the possible protective role of estrogen against the development of cardiovascular disease in women. However before accepting these conclusions we need first to answer the unsolved questions reported in Table 2 and second to wait for the outcome of ongoing controlled trials, such as the Heart and Estrogen/Progestin Replacement Study (HERS) and the Women's Health Initiative, on the possible effect of ERT on primary and secondary prevention of cardiovascular events in postmenopausal women [47].

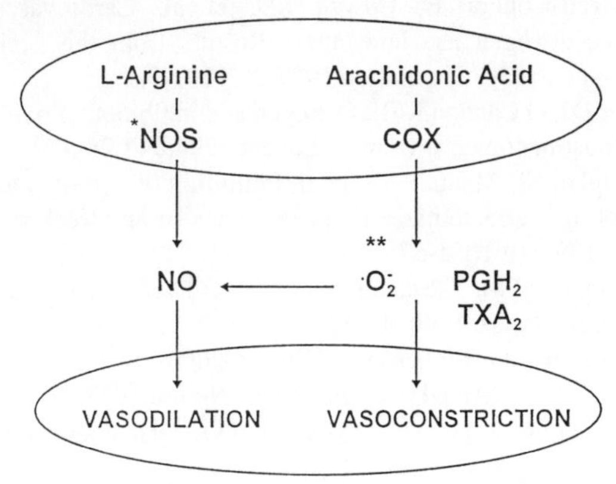

Figure 1. The protective effect of estrogen on endothelial function in normotensive women seems to be mediated both by the activation of L-arginine-NO pathway and by the inhibition of oxidative stress.

Table 2. Estrogens and Endothelial Function: Unsolved Questions

Estrogens and Endothelial Function: Unsolved Questions
• How long after menopause can estrogen restore endothelial function and how long is this effect detectable?
• Does the restoring action of estrogen on endothelial function differ according to the presence of different and/or various cardiovascular risk factors?
• Is there in the latter women a time and dose-dependent effect of estrogen on endothelial function?
• Can the action of different progestins influence the effect of estrogen on endothelial function?
• Do selective estrogen receptor modulators (SERMs) influence endothelial function?

References

1. Gerhard M, Ganz P. How do we explain the clinical benefits of estrogen? From bedside to bench. Circulation 1995;92:5-8
2. Grodstein F, Stampfer MJ, Manson JE, et al. Postmenopausal estrogen and progestin use and the risk of cardiovascular disease. N Engl J Med 1996;335:453-61.
3. Bush TL, Barrett-Connor E, Cowan LD, et al. Cardiovascular mortality and noncontraceptive estrogen use in women: Results from the Lipid Research Clinics Program Follow-up Study. Circulation 1987;75:1102-9.
4. Sack MN, Rader DJ, O Cannon R III. Ostrogen and inhibition of oxidation of low-density lipoproteins in postmenopausal women. Lancet 1994;343:269-70.
5. Adams MR, Kaplan JR, Manuck SB, et al. Inhibition of coronary artery atherosclerosis by 17β-estradiol in ovariectomized monkeys: Lack of an effect of added progesterone. Arteriosclerosis 1990;10:1051-57.
6. Luscher TF, Vanhoutte PM. The endothelium: Modulator of cardiovascular function. Boca Raton, Florida: CRC Press, 1990: 1-215.
7. Palmer RMJ, Ferrige AG, Moncada S. Nitric oxide release accounts for the biological activity of endothelium-derived relaxing factor. Nature 1987;327:424-526.
8. Furchgott RF. Role of endothelium in response of vascular smooth muscle. Circ Res 1983; 53:557-73.
9. De Mey JG, Vanhoutte PM. Role of the intima in cholinergic and purinergic relaxation of isolated canine femoral arteries. J Physiol 1981;316:347-55.
10. Luscher TF, Cooke JP, Houston DS, Neves R, Vanhoutte PM. Endothelium-dependent relaxations in human arteries. Mayo Clin Proc 1987;62:601-6.
11. Luscher TF, Diederich D, Siebenmann R, Lehmann K, et al. Difference between endothelium-dependent relaxation in arterial and in venous coronary bypass grafts. N Engl J Med 1988;319:462-67.
12. Luscher TF, Vanhoutte PM. Endothelium-dependent contractions to acetylcholine in the aorta of spontaneously hypertensive rats. Hypertension 1986;8:344-48.
13. Diederich D, Yang Z, Buhler FR, Luscher TF. Impaired endothelium-dependent relaxations in hypertensive resistance arteries involve the cyclooxygenase pathway. Am J Physiol 1990;258:H445-H451.
14. Shirahase H, Fujiwara M, Usui H, Kurahashi K. A possible role of thromboxane A2 in endothelium in maintaining resting tone and producing contractile response to acetylcholine and arachidonic acid in canine cerebral artery. Blood Vessels 1987;24:117-19.
15. Kato T, Iwama Y, Okumura K, Hashimoto H, Ito T, Satake T. Prostaglandin H2 may be the endothelium-derived contracting factor released by acetylcholine in the aorta of the rat. Hypertension 1990;15:475-82.
16. Katusic ZS, Vanhoutte PM. Superoxide anion is an endothelium-derived contracting factor. Am J Physiol 1989;257:H33-H37.
17. Taddei S, Virdis A, Ghiadoni L, Salvetti A. Hypertension and endothelial dysfunction. Cardiovasc Risk Factors 1997;7:76-87.
18. Rosselli M, Imthurn B, Macas E, Keller PJ, Dubey RK. Circulating nitrite/nitrate levels increase with follicular development: Indirect evidence for estradiol mediated NO-release. Biochem Biophys Res Commun 1994;202:1543-52.

19. Hashimoto M, Akishita M, Eto M, et al. Modulation of endothelium-dependent flow-mediated dilatation of the brachial artery by sex and menstrual cycle. Circulation 1995; 92:3431-35.
20. Casino PR, Kilcoyne CM, Quyyumi AA, Hoeg JM, Panza JA. The role of nitric-oxide in endothelium-dependent vasodilation of hypercholesterolemic patients. Circulation 1993; 88:2541-47.
21. Taddei S, Virdis A, Mattei P, et al. Aging and endothelial function in normotensive subjects and essential hypertensive patients. Circulation 1995;91:1981-87.
22. Taddei S, Virdis A, Mattei P, et al. Hypertension causes premature aging of endothelial function in humans. Hypertension 1997;29:736-43.
23. Chowienczyk PJ, Watts GF, Cockcroft JR, et al. Sex differences in endothelial function in normal and hypercholesterolemic subjects. Lancet 1994;344:305-6.
24. Furchgott RF, Zawadzky JV. The obligatory role of endothelial cells in the relaxation of arterial smooth muscle by acetylcholine. Nature 1980;288:373-76.
25. Celermajer DS, Sorensen, KE, Spielgelhalter DJ, Geogakopoulos D, Robinson J, Deanfield JE. Aging is associated with endothelial dysfunction in healthy men years before the age related decline in women. J Am Coll Cardiol 1994;24:471-76.
26. Taddei S, Virdis A, Ghiadoni L, et al. Menopause is associated with endothelial dysfunction in women. Hypertension 1996;28:576-82.
27. Pinto S, Virdis A, Ghiadoni L, et al. Endogenous estrogen and acetylcholine-induced vasodilation in normotensive women. Hypertension 1997;29:576-82.
28. Sorensen KE, Dorup I, Hermann AP, Mosekilde L. Combined hormone replacement therapy does not protect women against the age-related decline in endothelium-dependent vasomotor function. Circulation 1998;97:1234-38.
29. McCrohon JA, Adams MR, McCredie RJ, et al. Hormone replacement therapy is associated with improved arterial physiology in healthy post-menopausal women. Clin Endocrinol 1996;45:435-41.
30. Gilligan DM, Quyyumi AA, Cannon III RO. Effects of physiological levels of estrogen on coronary vasomotor function in postmenopausal women. Circulation 1994;89:2545-51.
31. Gilligan DM, Badar DM, Panza JA, Quyyumi AA, Cannon III RO. Acute vascular effects of estrogen in postmenopausal women. Circulation 1994;90:786-91.
32. Tagawa H, Shimokawa H, Tagawa T, Matsumoto MK, Hirooka Y, Takeshita A. Short-term estrogen augments both nitric oxide-mediated and non-nitric oxide-mediated endothelium-dependent forearm vasodilation in postmenopausal women. J Cardiovasc Pharmacol 1997;30:481-88.
33. Herrington DM, Braden GA, Williams JK, Morgan TM. Endothelial-dependent coronary vasomotor responsiveness in postmenopausal women with and without estrogen replacement therapy. Am J Cardiol 1994;73:951-52.
34. Lieberman EH, Gerhard MD, Uehata A, et al. Estrogen improves endothelium-dependent, flow-mediated vasodilation in postmenopausal women. Ann Intern Med 1994;121:936-41.
35. Sudhir K, Jennings GL, Funder JW, Komesaroff PA. Estrogen enhances basal nitric oxide release in the forearm vasculature in perimenopausal women. Hypertension 1996;28:330-34.
36. Lyons D, Webster J, Benjamin N. The effect of antihypertensive therapy on responsiveness to local intra-arterial N^G-monomethyl-L-arginine in patients with essential hypertension. J Hypertens 1994;12:1047-52.

37. Gilligan DM, Badar DM, Panza JA, Quyyumi AA, Cannon III RO. Effects of estrogen replacement therapy on peripheral vasomotor function in postmenopausal women. Am J Cardiol 1995;75:264-68.
38. Rosselli M, Imthurn B, Keller PJ, Jackson EK, Dubey RK. Circulating nitric oxide (nitrite/nitrate) levels in postmenopausal women substituted with 17 β-estradiol and norethisterone acetate. Hypertension 1995;25[part2]:848-53.
39. Reis SE, Gloth ST, Blumenthal RS, et al. Ethinyl estradiol acutely attenuates abnormal coronary vasomotor responses to acetylcholine in postmenopausal women. Circulation 1994;89:52-60.
40. Kawano H, Motoyama T, Kugiyama K, et al. Gender difference in improvement of endothelium-dependent vasodilation after estrogen supplementation. J Am Coll Cardiol 1997;30:914-19.
41. New G, Timmins K, Duffy S, et al. Long-term estrogen therapy improves vascular function in male to female transsexuals. J Am Coll Cardiol 1997;29:1437-44.
42. McCrohon JA, Walters W, Robinson J, et al. Arterial reactivity is enhanced in genetic males taking high dose estrogens. J Am Coll Cardiol 1997;29:1432-36.
43. Weiner CP, Lizasoain I, Baylis SA, Knowles RG, Charles IG, Moncada S. Induction of calcium-dependent nitric oxide synthases by sex hormones. Proc Natl Acad Sci USA 1994;91:5212-16.
44. Keaney JF, Shwaery GT, Xu A, et al. 17β-estradiol preserves endothelial vasodilator function and limits low-density lipoprotein oxidation in hypercholesterolemic swine. Circulation 1994;89:2251-59.
45. Pinto S, Virdis A, Ghiadoni L, et al. Mechanisms responsible for endothelial dysfunction associated to acute estrogen deprivation in normotensive women (abstract). Eighth European Meeting on Hypertension, Milan, 1997.
46. Taddei S, Virdis A, Ghiadoni L, Magagna A, Salvetti A. Vitamin C improves endothelium-dependent vasodilation by restoring nitric oxide activity in essential hypertension. Circulation 1998;97:2222-29.
47. Vogel RA and Corretti MC. Estrogens, progestins, and heart disease. Circulation 1998;97: 1223-26.

ACTION OF SPECIFIC ESTROGENS ON THE CORONARY ARTERY: EFFECTS ON LIPOPROTEINS, COAGULATION, AND FIBRINOLYSIS

Jay M. Sullivan

Lipoprotein-Related Mechanisms of Cardioprotection by Estrogen Replacement Therapy

Among the known cardiovascular risk factors, high low-density lipoprotein (LDL) cholesterol and low high-density lipoprotein (HDL) cholesterol are important predictors of atherosclerotic disease. Analysis of data from observational studies initially suggested that the cardioprotective effect of estrogen was due to estrogen-induced changes in serum lipoproteins [1]. After menopause, LDL levels rise while HDL levels decline [2], thus worsening the LDL-to-HDL ratio and accelerating the development of atherosclerotic disease. Estrogen replacement results in lower levels of total cholesterol, decreased levels of LDL cholesterol, increased levels of HDL cholesterol, and increased levels of triglycerides [3,4]. Estrogen stimulates the liver to increase syntheses of VLDL which results in higher blood triglyceride levels. However, estrogen also stimulates hepatic receptor-mediated clearance of highly atherogenic LDL cholesterol. Another beneficial effect of estrogen is increased synthesis of apolipoprotein A1 and A2 which results in higher blood levels of HDL cholesterol. This effect is also due to decreased clearance of HDL cholesterol, leading to increased blood HDL levels and increased clearance of tissue cholesterol by the reverse cholesterol transport pathway.

Many observational studies have reported lower LDL and higher HDL concentrations in current users of estrogen when compared to women who do not use estrogen. Two randomized controlled trials have studied the effect of various hormone replacement regimens on serum lipoproteins. Walsh et al. [3] randomized 40 women to receive either placebo; conjugated estrogen in doses of 0.625 and 1.25 mg daily; oral micronized estrogen, 2 mg daily; or transdermal estradiol, 0.1 mg twice a week for six weeks. They found a dose-dependent decrease in LDL and increase in HDL and VLDL with conjugated estrogen. Oral estradiol had similar effects but no effect on lipoproteins was seen with transdermal estradiol.

The 3-year PEPI Study [5] randomized 875 postmenopausal women to therapy with placebo; conjugated equine estrogen (CEE) 0.625 mg daily without and with either medroxyprogesterone acetate (MPA) 10 mg/dl for 12 days a month, or MPA, 2.5 mg daily; or micronized progesterone 200 mg daily for 12 days a month. All regimens reduced LDL

R. Paoletti et al. (eds.), Women's Health and Menopause, 99–104.

by 14.5 to 17.7 mg/dl. CEE alone raised HDL by 5.6 mg/dl, with MPA by 1.2 to 1.6 mg/dl and with micronized progesterone by 4.1 mg/dl. This confirmed the LDL lowering effect of estrogen and showed that micronized progesterone attenuated the rise in HDL less than MPA.

It is interesting to compare the effects of estrogen on lipoproteins with the effects of an HMG-CoA reductase inhibitor, or statin. Darling et al. [6] studied 58 women with total cholesterol exceeding 250 mg/dl in a randomized, crossover trial comparing simvastatin, 10 mg daily with CEE 1.25 mg and MPA 5 mg daily. They found that both regimens raised HDL equally while simvastatin resulted in greater reduction of total and LDL cholesterol and of triglycerides. However statin did not lower Lp(a) while CEE plus MPA did. This suggests that certain women with dyslipidemia may require addition of a statin to a program of hormone replacement.

Davidson et al. [7] compared pravastatin to CEE and found that CEE increased HDL more than pravastatin but lowered LDL less. The addition of a statin to CEE had no greater effect on HDL than CEE alone. Similarly CEE plus statin did not lower LDL more than statin alone.

Raloxifene, a selective estrogen receptor modulator, has been found to lower LDL, although less than estrogen. Raloxifene has no effect on total HDL but increases the HDL_2 subfraction [8].

However, recent epidemiologic estimates and experimental results suggest that no more than 25-35% of the cardioprotective effect of estrogen replacement therapy can be explained by changes in serum lipids.

Other mechanisms that have been considered have been the effect of ERT on lipoprotein (a) levels, which climb after menopause and which are lowered towards normal by estrogen or hormone replacement therapy [9]. ERT has also been found to impede the oxidation of LDL cholesterol [10] which in turn could result in less cholesterol accumulating in arterial walls. In the nonhuman primate, ERT decreases the rate at which LDL is taken up by the blood vessel wall in postmenopausal animals fed a high-fat diet. Postmortem studies have demonstrated less extensive atherosclerosis in estrogen-treated postmenopausal monkeys, even when serum lipid levels do not change substantially [11].

Estrogen lowers blood pressure, although perhaps not sufficiently to have clinically important effects on the development of atherosclerosis [12].

Effect of Estrogen and Progestins on the Coagulation Cascade and Fibrinolytic Mechanisms

Studies of the effect of hormone replacement on fibrinolysis and components of the coagulation system have yielded inconsistent results. For example, Gordon et al. [13] studied a group of 74 women who were randomly assigned to receive 5, 10, or 20 micrograms of ethinyl estradiol and varying doses of norethindrone acetate. After a year of observation, they found no changes in prothrombin time, Factor X, fibrinogen, Factor VII, or fibrinopeptide A. The partial thrombopastin time shortened in all groups and Factor

XII climbed. A dose of ethinyl estradiol of 20 micrograms daily resulted in a lowered titer of antithrombin III; however, this effect was not seen at lower dose levels. There were no thromboembolic episodes.

Alkjaersig et al. [14] studied 23 women who received therapy with conjugated oral estrogens or transdermal estradiol in random order. Fibrinopeptide A and fibrinogen did not change. Alpha I antitrypsin and plasminogen concentrations increased significantly at a dose of oral estrogens of 1.25 mg daily but not with transdermally administered estrogen. Bar et al.[15] studied the effect of hormone replacement on platelet function in 51 women and found that replacement of estrogen and progestins resulted in a significant decrease in platelet aggregation induced by adrenaline. The addition of medroxyprogesterone acetate to estrogen resulted in less reduction of platelet aggregation than with estrogen alone. Gilabert et al. [16] studied the effect of estrogen replacement therapy and progestins on the fibrinolytic system and coagulation inhibitors. Seventy-five women were studied before and 3-12 months after hormone therapy. They found an increase in plasma fibrinolytic activity due to a significant increase in tissue type plasminogen activator and a decrease in plasminogen activator inhibitor Type I. Protein S declined significantly however. There were no changes in activator protein C/Alpha I antitrypsin complexes, plasminogen, or antithrombin III. In a study of the Framingham offspring by Gebara et al. [17] premenopausal women and women on hormone replacement therapy were compared to males and postmenopausal women not receiving hormone replacement therapy. Those individuals with high estrogen levels were found to have greater fibrinolytic potential with lower PAI I levels. Similarly, Koh et al. [18] in a study of 30 postmenopausal women found that conjugated estrogen, either by itself or with combined progestin replacement, resulted in lower PAI I levels by about 50%. D-dimer levels climbed which is consistent with increased fibrinolysis. Transdermal estradiol had no effect on PAI I levels.

Chetkowski et al. [19] compared the effects of conjugated equine estrogens at doses of 0.625 and 1.25 with transdermal estradiol at doses of 25, 50, 100, and 200 micrograms per 24 hours. They measured fibrinopeptide A, fibrinogen, and antithrombin III levels and activity and found no significant change with either transdermal or conjugated oral estrogens.

Poller et al. [20], in a study of 21 women, found that conjugated estrogens resulted in increased levels of Factor VII and X, accelerated prothrombin time and increased platelet aggregation. Stangel et al. [21] found levels of thrombin and heparin antithrombin activity suggesting a hyper-coagulable state in 57.2% of a group of women taking hormone replacement but in only 14.7% of women not receiving replacement.

Epidemiologic studies such as the Framingham Study have also found that fibrinogen levels are an independent risk factor for the development of cardiovascular disease. They also found that fibrinogen levels rise after menopause.

In the Atherosclerosis Risk in Community Study, Nabulsi et al. [22] studied 15,800 subjects, comparing estrogen users with estrogen nonusers. They found that the estrogen users had lower levels of fibrinogen and antithrombin III than nonusers. However, Factor VII and protein C levels were higher in estrogen users.

The PEPI Trial [5] of 875 women who were randomized to receive either placebo, or conjugated equine estrogens with either medroxyprogesterone acetate serially or sequentially or micronized progesterone cyclically found that fibrinogen levels rose in the placebo group but remained stable in estrogen users. The addition of various doses and types of progestins did not significantly change this observation.

Part of the effect of hormone replacement on platelet behavior is probably due to an effect on arachidonic acid metabolism. Arachidonic acid can be metabolized into prostaglandin I_2 which causes vasodilatation and inhibits platelet aggregation or to thromboxane A_2 which causes vasoconstriction and stimulates platelet aggregation. Studies in the rat and in the human have both demonstrated that hormone replacement therapy results in higher levels of the metabolites of prostaglandin I2 and reduction in the metabolites of thromboxane A_2 [23].

With regard to outcome, three recent studies have reported an increased incidence of thromboembolism in women receiving hormone replacement therapy. Daly et al. [24], in a case-control British Study of women admitted with a suspected diagnosis of venous thromboembolism, found that the risk of venous thromboembolism was increased over threefold in current users of hormone replacement therapy. This risk was highest among short-term current users. Jick et al. [25], in a study of participants in The Group Health Cooperative of Puget Sound compared cases of women discharged from the hospital with a diagnosis of deep vein thrombosis or pulmonary embolism with control subjects and found that 49% of the women with thromboembolism were current users of estrogen while only 26% of the control group were estrogen users. The relative risk, comparing current estrogen users with nonusers, was 3.2 (95% CI 1.5-6.8). The Nurses' Health Study [26], which involved 112,593 women, documented 123 cases of primary pulmonary embolism. This study found that current postmenopausal hormone users had an increased risk of primary pulmonary embolism of 2.1 (95% CI 1.2-3.8).

References

1. Bush TL. The epidemiology of cardiovascular disease in postmenopausal women. Ann NY Acad Sci 1990;592:263-271.
2. Sacks FM, Walsh BW. The effect of reproductive hormones on serum lipoproteins: Unresolved issues in biology and clinical practice. Ann NY Acad Sci 1990;592:272-285.
3. Walsh BW, Schiff I, Rosner B, et al. Effects of postmenopausal estrogen replacement on the concentrations and metabolism of plasma lipoproteins. N Engl J Med 1991;325:1196-1204.
4. Colvin PL, Auerbach BJ, Case LD et al. A dose-response relationship between sex hormone-induced change in hepatic triglyceride lipase and high-density lipoprotein cholesterol in postmenopausal women. Metabolism 1991;40:1052-56.
5. The Writing Group for the PEPI Trial. Effects of estrogen or estrogen/progestin regimens on heart disease risk factors in postmenopausal women. JAMA 1995;273:199-208.
6. Darling GM, Johns JA, McCloud PI, Davis SR. Estrogen and progestin compared with simvastatin for hypercholesterolemia in postmenopausal women. N Engl J Med 1997;337:

595-601.
7. Davidson MH, Testolin LM, Maki KC, vonDuvillard S, Drennan KB. A comparison of estrogen replacement, pravastatin, and combined treatment for the management of hypercholesterolemia in postmenopausal women. Arch Intern Med 1997;157:1186-92.
8. Clarkson TB, Anthony MS, Jerome CP. Lack of effect raloxifene on coronary artery atherosclerosis of postmenopausal monkeys. J Clin Endocrinol Metab 1998;83:721-26.
9. Sacks FM, McPherson R, Walsh BW, et al. Effect of postmenopausal estrogen replacement on plasma Lp(a) lipoprotein concentrations. Arch Intern Med 1994;154:1106-10.
10. Sacks MN, Rader DJ, Cannon RO, et al. Oestrogen and inhibition of oxidation of low-density lipoproteins in postmenopausal women. Lancet 1994;343:269-70.
11. Adams MR, Williams JK, Clarkson TN, et al. Effects of oestrogens and progestogens on coronary atherosclerosis and osteoporosis of monkeys. Bailliere's Clinical Obstetrics and Gynaecology 1991;5:915-34.
12. Fowlkes L, Sullivan JM. Estrogens, blood pressure and cardiovascular disease. Cardiol Rev 1995;3(2):106-14.
13. Gordon EM, Williams SR, Frenchek B, Mazur CA, Speroff L. Dose-dependent effects of postmenopausal estrogen and progestin on antithrombin III and factor XII. J Lab Clin Med 1988;111:52-56.
14. Alkjaersig N, Fletcher AP, De Ziegler D, Steingold KA, Meldrum DR, Judd HL. Blood coagulation in postmenopausal women given estrogen treatment: Comparison of transdermal and oral administration. J Lab Clin Med 1988;111:224-28.
15. Bar J, Tepper R, Fuchs J, Pardo J, Goldberger S, Ovadia J. The effect of estrogen replacement therapy on platelet aggregation and adenosine triphosphate release in postmenopausal women. Obstet Gynecol 1993;81:261-64.
16. Gilabert J, Estelles A, Cano A, et al. The effect of estrogen replacement therapy with or without progestogen on the fibrinolytic system and coagulation inhibitors in postmenopausal status. Am J Obstet Gynecol 1995;173(6):1849-54.
17. Gebara OCE, Mittleman MA, Sutherland P, et al. Association between increased estrogen status and increased fibrinolytic potential in the Framingham Offspring Study. Circulation 1995;91:1952-58.
18. Koh KK, Mincemoyer R, Bui MN, et al. Effects of hormone-replacement therapy on fibrinolysis in postmenopausal women. N Engl J Med 1997;336:683-90.
19. Chetkowski RJ, Meldrum DR, Steingold KA, et al. Biologic effects of transdermal estradiol. N Engl J Med 1986;314:1615-20.
20. Poller L, Thomson JM, Coope J. Conjugated equine estrogens and blood clotting: A followup report. Br Med J 1977;1:935-36.
21. Stangel JJ, Innerfield I, Reyniak J, Stone ML. The effect of conjugated estrogens on coagulability in menopausal women. Obstet Gynecol 1977;49:314-16.
22. Nabulsi AA, Ch B, Folsom AR, et al. Association of hormone-replacement therapy with various cardiovascular risk factors in postmenopausal women. N Engl J Med 1993;328:1069-75.
23. Fogelberg M, Vesterqvist O, Dicfalusy U, et al. Experimental atherosclerosis: Effects of oestrogen and atherosclerosis on thromboxane and prostacyclin formation. Eur J Clin Invest 1990;20:105-10.
24. Daly E, Vessey MP, Hawkins MM, Carson JL, Gough P, Marsh S. Increased risk of

venous thromboembolism in hormone replacement therapy users. Lancet 1996;348:977-80.

25. Jick H, Jick SS, Gurewich V, Myers MW, Vasilakis C. Risk of idiopathic venous thromboembolism among users of postmenopausal estrogens. Lancet 1996;348:981-82.
26. Grodstein F, Stampfer MJ, Goldhaber SZ, et al. Prospective study of exogenous hormones and risk of pulmonary embolism in women. Lancet 1996;348:983-87.

CORONARY HEART DISEASE IN WOMEN: STATUS 1998

Nanette Kass Wenger

Despite major recent advances in the recognition and management of coronary heart disease as a preeminent health problem for women, its high prevalence and lethality persist. Coronary heart disease is the major cause of mortality for U.S. adult women, accounting for about 45% of all female deaths [1,2]. The total mortality from myocardial infarction, stroke, and other cardiovascular diseases has been greater for U.S. women than for men each year since 1984, because the U.S. population of older women substantially exceeds in size that of older men. Prominent age-dependency is a gender-specific characteristic of coronary heart disease in women in that 1 of 8 or 9 women aged 45-64 years has clinical evidence of coronary disease, as compared with 1 of 3 women older than 65 years of age. [1] Any initial clinical manifestation of coronary heart disease occurs about 10 years later for women than for men, with myocardial infarction sustained as much as 20 years later; explanations for this delayed onset remain poorly understood. Once coronary heart disease becomes clinically evident, it results in substantial morbidity and disability for women, compromising the functional status and life quality, particularly for older women; as well, coronary heart disease contributes prominently to hospital admissions and physicians visits for women.

Coronary risk factors are highly prevalent among women [3,4] of all racial and ethnic groups in the U.S. Only 30% of U.S. women are free of at least 1 major coronary risk factor, with even this small percentage decreasing in women of older age. Female-male crossover with aging is evident for several major coronary risk factors; whereas hypertension, hypercholesterolemia and diabetes predominate in men of young-to-middle age, these risk attributes escalate in women at older age to exceed the prevalence for men. The cornerstones of coronary risk reduction [5] include smoking cessation, nonpharmacologic and pharmacologic control of lipid abnormalities and of hypertension, regular moderate-intensity physical activity, a healthy diet, and weight control. Continued investigation is warranted regarding the role of a risk intervention unique to women–the use of postmenopausal hormone therapy, [6] as well as of selective estrogen receptor modulator use. [7] Data from randomized clinical trials currently in progress should better guide the application of these therapies.

Angina pectoris is the predominant initial and subsequent clinical manifestation of coronary heart disease in women, in contrast to myocardial infarction and sudden death for men. [8] Although women who present with angina pectoris are older than men and more

R. Paoletti et al. (eds.), Women's Health and Menopause, 105–108.

frequently have comorbid problems, they less often have had either prior myocardial infarction or myocardial revascularization. [9] Since women with an initial myocardial infarction are more likely than men to have had antecedent stable angina pectoris, [10] it must be ascertained whether earlier noninvasive risk stratification and resultant appropriate interventions may avert myocardial infarction and improve prognosis. As knowledge about the adverse outcomes of coronary heart disease in women has evolved, U.S. physicians currently more frequently undertake objective testing earlier in women with chest pain syndromes. [11] Coronary arteriography appears to be the most important determinant of access to myocardial revascularization procedures; with comparable coronary arteriographic obstruction, comparable myocardial revascularization is now undertaken in women and men, with no gender differences in event rates evident during followup. [12])

Following both myocardial infarction and myocardial revascularization procedures, clinical outcomes are less favorable for women than for men. Women are more likely to die during an episode of myocardial infarction than are men, and first year postinfarction mortality is comparably greater for women; earlier and more frequent reinfarction occurs among female than male survivors [13]. The contribution of gender *per se* versus that of excess comorbidity, particularly diabetes and hypertension, and of the older age of women at infarction remains to be ascertained. Equally pivotal is the less frequent application of acute coronary catheterization, coronary thrombolysis, use of beta blocking drugs and of aspirin [14], postinfarction risk stratification procedures, coronary revascularization procedures, and referral to cardiac rehabilitation. Women have a doubled mortality from coronary artery bypass graft surgery [15,16] compared with men as well as less graft patency, less symptomatic relief, greater perioperative infarction and heart failure, and more frequent reoperation within the initial postoperative 5 years. However survival is improved among operated women compared with those who are surgical candidates and refuse or are not referred for operation [17,18]. Despite comparable initial favorable outcomes of percutaneous transluminal coronary angioplasty to their male counterparts, women have less long-term symptomatic relief and decreased long-term survival rates, the latter reflecting their older age. Data from the NACI registry [19] highlight comparable favorable outcomes by gender with new intravascular devices, including coronary stents, despite more frequent in-hospital complications for women. Again, the contribution to the unfavorable outcome of myocardial revascularization procedures of gender versus comorbidity versus older age versus the greater likelihood of severe and unstable angina mandating more frequent urgent or emergency revascularization must be carefully examined. As well, the impact of contemporary changes in clinical practice must be carefully reviewed. During the past decade, in part reflecting the doubling of coronary arteriography among U.S. women and in part reflecting the increased application of myocardial revascularization at elderly age, there has been an almost threefold increase in the rates of coronary artery bypass graft surgery and transcatheter revascularization procedures among women. The challenge in comparing their outcomes to those of women in prior years is that older and sicker women currently receive myocardial revascularization. [20]

These data underscore the need for adoption of a healthy lifestyle and for other aspects of coronary prevention for women across their lifespan. Research needs include the optimal methods of effecting and maintaining coronary risk reduction among women. As well, current gender-specific evaluations of diagnostic and therapeutic strategies offer promise to identify approaches most likely to be associated with an improved prognosis for women. Nonetheless, unless women and their health care providers appreciate the vulnerability of women to coronary heart disease, coronary preventive care and prompt clinical recognition of coronary heart disease and appropriate therapeutic interventions are unlikely to be undertaken.

References

1. Wenger NK. Coronary heart disease in women: Evolving knowledge is dramatically changing clinical care. In: Julian DG, Wenger NK, editors. Women and heart disease. London: Martin Duntiz Publishers, 1997:21-38.
2. Wenger NK, Speroff L, Packard B. Cardiovascular health and disease in women. N Engl J Med 1993;329:247-56.
3. Eaker ED, Chesebro JH, Sacks FM, Wenger NK, Whisnant JP, Winston M. Cardiovascular disease in women. Circulation 1993;88(part 1):1999-2009.
4. National Center for Health Statistics. Health: United States, 1990. U.S. Public Health Services. Hyattsville, Maryland, 1991, Centers for Disease Control.
5. Rich-Edwards JW, Manson JE, Hennekens CH, Buring JE. The primary prevention of coronary heart disease in women. N Engl J Med 1995;332:1758-66.
6. Wenger NK. Postmenopausal hormone therapy: Is it useful for coronary prevention? Cardiol Clin 1998;16:17-25.
7. Barrett-Connor E, Wenger NK, Grady D, et al. Hormonal and nonhormonal therapy for the maintenance of postmenopausal health: The need for randomized controlled trials of estrogen and raloxifene. J Women's Health 1998; in press.
8. Lerner DJ, Kannel WB. Patterns of coronary heart disease morbidity and mortality in the sexes: A 26-year follow-up of the Framingham population. Am Heart J 1986;111:383-90.
9. Pepine CJ, Abrams J, Marks RG, Morris JJ, Scheidt SS, Handberg E, for the TIDES Investigators. Characteristics of a contemporary population with angina pectoris. Am J Cardiol 1994;74:226-31.
10. Goldberg RJ, Gorak EJ, Yarzebski J, et al. A communitywide perspective of sex differences and temporal trends in the incidence and survival rates after acute myocardial infarction and out-of-hospital deaths caused by coronary heart disease. Circulation 1993;87:1947-53.
11. Lauer MS, Pashkow FJ, Snader CE, Harvey SA, Thomas JD, Marwick TH. Gender and referral for coronary angiography after treadmill thallium testing. Am J Cardiol 1996;78:278-83.
12. Sullivan AK, Holdright DR, Wright CA, Sparrow JL, Cunningham D, Fox KM. Chest pain in women: Clinical, investigative, and prognostic features. BMJ 1994;308:883-86.
13. Kudenchuk PJ, Maynard C, Martin JS, Wirkus M, Weaver WD, for the MITI Project Investigators. Comparison of presentation, treatment, and outcome of acute myocardial infarction in men versus women (The Myocardial Infarction Triage and Intervention

Registry). Am J Cardiol 1996;78:9-14.
14. McLaughlin TJ, Soumerai SB, Willison DJ, et al. Adherence to national guidelines for drug treatment of suspected acute myocardial infarction. Evidence for undertreatment in women and the elderly. Arch Intern Med 1996;156:799-805.
15. Maynard C, Weaver WD. Treatment of women with acute MI: New findings from the MITI Registry. J Myocard Ischemia 1992;4:27-37.
16. O'Connor GT, Morton JR, Diehl MJ, et al for the Northern New England Cardiovascular Disease Study Group. Differences between men and women in hospital mortality associated with coronary artery bypass graft surgery. Circulation 1993;88(part 1):2104-10.
17. Shaw LJ, Miller DD, Romeis JC, Kargl D, Younis LT, Chaitman BR. Gender differences in the noninvasive evaluation and management of patients with suspected coronary artery disease. Ann Intern Med 1994;120:559-66.
18. Davis KB, Chaitman B, Ryan T, Bittner V, Kennedy JW. Comparison of 15-year survival for men and women after initial medical or surgical treatment for coronary artery disease: A CASS Registry Study. J Am Coll Cardiol 1995;25:1000-1009.
19. Robertson T, Kennard ED, Mehta S, et al, for the NACI Investigators. Influence of gender on in-hospital clinical and angiographic outcomes and on one-year follow-up in the New Approaches to Coronary Intervention (NACI) Registry. Am J Cardiol 1997;80(10A):26K-39K.
20. Jacobs AK, Kelsey SF, Yeh W, et al. Documentation of decline in morbidity in women undergoing coronary angioplasty (A report from the 1993-94 NHLBI Percutaneous Transluminal Coronary Angioplasty Registry). Am J Cardiol 1997;80:979-84.

INTRODUCTION TO OSTEOPOROSIS

Claus Christiansen

Introduction

Osteoporosis is a major and growing health problem worldwide. It affects an estimated 75 million people in the United States, Europe, and Japan combined, including one in three postmenopausal women and a majority of the elderly, including a substantial number of men.

The currently accepted conceptual definition of osteoporosis is that it is a systemic skeletal disease characterized by low bone mass and microarchitectural deterioration of bone tissue, with a consequent increase in bone fragility and susceptibility to fractures. Bone mineral can be measured with acceptable accuracy and precision and forms the basis for an operational definition of osteoporosis with better clinical utility. A study group of the World Health Organization (WHO) has proposed diagnostic guidelines for the interpretation of a bone mass measurement in Caucasian women [1]:

1. Severe osteoporosis: Bone mineral density (BMD) more than 2.5 standard deviations (SD) below the mean value of peak bone mass in young normal women and the presence of fractures.
2. Osteoporosis: BMD more than 2.5 SD below the mean value of peak bone mass in young normal women.
3. Low bone mass (or osteopenia): BMD within -1 SD and -2.5 SD of the mean value of peak bone mass in young normal women.
4. Normal: BMD not more than 1 SD below the mean value of peak bone mass in young normal women.

This operational definition is practical but clearly not perfect and as in the diagnosis and assessment of most disorders, history, physical, and biochemical examination are important features which should be included in the diagnostic and therapeutic considerations.

Bone Mass Measurements

A number of techniques are available to evaluate bone density in the peripheral, central, or entire skeleton as well as the cancellous or cortical bone envelopes. These include radiographic absorptiometry (RA), single X-ray absorptiometry (SXA), dual X-ray

R. Paoletti et al. (eds.), Women's Health and Menopause, 109–116.

absorptiometry (DXA), spinal and peripheral quantitative computed tomography (QCT/pQCT), and quantitative ultrasound (QUS).

Techniques relying on the absorption of X-ray all work on the same principle but differ from ultrasound. In order to be useful, such methods need to be accurate, precise, rapid, reliable, inexpensive, and expose patients to minimal radiation.

The most widely used methods for bone density measurement are single and dual X-ray absorptiometry for peripheral (forearm, heel) or axial (hip, spine) measurements. Several studies have shown that bone mineral density (BMD) measured by these methods predicts subsequent fracture occurrence.

Recently, a meta-analysis of prospective cohort studies published between 1985 and 1994 was completed by Marshall et al. [2]. A summary of the meta-analysis is shown in Table 1.

Table 1. Meta-analysis of Bone Mass Measurements to Predict the Risk of Fracture [2]

Site of measurement of bone density	Fracture Type			
	Forearm	Hip	Vertebral	All
Measurement by methods other than ultrasound				
Proximal radius	1.8 (1.5-2.1)	2.1 (1.6-2.7)	2.2 (1.7-2.6)	1.5 (1.3-1.6)
Distal radius	1.7 (1.4-2.0)	1.8 (1.4-2.2)	1.7 (1.4-2.1)	1.4 (1.3-1.6)
Hip	1.4 (1.4-1.6)	2.6 (2.0-3.5)	1.8 (1.1-2.7)	1.6 (1.4-1.8)
Lumbar spine	1.5 (1.3-1.8)	1.6 (1.2-2.2)	2.3 (1.9-2.8)	1.5 (1.4-1.7)
Calcaneous	1.6 (1.4-1.8)	2.0 (1.5-2.7)	2.4 (1.8-3.2)	1.5 (1.3-1.8)
All	1.6 (1.5-1.7)	2.0 (1.7-2.4)	2.1 (1.9-2.3)	1.5 (1.4-1.6)
Measurement by ultrasound				
Calcaneous		2.2 (1.8-2.7)	1.8 (1.5-2.2)	1.5 (1.4-1.7)

As it appears, every measurement site has the same capability to predict the overall risk fracture (including the risk of hip and spinal fracture). Probably, the patient who receives a bone mass measurement is interested in knowing the overall fracture risk. With this information, the patient and the physician are able to discuss and decide whether or not a treatment should be instituted.

Quantitative ultrasound is a diagnostic tool with possibility for clinical use, related to its lack of radiation exposure and low cost. Most systems measure the calcaneus, but other sites such as patella, tibia, and phalanges are being investigated. Providing a specific area of interest and knowing the size of the bone may be important to improve precision. Ultrasound predicts the risk of fractures, but it is unclear if it can replace existing bone density measurements or augment the information obtained from them. Criteria for clinical use of ultrasound in prognosis of osteoporosis are probably similar to those for bone density measurements.

The assessment of bone density forms the corner-stone for the diagnosis of osteoporosis, however, bone mineral measurements should only be obtained when decisions regarding treatment are dependent on the results. Diagnostic thresholds should not be confused with interventional thresholds which may depend on factors other than bone density. For example previous fracture, menopausal status, and age may modify the decision to treat.

Bone densitometry may also be used to evaluate response to treatment. Although there is limited information upon which to base a recommendation for repeat measurements, the current consensus is that bone mass should be measured at least twice following initiation of treatment at intervals annually or greater.

Treatment options which have their only pharmacological effect on the skeleton should not be utilized without measurement of bone density, except in very specific subgroups, e.g. those who present with multiple osteoporotic fractures.

Estrogen and Estrogen-Progestogen Replacement Therapy

The metabolism of calcium changes dramatically when estrogen production declines in women. The main characteristic is increased bone remodeling, which is reflected in high serum concentrations of biochemical estimates of bone resorption and bone formation. This increased bone turnover is maintained throughout life and is also seen in women with symptomatic postmenopausal osteoporosis [3].

All conditions of estrogen deprivation result in loss of bone. This includes natural or surgical menopause and drugs that inhibit estrogen production or its effect (luteinizing hormone releasing hormone agonists, antagonists, or antiestrogens). Strenuous exercise or other conditions that provoke anovulation also may result in bone loss. Withdrawal of estrogen primarily affects bone resorption, as shown by a rapid increase in the biochemical estimates of bone resorption. Owing to the coupling of bone resorption and formation, bone formation will show a secondary increase, reflected by a delayed elevation in the biochemical estimates of bone formation [4].

In early postmenopausal women, numerous studies have demonstrated that oral estrogen and estrogen-progestogen regimens stop bone loss. This is true for use of both synthetic and nonsynthetic estrogens, keeping in mind that nonsynthetic estrogens should be preferred for postmenopausal therapy. The estrogenic effect on bone is dose dependent; if a sufficient serum concentration of estrogen is not obtained, bone loss will not be arrested completely. For oral estrogen and estrogen-progestogen therapy, studies have demonstrated that the doses of 0.625 mg of conjugated estrogens and 2 mg of 17β-estradiol prevents early postmenopausal bone loss [5-7]. However, recent studies [8-12] suggest that lower doses may be sufficient to prevent bone loss which is important since adverse effects are dose-dependent.

In addition to oral HRT, other delivery systems are available. Percutaneous 17β-estradiol, that is, estradiol given in a gel applied every day on the skin, prevents skeletal bone loss as effectively as oral HRT [13]. The same is the case with transcutaneous estradiol, that is, estradiol given in a patch that is changed two to three times per week. Studies have demonstrated that this delivery system also prevents bone loss [14].

When HRT is stopped, bone loss recurs. The rate of bone loss after stopping HRT has been shown to be accelerated compared to average postmenopausal bone loss [15]. However, others have shown that the rate of bone loss after cessation of HRT is similar to average bone loss [16]. Several epidemiological studies have shown that HRT provides protection against fracture [17,18]. However, it is obvious that the treatment has to be continued for some years to be effective. It is difficult to suggest a definite treatment period of time, but 5 to 10 years is what may be necessary before this fracture protection can be observed, if estrogens are prescribed in the immediate postmenopausal period. A number of case-control studies have demonstrated that HRT given for at least 6 years reduces the risk of hip fractures and Colles' fractures by 50%. Cohort studies have also shown that long-term HRT reduces the incidence of vertebral deformities in postmenopausal women by about 90% [3].

Estrogen Analogs

ANTIESTROGENS AND SELECTIVE ESTROGEN RECEPTOR MODULATORS (SERMS)

In 1973, McGuire and Chamness (In: O'Malley BW, Means AR, editors. Receptors for Reproductive Hormones, Plenum Press), summarized their work on the estrogen receptor in animal and human breast tumors, and in so doing described a target for therapeutic intervention. At that time there were no clinically useful antiestrogens, but the subsequent development of tamoxifen for breast cancer therapy has revolutionized the approach to treatment. In addition to breast cancer therapy with antiestrogens a new strategy is being developed to exploit the target site for the beneficial actions of estrogen, i.e. prevention of bone loss and cardiovascular disease [19].

Today, three so-called selective estrogen receptor modulators (SERMs) are evaluated as a treatment for osteoporosis, i.e. raloxifene, levomeloxifene, and droloxifene.

Raloxifene is the compound which is in the most advanced stage of development, and the first interim analysis of the first phase three studies has just been published [20]. Raloxifene is a selective estrogen receptor modulator that in animal models act as an estrogen receptor antagonist in breast and endometrial tissue, but as an estrogen agonist in the skeletal and cardiovascular systems. The clinical trials have demonstrated that raloxifene is well tolerated and normalizes bone turnover and slows the bone loss in healthy early postmenopausal women without stimulating uterus and breast tissues. In contrast, preliminary data suggests that raloxifene may even decrease the risk of breast cancer [20].

In elderly women with osteoporosis, defined as more than one prevalent vertebral fracture, the same effects were seen as in the younger women, i.e. a normalization of bone turnover and reduction and even increase in bone mass [21]. In this study, which was a preliminary fracture study, the effect on fracture incidence was of borderline statistical significance. There were overall multiple beneficial effects on bone, although, under the conditions of that study, they appear to be less prominent than have been reported with estrogen [21].

Raloxifene has been shown significantly to reduce aortic atherosclerosis in cholesterol-fed rabbits [22]. In the clinical trials reported so far, raloxifene accordingly lowers serum cholesterol and LDL-cholesterol significantly, both in early and late postmenopausal women [21]. Raloxifene is expected to be registered in the first countries within the next year or so.

Levomeloxifene, another SERM compound, has previously been shown to inhibit bone loss and arterial cholesterol accumulation in estrogen deplete animal models without stimulation of the endometrial glands and the epithelium. Different doses of levomeloxifene decrease bone turnover parameters and serum cholesterol and LDL-cholesterol in postmenopausal women [23].

Droloxifene, a third SERM compound, has been shown to decrease bone turnover, prevent bone loss and reduce total serum cholesterol in ovariectomized rats without stimulating the endometrium [24].

Bisphosphonates

The bisphosphonates are a class of drugs developed in the past two decades for use in various diseases of bone and teeth, as well as in calcium metabolism. Bisphosphonates are compounds characterized by two carbon-phosphorus bonds. They are analogous with the physiologically occurring inorganic pyrophosphate in which an oxygen atom has been replaced by a carbon atom. The structure of bisphosphonates allows for a great number of variations. Each bisphosphonate has its own physicochemical and biological characteristics. One should thus not speak generally about the effect of "the bisphosphonates" but rather consider each bisphosphonate on its own.

It is known that the pyrophosphate impairs the crystallization as well as the dissolution of calcium phosphate crystals. This effect is apparently related to the strong surface absorption of pyrophosphate on solid-phase calcium phosphate. Bisphosphonates act in a similar way and thus have the ability both to block the production of apatite crystals and to delay the dissolution of the crystals. Bisphosphonates, when given both parenterally and orally, have been found to inhibit experimentally induced calcification of soft tissues, such as arteries, kidney, skin, and heart. If administered in sufficiently high doses, certain bisphosphonates can also inhibit the normal calcification that occurs in bone, cartilage, and teeth. Bisphosphonates are resorption inhibitors, as are estrogen and calcitonin; that is, they decrease the rate of bone resorption.

Bisphosphonates have undergone a revolutionary development during the past 5 years and they are now a reality as an option for osteoporosis prevention and treatment. In the Unites States, and in many other countries, alendronate (Fosamax) is now approved for both prevention and treatment of osteoporosis. The approval was based on a number of very large double-blind studies in both early and late postmenopausal women as well as in osteoporotic women.

Prevention of Bone Loss

For prevention of bone loss a dose of 5 mg/day of alendronate is approved [25]. This dose has now been studied for 5 years [26] and shows that in the spine there is an increase in bone density within the first two years of treatment by a total of approximately 4%. Thereafter a plateau is reached, as would be expected by an antiresorptive agent (see above), and bone mass is hereafter kept constant. The same pattern is seen in the hip, forearm, and total body.

Other bisphosphonates are under development for prevention of bone loss including ibandronate [27]).

Treatment of Osteoporosis and Prevention of Fracture

For treatment of women who already have osteoporosis a dose of 10 mg/day of alendronate is approved. A very large study [28] showed that alendronate reduced the risk of the most common forms of osteoporotic fracture (spine, hip, wrist) by approximately 50% in women with low hip bone density and previous fracture. Also women with no prevalent fracture at the enrollment into the study had fracture incidence decreased by approximately 50% [29].

In many countries, but not in the United States, etidronate is approved for treatment. In two double-blind, placebo-controlled studies [30,31], etidronate was given orally at 400 mg/day for 2 weeks, followed by 10 to 13 weeks of calcium supplementation (a cyclical regimen). After 2 to 3 years of study, vertebral fractures in cyclical etidronate-treated patients were reduced by 50% compared to patients receiving calcium. The treatment regimen also produced increases in vertebral bone mass of 4% to 8% versus control over the 2 to 3 years. Furthermore, cyclic etidronate maintained cortical bone at the hip and wrist.

References

1. World Health Organization. Assessment of fracture risk and its application to screening for postmenopausal osteoporosis. WHO Technical Report Series 843. Geneva:WHO, 1994.
2. Marshall D, Johnell O, Wedel H. Meta-analysis of how well measures of bone mineral density predict occurrence of osteoporotic fractures. BMJ 1996;312:1254-59.
3. Riggs BL, Mann KG. Assessment of bone turnover in osteoporosis using biochemical marker. In: Christiansen C, et al, editors. Osteoporosis, vol 2. Copenhagen: Osteopress

ApS, 1987: 672-76.
4 Christiansen C, Roedbro P, Tjellesen L. Serum alkaline phosphatase during hormone treatment in early postmenopausal women. Acta Med Scand 1984;216:11-17.
5. Horsman A, Jones M, Francis R, Nordin BBC. The effect of estrogen dose on postmenopausal bone loss. N Engl J Med 1983;309:1405-7.
6. Christensen MS, Hagen C, Christiansen C, Transbol I. Dose response evaluation of cyclic estrogen/gestagen in postmenopausal women. Placebo-controlled trial of its gynecologic and metabolic actions. Am J Obstet Gynecol 1982;144:873-79.
7. Lindsay R, Hart CM, Clark DM. The minimum effective dose of estrogen for prevention of postmenopausal bone loss. Obstet Gynecol 1983;63:759-63.
8. Alexandersen P, Hassager C, Sandholdt I, Riis BJ, Christiansen C. The effect of cyclophasic hormone therapy on postmenopausal bone mass and bone turnover. J Bone Min Res 1997;12 (Suppl.1):S499
9. Bjarnason NH, Hassager C, Christiansen C. 17β oestradiol 1 mg and 2 mg in combinations with a new gestagen, Gestodene are equally preventive on bone loss in early postmenopausal women. Bone 1997;20,93S.
10. Bjarnason NH, Hassager C, Christiansen C. Profile of a new substitution principle: Low dose 17β estradiol and gestodene. Acta Obstet Gynecol Scand 1997;76(Suppl.167):S56.
11. Bjarnason NH, Bjarnason K, Haarbo J, Rosenquist C, Christiansen C. Tibolone: Prevention of bone loss in late postmenopausal women. J Clin Endocrinol Metab 1996;81:2419-22.
12. Stadberg E, Mattson L-C, Uvebrant M. Low doses 17-beta-estradiol and norethisterone acetate as continuous combined hormone replacement therapy in postmenopausal women: Lipid metabolic effects. Menopause 1996;3:90-96.
13. Christiansen C, Nilas L, Riis BJ, Rødbro P, Deftos L. Uncoupling of bone formation and resorption by combined oestrogen and progestogen therapy in postmenopausal osteoporosis. Lancet 1985,ii:800-801.
14. Ribot C, Tremollieraaes F, Pouillaaes JM. Cyclic Estraderm TTS 50 plus oral progestogen in the prevention of postmenopausal bone loss over 24 months. In: Christiansen C, Overgaard K, editors. Osteoporosis. Vol 2. Copenhagen: Osteopress ApS, 1990:1979-84.
15. Lindsay R, Hart DM, Maclean A, Clark AC, Kraszewski A, Garnwood J. Bone response to termination of estrogen treatment. Lancet 1978;1:1321-27.
16. Christiansen C, Christensen MS, Transbøl I. Bone mass in postmenopausal women after withdrawal of estrogen/gestagen replacement therapy. Lancet 1981;1:459-61.
17. Weiss NS, Ure CL, Ballard JH, Williams AR, Dalin JR. Decreased risk of fractures of the hip and lower forearm with postmenopausal use of estrogen. N Engl J Med 1980;303:1195-98.
18. Kiel DP, Felson DT, Andersen JJ, Wilson PWF, Moskowitz MA. Hip fracture and the use of estrogens in postmenopausal women. The Framingham study. N Engl J Med 1987;317:1169-74.
19. Jordan VC. Third annual William L. McGuire memorial lecture. "Studies on the estrogen receptor in breast cancer" - 20 years as a target for the treatment and prevention of cancer. Breast Cancer Res Treat 1995;36:367-85.
20. Delmas PD, Bjarnason NH, Mitlak BH, et al. The effects of raloxifene on bone mineral density, serum cholesterol, and uterine endometrium in postmenopausal women. N Engl J Med 1997;337:1641-47.

21. Lufkin EG, Whitaker R, Argueta R, Caplan RH, Nickelsen T, Riggs BL. Raloxifene treatment of postmenopausal osteoporosis. J Bone Miner Res 1997;12(Suppl.1):S150.
22. Bjarnason NH, Haarbo J, Byrjalsen I, Kauffman RF, Christiansen C. Raloxifene inhibits aortic accumulation of cholesterol in ovariectomized, cholesterol-fed rabbits. Circulation 1997;96: 1964-69.
23. Bjarnason K, Skrumsager BK, Kiehr B. Levomeloxifene, a new partial estrogen receptor agonist, demonstrates anti-resorptive and anti-atherogenic properties in postmenopausal women. J Bone Miner Res 1997;12(Suppl.1):S346.
24. Ke HZ, Chidsey-Frink KL, Oi H, et al. Droloxifene increases bone mass in ovariectomized rats with established osteopenia. J Bone Miner Res 1997;12(Suppl.1):S349.
25. Hosking D, Chilvers CED, Christiansen C, et al. Prevention of bone loss with alendronate in postmenopausal women under age 60 years of age. N Engl J Med 1998;338:485-92.
26. Weiss S, McClung M, Gilschrist N, et al. Five-years efficacy and safety of oral alendronate for prevention of osteoporosis in early postmenopausal women. J Bone Miner Res 1997;12 (Suppl.1):S144.
27. Ravn P, Clemmesen B, Riis BJ, Christiansen C. The effect on bone mass and bone markers of different doses of ibandronate: A new bisphosphonate for prevention and treatment of osteoporosis: a 1-year, randomized, double-blind, placebo-controlled dose-finding study. Bone 1996;5:527-33.
28. Black DM, Cummings SR, Karpf DB, et al. Randomised trial of effect of alendronate on risk of fracture in women with existing vertebral fracture. Fracture intervention trial research group. Lancet 1996;348:1535-41.
29. Cummings SR, Black DM, Thompson DE for the FIT research group. Alendronate reduces the risk of vertebral fractures in women without pre-existing vertebral fractures: Results of the fracture intervention trial. J Bone Miner Res 1997;12(Suppl.1):S149.
30. Storm TM, Thamsborg G, Steiniche T, Genant HK, Storensen OM. Effect of intermittent cyclical etidronate therapy on bone mass and fracture rate in women with postmenopausal osteoporosis. N Engl J Med 1990;322:1265-71.
31. Nelson BW, Harris ST, Genant HK, et al. Intermittent cyclical etidronate treatment of postmenopausal osteoporosis. N Engl J Med 1990;323:73-79.

Genetics of Osteoporosis

Laura Masi, Luigi Gennari, Alberto Falchetti, and Maria Luisa Brandi

Introduction

Understanding the genetic base of multifactorial diseases represents the future task for scientists in order to explain new physiopathological aspects of complex traits. One of these diseases is represented by osteoporosis (OP).

OP is a common disorder associated with reduced bone mineral density, affecting up to 40% of women and 12% of men at same point during life. Osteoporotic fractures are an increasing health care burden in all aging communities. A major determinant of fracture risk is represented by bone mineral density (BMD), independent of other factors such as aging *per se* and falls [1]. BMD depends upon both the peak bone mass achieved in adolescence and the subsequent bone loss. However, peak bone mass is the major determinant of bone mineral density for up to 10-20 years after menopause, until age-related factors become relatively more important in determining bone mass loss. Although OP is a multifactorial trait, genetic factors play an important effects on peak bone mass and in the pathogenesis of OP. The development in advanced techniques for measuring BMD made possible to have available a quantitative trait for segregation analysis. In fact, quantitative traits are defined as characters that everyone has, that are measurable and that exhibit a normal distribution in the population. Up to 75% of variation in BMD has been suggested to be under genetic influences [2]. However, the inclusion of OP in the list of genetic disorders is still debatable. Twin studies have shown a strong genetic effect of BMD at both peripheral and axial sites [2]. The largest genetic influence was observed at sites of high trabecular bone content. Although twin studies have been powerful tools for studying genetic effects, they show some limitations and can only imply but not prove genetic influence. A variety of experimental design appropriate models for establishing genetic background of OP have been proposed, such as linkage analysis, allele sharing methods, association studies, and experimental crosses [3] (Figure 1).

Family studies suggest a significant effect of genetic factors on peak bone mass. For example, using the early approach of metacarpal/cortical bone thickness, parent-offspring correlations indicated that bone mass was for a large portion genetically determined [4]. In addition sib-pair studies, in premenopausal daughters of women with OP, have also shown modest but significant reductions in lumbar spine, femoral neck, and femoral shaft BMD compared to premenopausal women without a family history of OP [5].

R. Paoletti et al. (eds.), Women's Health and Menopause, 117–123.

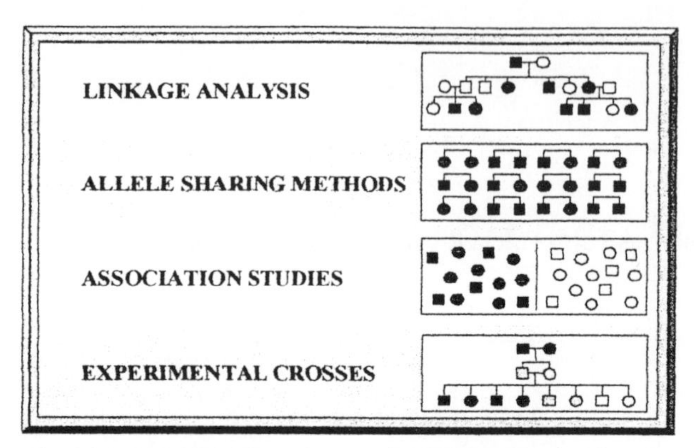

Figure 1. Genetic dissection of complex traits.

In the last five years association studies have provided new and contoversial information. The studies have compared allele frequencies for a particular polymorphism or candidate gene in disease population with that in nondisease population. Despite linkage analysis, where a physical connection between trait and marker locus must exist, an allele can be considered associated with a trait when it occurs more frequently in individuals with the trait than those without the trait. It confers an increased risk for disease to individuals carrying the associated allele (Figure 2). These studies have used the candidate gene approach and, given the number of factors that are likely to be involved, there is a seemingly unlimited supply of candidate genes for OP.

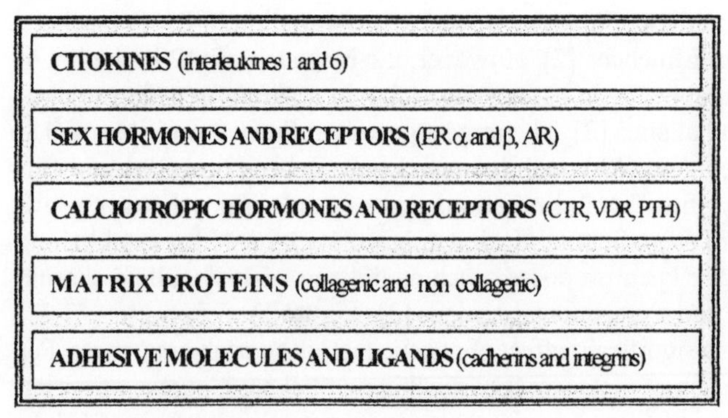

Figure 2. Candidate genes for osteoporosis.

Discrepancy among studies can be explained on the basis of the quantitative polygenic nature of this disorder, where the effect of a given gene can easily be modified by epistatic and/or pleiotropic effects of other genes. It is likely that interactions between different genes could, at least in part, explain the discrepancy among the studies. To date, association studies have represented the more frequently used approach for the identification of genetic effects in OP.

Candidate Genes Associated with Osteoporosis

VITAMIN D RECEPTOR GENE

Vitamin D exerts a central role in bone and calcium homeostasis, since, through the binding with its receptor, it influences bone mineralization, intestinal calcium absorption, and parathyroid hormone secretion. The vitamin D receptor gene (VDR gene) has been postulated as a major locus for genetic influences on bone mass and restriction fragment length polymorphisms at the 3' end of VDR gene, identified by Apa I, Bsm I, and Taq I endonucleases, have been originally correlated to variation in BMD and rates of bone loss in an Australian population [6]. The subjects homozygous for the presence of the Bsm I restriction endonuclease site were reported to have a bone mass about 15% higher than that of subjects homozygous for the presence of the site [6]. However, agreement on this relationship is not universal among different populations. In fact, successive reports confirmed the role of the VDR allelic variant 1) on determination of BMD at multiple sites [7-9], 2) on intestinal calcium absorption efficiency [10-12], and 3) on bone turnover rate [13]. Conversely, other studies failed to demonstrate a significant effect [14-17]. The cause of these discrepancies remain to be determined and in part it may be caused by the limited sample size of many studies. Indeed relationship between VDR gene alleles and osteoporotic fracture risk, mediated through differences in BMD, are unlikely to be observed in relatively small samples. Potential confusion in some of these studies may also be given from the health-based selection bias, with the tendency to exclude osteoporotic women. Heterogeneity is also likely, with different major genes segregating in different patient samples. In this regard, other polymorphic genes, such as the one encoding for estrogen receptor α, have been shown to modulate VDR gene effect in the determination of BMD [18]. In addition, environmental factors could also reciprocally interact with genetic factors. Recently, dietary calcium intake has been reported to contribute to the expression of VDR gene effect both on BM [19] and intestinal calcium absorption [10], raising the possibility that the influence of VDR genotypes on bone metabolism could be observed only among populations with a relatively low calcium intake.

ESTROGEN RECEPTOR GENE

Osteoblast, osteoclasts, and bone marrow stromal cells bear estrogen receptors (ER) and are modulated by estrogen. Recently, the clinical observation of an osteoporotic phenotype in a man with a disruptive mutation of the ER gene, indicated the ER as a candidate gene for osteoporosis [20]. It is possible that common allelic variants of the ER, causing milder estrogen resistance, exist and that compensatory hyperestrogenism can initially overcame the resistance resulting in normal phenotype. This compensatory balance could be altered later on by aging or by a condition like menopause, leading to clinical disorders such as osteoporosis. Both intronic polymorphisms (recognized by the restriction endonucleases Pvu II and Xba I) and polymorphic variable number of (TA)n repeat upstream the ERα, have been associated with BMD in the Japanese population [21,22]. The relationship between these polymorphisms and BMD has not been widely studied in larger samples and similar studies in other populations failed to confirm these data [23].

Calcitonin Receptor Gene

One of the hormones involved in bone metabolism is calcitonin (CT), a polypeptide hormone secreted by parafollicular cells of the thyroid gland [24] able to inhibit osteoclastic bone resorption and to stimulate urinary calcium excretion [25]. The human CT receptor (CTR) was cloned from an ovarian carcinoma cell line (BIN-67) [26] and recognized to be a prototypic member of a distinct family of G-protein-coupled receptors with seven spanning domains [27]. Recently, a previously unrecognized restriction fragment length polymorphism (RFLP) at the human CTR gene was detected by the restriction endonuclease Taq I. In that study the gene encoding for a peptide hormone receptor, the CTR, has been evaluated for its potential role in BMD determination in a population of postmenopausal women. The authors demonstrated a significant correlation between the CTR genotypes and lumbar BMD [28] . Recently, using SSCP analysis two alleles of human CTR gene have been identified in the French population but no association between the polymorphism recovered and development of osteoporosis was found [29]. In addition, Nakamura et al. [30] have described an Alu I RFLP at the CTR gene locus in the Japanese population. Significant differences on LS-BMD among various Alu I CTR allelic variants were recently described [31]. No significant variations of BMD among the CTR genotypes were observed in the femoral neck. The CTR gene becomes a new actor in the scenario of genes influencing BMD determination. As for other described genetic influences the history is just at the beginning and it is prudent to remain cautious about the significance of this correlation.

Collagen Type I Genes

Collagen type I is the major protein constituent of bone matrix, and therefore, it can easily be recognized as a candidate gene in determination of bone mass. Indeed, the osteoporotic phenotype of osteogenesis imperfecta is due to mutations that affect the coding regions of the collagen type I genes (COLIA1 and COLA2) [32]. Recently, Grant et al. showed a G/T polymorphism in the first intron strongly associated with BMD [33]. Additional data in other populations support these findings, although the mechanisms by which the described COLIA1 polymorphism associates with bone mass are currently unclear.

Final Consideration

The potential application of the genetic markers in the assessment of osteoporosis risk is of great interest, due to the stability of DNA. The potential use of genetic polymorphisms as predictive tools of osteoporotic risk becomes an attractive proposition. It is clear that because osteoporosis is a multifactorial disorder with highly variable and complex phenotype, genetic studies of osteoporosis/bone mass should be interpreted with caution. However, the use of this approach in the clinical practices once evaluated, could allow prediction of peak bone mass, responsiveness to treatment, and ultimately fracture risk. The area to be explored is enormous and the questions exceed the answers.

References

1. Hui SL, Slemanda CW, Johston CC. Age and bone mass as predictors of fracture in prospective study. J Clin Invest 1988;81:1804-9.
2. Pocock NA, Lisman IA, Hopper I, Yeates MG, Sambrook PN, Eberl S. Genetic determinants of bone mass in adults. J Clin Invest 1987;80:706-10.
3. Lander ES, Schork NJ. Genetic dissection of complex traits. Science 1994;265:2037-48.
4. Sambrook PN, Kelly PJ, White NA, Morrison NA, Eisman JA. Genetic determinants of bone mass. In: Marcus R, Feldman D, Kelsey J, editors. Osteoporosis. Academic Press, 1995: 477-82.
5. Seeman E, Hopper JL, Bach LA, et al. Reduced bone mass in daughters of women with osteoporosis. N Engl J Med 1989;320:554-58.
6. Morrison NA, Cheng JQI, Akifumi T, et al. Prediction of bone density from vitamin D receptor alleles. Nature 1994;367:284-87.
7. Matsuyama T, Ishii S, Yabuta K, Yamamori S, Morrison NA, Eisman JA. Vitamin D receptors and bone mineral density. Lancet 1995;345:1239-40.
8. Fleet JC, Harris SS, Wood RJ, Dawson-Hughes B. The Bsm I vitamin D receptor restriction fragment length polymorphism (BB) predicts low bone density in premenopausal black and white women. J Bone Miner Res 1995;10(6):985-90.
9. Riggs BL, Nguyen TV, Melton LJ, et al. The contribution of vitamin D receptor gene alleles to the determination of bone mineral density in normal and osteoporotic women. J Bone Miner Res 1995;10(6):991-96.
10. Dawson-Hughes B, Harris SS, Finneran S. Calcium absorption on high and low calcium intakes in relation to vitamin D receptor genotype. J Clin Endocrinol Metab 1995;80:3657-61.
11. Wishart JM, Horowitz M, Need AG, et al. Relations between calcium intake, calcitriol, polymorphisms of the vitamin D receptor gene, and calcium absorption in premenopausal women. Am J Clin Nutr 1997;65:798-802.
12. Gennari L, Becherini L, Masi L, et al. Vitamin D receptor genotypes and intestinal calcium absorption in postmenopausal women. Calcif Tissue Int 1997;61:460-63.
13. Ferrari S, Rizzoli R, Chevalley T, Slosman D, Eisman JA, Bonjour JP. Vitamin D receptor gene polymorphisms and change in lumbar spine bone mineral density. Lancet 1995;345: 423-24.
14. Looney JE, Yoon HK, Fischer M, et al. Lack of a high prevalence of the BB vitamin D receptor genotype in severely osteoporotic women. J Clin Endocrinol Metab 1995;80:

2158-62.
15. Lim SK, Park YS, Park JM, et al. Lack of association between vitamin D receptor genotypes and osteoporosis in Korean. J Clin Endocrinol Metab 1995;80:3677-81.
16. Garnero P, Borel P, Sornay-Rendu E, Arlot ME, Delmas PD. Vitamin D receptor gene polymorphisms are not related to bone turnover, rates of bone loss and bone mass in postmenopausal women: the OFELY study. J Bone Miner Res 1996;11:827-34.
17. Vandevyver C, Wylin T, Cassiman JJ, Raus J, Geusens P. Influence of the vitamin D receptor gene alleles on bone mineral density in postmenopausal and osteoporotic women. J Bone Miner Res 1997;12:241-47.
18. Gennari L, Becherini L, Masi L, et al. Vitamin D and estrogen receptor allelic variants in postmenopausal women: Evidence of multiple gene contribution on bone mineral density. J Clin Endocrinol Metab 1998;83:939-44.
19. Krall E, Parry P, Lichter JB, Dawson-Hughes B. Vitamin D receptor alleles and rates of bone loss : Influences of years since menopause and calcium intake. J Bone Miner Res 1995;10: 978-84.
20. Smith EP, Boyd J, Frank GR, et al. Estrogen resistance caused by a mutation in the estrogen-receptor gene in a men. N Engl J Med 1994;331:1056-61.
21. Sano M, Inoue S, Hosoi T, et al. Association of estrogen receptor dinucleotide repeat polymorphism with osteoporosis. Biochem Biophys Res Commun 1995;217:378-83.
22. Kobayashi S, Inoue S, Hosoi T, Ouchi Y, Shiraki M, Orimo H. Association of bone mineral density with polymorphism of the estrogen receptor gene. J Bone Miner Res 1996;11:306-11.
23. Han KO. Nonassociation of estrogen receptor genotypes with bone mineral density and estrogen responsiveness to hormone replacement therapy in Korean postmenopausal women. J Clin Endocrinol Metab 1997;82:991-95.
24. Bussolati G, Pearse AGE. Immunofluorescent localization of calcitonin in the "C cells" of pig and dog thyroid. J Endocrinol 1987;37:205-9.
25. Bijovet OLM, Van der Sluys Veer J, De Vries HR, Van Koppen ATJ. Natriuretic effect of calcitonin in man. N Engl J Med 1971;284:681-88.
26. Gorn AH, Lin HY, Yamin M, et al. Cloning, characterization and expression of a human calcitonin receptor from ovarian carcinoma cell line. J Clin Invest 1992;90:1726-35.
27. Segre GV, Goldring SR. Receptors for secretin, calcitonin, parathyroid hormone PTH/PTH-related peptide, vasoactive intestinal peptide, glucagonlike peptide 1, growth hormone-releasing hormone, and glucagon belong to a newly discovered G-protein-linked receptor family. Trends Endocrinol Metab 1993;4:309-14.
28. Masi L, Becherini L, Colli E, et al. Polymorphisms of the calcitonin receptor gene are associated with bone mineral density in postmenopausal women. Biop Bioch Res Comm 1998; in press.
29. Taboultet J, Frendoc JL, Delage-Mourroux R, Pichaud F, de Vernejoul MC. Evidence for two allelic forms of calcitonin receptor gene: distribution in normal and osteoporotic women. J Bone Min Res 1996;11(S1):S456.
30. Nakamura M, Zhang Z, Shan L, et al. Allelic variants of human calcitonin receptor in the Japanese population Hum Genet 1997;99:38-41.
31. Masi L, Becherini L, Gennari L, et al. Allelic variant of human calcitonin receptor: Distribution and association with bone mass in postmenopausal Italian women. Bioch Bioph Res Commun 1998;245:622-26.
32. Sykes B. Bone disease cracks genetics. Nature 1990;348:18-20.

33. Grant SFA, Reid DM, Blake G, Herd R, Fogelmaan I, Ralston SH. Reduced bone density and osteoporosis associated with polymorphic Sp1 site in the collagen type I alpha 1 gene. Nat Genet 1996;14:203-5.

The Clinical Management of Osteoporosis

Socrates E. Papapoulos

Introduction

The clinical presentation of a patient with osteoporosis will determine the diagnostic work-up and attention should be given to excluding causes of secondary osteoporosis especially in patients with vertebral fractures. This will not be discussed further and this short review will focus on the management of the patient in whom the diagnosis primary osteoporosis has been made. The overall management consists of an integrated approach including general measures, nonpharmacological and pharmacological interventions aiming at preventing osteoporotic fractures in patients who have not fractured yet or in those who have already sustained a fragility fracture. This subject has been recently reviewed elsewhere [1,2].

General Measures

In patients with fractures, early mobilization is essential because bed rest aggravates bone loss. Acute pain due to a recent vertebral fracture responds to bed rest, analgesics, heat, or transcutaneous electrical nerve stimulation (TENS) of paravertebral muscles to alleviate the spasm. Total bed rest should not exceed a few days and progressive mobilization should be recommended. Physiotherapy and an exercise program to maintain flexibility of the spine and to strengthen muscles are helpful. Lifting of heavy objects should be avoided. The use of a walker provides stability and confidence for the patients, especially following a hip fracture and psychological support is essential. All medications taken by the patient should be reviewed and those which predispose to falls or adversely affect bone mass and bone turnover should be discontinued or reduced to the lowest possible effective dose. Visual impairment is very common above the age of 65 years and should be corrected. Finally advice about potential indoor and environmental hazards should be given. Nutritional advice should be provided to all patients, especially elderly subjects living indoors. This includes advice about protein intake and most importantly about calcium and vitamin D. These are essential for bone health at all stages of life, but calcium and vitamin D deficiency or insufficiency are important risk factors for fractures, especially hip fractures in institutionalized elderly subjects. This has been demonstrated in a large French study of institutionalized women of a mean age of 84 years who were given either placebo or

R. Paoletti et al. (eds.), Women's Health and Menopause, 125–130.

calcium and vitamin D. After 3 years of treatment the risk of new fractures including those of the hip was decreased by about 30% in the subjects receiving active treatment [3,4]. In another study younger women (mean age 73.6 years) with a low dietary calcium intake (mean 431 mg/day) were treated with calcium 1,200 mg/day or placebo for 4.3 years [5]. At the end of the study there was a significant decrease in the incidence of new vertebral fractures in the women who had prevalent vertebral fractures and were treated with calcium as compared to those who received placebo. Finally, the value of optimal calcium and vitamin D intake was also recently shown in a study of healthy elderly men and women in which significantly fewer peripheral fractures were recorded in the subjects receiving calcium and vitamin D [6]. Thus, independently of any further advice, osteoporotic subjects should receive instructions about adequate calcium and vitamin D intake.

Nonpharmacological Interventions

Nonpharmacological interventions aim mainly at reducing the risk and impact of falls. Falls are very common in the elderly and approximately 30% of community dwelling men and women over 65 years of age fall each year. Most hip and wrist fractures are due to falls. Prevention of falls, therefore, seems a rational approach to preventing fractures. It should be mentioned, however, that only a very small proportion of the subjects who fall fractures. This makes the design of controlled trials of fall prevention with endpoint the prevention of fracture difficult. Thus, although a number of controlled studies has recently shown that the incidence of falls can be reduced by a variety of interventions, none has documented yet that this is associated with a reduction in the risk of fractures. Mobility impairment is one of the most important and potentially modifiable risk factors for falls and there is a decline in muscular skeletal function with aging. A number of controlled studies suggests that exercise can improve muscular skeletal function in elderly people and can reduce the risk of falls by an average of 10% although marked improvement has been reported in studies employing special forms of exercise [7]. It should be noted that in elderly subjects exercise has a minimal effect on bone mass and its beneficial effect is mostly exerted through improvement of the muscular function. Furthermore, this should not be intense as this can lead to opposite effects in frail individuals who have not been used to regular exercise. Much more promising and of direct clinical relevance have been studies of measures to reduce the impact of a fall. The risk of a hip fracture is increased for falls directly on the trochanteric area. Hip protectors may attenuate the impact for sustaining a fall on the hip by either absorbing the impact energy or shunting the energy away from the trochanter into the surrounding tissues or both. In a randomized controlled trial of elderly nursing home residents, aged 70 years and over, Lauritzen et al. [8] reported a 53% reduction in hip fractures in the group who were assigned to wear external hip protectors. These data have been recently confirmed by other groups. A major problem in the use of hip protectors is the poor compliance of the subjects and there is currently a great effort at improving their design and fit. Hip protectors are relatively inexpensive and are particularly indicated in patients who have already sustained one hip fracture and those living in nursing homes.

Pharmacological Interventions

Pharmacological interventions for the prevention of fractures in patients with osteoporosis aim mainly at correcting the bone remodelling imbalance by either reducing bone resorption and bone turnover or stimulating bone formation. A number of agents is currently available but only a few have been evaluated under controlled conditions in clinical trials with primary efficacy endpoint the prevention of fractures. When assessing the efficacy of pharmacological intervention in the prevention of osteoporotic fractures certain issues need to be considered. Methodologically, the best approach is the prospective, randomized, placebo-controlled trial (RCT). Randomization ensures that potential confounders are distributed equally between treatment and control groups and reduces the risk of bias in treatment allocation. If properly conducted, differences in outcome between groups can be more confidently attributed to the effect of the intervention. Because of its advantage over other types of study evidence (e.g. cross-sectional, case-cohort studies), the RCT constitutes a prerequisite for the approval of new drugs for the treatment of osteoporosis by most regulatory authorities. Clinical trials should include adequate number of patients to allow statistically valid and clinically relevant conclusions and should be planned for periods sufficient to detect differences between the test and the treatment groups. These requirements are particularly relevant if the incidence of fractures is expected to be low (for example, a higher incidence of fractures is anticipated in patients with low bone mass and prevalent fractures than in patients with low bone mass only). In addition, the natural history of fractures should be taken into account; vertebral fractures, for example, are events which usually occur in cycles. A study period of 3 years is considered sufficient for a clinical trial aiming at detecting differences in incident fractures and is generally required by regulatory agencies for the approval of antiosteoporotic drugs. Hip fractures are events which can easily be registered because patients are admitted to hospital. This is not, however, the case with vertebral fractures about two-thirds of which do not come to clinical attention. The presence, however, of a vertebral fracture is indicative of structural failure of the skeleton and increases the risk of new fractures independently of the level of BMD. Documentation, therefore, of vertebral fractures in RCT's should be done by an objective, precise, and consistent method on serial X-rays of the spine without knowledge of the patient's assignment group. Several objective methods for assessing vertebral morphology are currently used, but it should be noted that there is no gold standard for defining a vertebral fracture and some morphometric methods may overestimate the incidence of deformities. Finally, deformities of previously normal vertebrae occurring during the trial rather than progression of pre-existing vertebral deformities should be counted as new events and the number of patients with new vertebral fractures and not the number of new fractures should be analyzed. With these background considerations the available evidence of antifracture efficacy of the various interventions varies markedly. The majority of data has been obtained with the use of inhibitors of bone turnover.

Estrogens are usually recommended as the first line of treatment of postmenopausal women with osteoporosis. A number of large observational studies have provided strong evidence of the antifracture effectiveness of estrogens. However, data from RCTs in women with osteoporosis are scarce. Only one RCT with transdermal estradiol of only one-year duration in a small number of women with osteoporosis has shown that estrogens can reduce the incidence of new vertebral fractures [9] and there is no information about their effect on hip fractures in prospective studies. The best available evidence about the antifracture efficacy of inhibitors of bone turnover has been obtained with the use of bisphosphonates. Earlier studies with the use of etidronate given intermittently (400 mg/day for two weeks every three months) showed that this treatment is effective in reducing the rates of new vertebral fractures in patients with severe osteoporosis, namely those with low bone mass and three or more prevalent fractures [10]. The most convincing data, however, have been obtained with the use of the nitrogen-containing bisphosphonate alendronate [11-13]. This is given daily and the recommended dose is 10 mg/day. In large prospective studies alendronate was shown to significantly increase bone mineral density at all skeletal sites up to 5 years of treatment and to reduce significantly the incidence of new vertebral, hip, and wrist fractures by about 50% in osteoporotic women. The effect was more pronounced in women with prevalent vertebral fractures [12]. Calcitonin has been used for many years in the treatment of patients with osteoporosis but convincing data about its antifracture efficacy are lacking. Only one study of intranasal calcitonin given in different doses for a period of two years to osteoporotic women showed an overall significant decrease in the incidence of new vertebral fractures [14]. A larger study has recently been completed but results are not yet available. Other agents which have been tested for their antifracture efficacy include active metabolites of vitamin D such as 1α-hydroxycholecalciferol and calcitriol but such efficacy has not been adequately documented. Currently large studies involving many thousands of patients are underway with other nitrogen-containing bisphosphonates such as risedronate and ibandronate and with the SERM raloxifen. Of the stimulators of bone formation only fluoride has been tested in prospective clinical studies with fracture prevention as endpoint. All studies, apart from one in which sodium fluoride was given in slow releasing tablets intermittently, failed to show any antifracture effect.

Analysis of the results of published clinical trials has revealed also a very important issue in the understanding of skeletal response to treatment in osteoporosis. As shown in the fluoride studies increases in bone mineral density are not necessarily associated with increased resistance of the skeleton to fractures and improvement of bone quality is very important for that. On the other hand it appears that the reduction in fracture frequency by inhibitors of bone turnover is much larger than would have been expected by the moderate changes in bone mass observed. The reasons for the additional beneficial effect of these treatments on the mechanical competence of bone are not clear. Possible explanations include the prevention of the perforation of the trabaculae and/or the reduction in the number and depth of resorption lacunae of the trabeculae by the induced decrease in bone turnover.

Conclusions

The management of the patient with osteoporosis, whether this is defined by the presence of fractures or by low bone mass or both, should comprise the integrated use of a number of pharmacological and nonpharmacological measures aiming at improving the quality of life of the patient and reducing the risk of fractures. The practicing clinician confronted with the individual patient is now in a much better position than only a few years ago when decisions were based exclusively on experience, intuition, and extrapolation of data. Today management decisions can be based on evidence obtained in well-performed large clinical studies. Particular attention should be paid, however, not only to the evidence and the efficacy of the available interventions but also to their side effects and the costs. It should not be forgotten that treatment of chronic diseases as osteoporosis is a major concern of societies confronted with the aging of the population and the financial burden of modern health care. These issues need to be considered for the benefit of both the individual and the society.

References

1. O'Neill T, Papapoulos S. Can we prevent fractures? Baillière's Clin Rheumatol 1997;11:565-82.
2. Papapoulos SE. Pharmacological management of osteoporosis. Aging 1998; in press.
3. Chapuy MC, Arlot ME, Delmas PD, Meunier PJ. Effect of calcium and cholecalciferol treatment for three years on hip fractures in elderly women. Br Med J 1994;308:1081-82.
4. Chapuy MC, Arlot ME, Duboef F, et al. Vitamin D3 and calcium to prevent hip fractures in elderly women. N Engl J Med 1992;327:1637-42.
5. Recker RR, Hinders S, Davies KM, et al. Correcting calcium nutritional deficiency prevents spine fractures in elderly women. J Bone Miner Res 1996;11:1961-66.
6. Dawson-Hughes B, Harris SS, Krall EA, Dallal GE. Effect of calcium and vitamin D supplementation on bone density in men and women 65 years of age and older. New Engl J Med 1997;337:670-76.
7. Province MA, Hadley EC, Hornbrook MC, et al. The effects of exercise on falls in elderly patients. A preplanned meta-analysis of the FICSIT trials. JAMA 1995;273:1341-47.
8. Lauritzen JB, Petersen MM, Luud B. Effect of external hip protectors on hip fractures. Lancet 1993;341:11-13.
9. Lufkin EG, Wahner HW, O'Fallon WM, et al. Treatment of postmenopausal osteoporosis with transdermal estrogen. Ann Int Med 1992;117:1-9.
10. Harris ST, Watts NB, Jackson RD, et al. Four-year study of intermittent cyclic etidronate treatment of postmenopausal osteoporosis: Three years of blinded therapy followed by one year of open therapy. Am J Med 1993;95:557-67.
11. Liberman UA, Weiss SR, Bröll J, et al. Effect of oral alendronate on bone mineral density and the incidence of fractures in postmenopausal osteoporosis. N Engl J Med 1995;333:1437-43.
12. Black DM, Cummings SR, Karpf DB, et al. Randomized trial of effect of alendronate on risk of fractures in women with existing vertebral fractures. Lancet 1996;348:1535-41.
13. Karpf DB, Shapiro DR, Seeman E. Prevention of non-vertebral fractures by alendronate:

A meta-analysis. JAMA 1997;227:1159-64.

14. Overgaard K, Hansen MA, Jensen SB, Christiansen C. Effect of calcitonin given intranasally on bone mass and fracture rates in established osteoporosis: A dose-response study. Br Med J 1992;305:556-61.

MENOPAUSE AND PSYCHOPHARMACOLOGY: SIGNS, SYMPTOMS, AND TREATMENT

Linda S. Kahn and Uriel Halbreich

Introduction

Approximately 30% of postmenopausal women in Western countries receive hormone replacement therapy (HRT), mostly utilizing estrogen and progesterone. This number is expected to substantially increase over the next 10 years.

HRT is most commonly prescribed for the alleviation of the physical symptoms of menopause (i.e. hot flashes, vaginal dryness), and for the prevention of long-term conditions, including osteoporosis [1,2] and cardiovascular disorders [3]. It is also indicated for the treatment of atrophic vaginitis [4] and menopausal vasomotor symptoms. There is growing evidence that HRT may improve mood and well-being of postmenopausal women. Recent data also suggest that HRT might enhance and prevent the deterioration of selective cognitive functions in postmenopausal women and decrease their risk of Alzheimer's dementia.

Estrogen and Mood

Estrogen increases serotonergic (5-HT) postsynaptic responsivity and increases both the number of serotonergic receptors and neurotransmitter uptake. Estrogen also increases 5-HT synthesis, and 5-HTIAA levels. It upregulates 5-HT_1 receptors and down-regulates 5-HT_2 receptors. It decreases MAO activity. The cumulative effect of estrogen on serotonergic function is as a 5-HT agonist [5].

Similarly, estrogen acts as a cholinergic agonist in selective brain regions. It increases activity of acetylcholine transferase in the preoptic area, amygdala, horizontal diagonal nucleus, frontal cortex, and area CA1 of the hippocampus [6]. It increases the number of muscarinic receptors in the medial, lateral, and ventromedial hypothalamus, but decreases their number in the medial preoptic area. Estrogen also increases electrical firing of neurons in the hypothalamus in response to acetylcholine [5].

Estrogen selectively increases norepinephrine (NE) activity in the brain. Its effect on the enzyme tyrosine hydroxylase is mixed. Estrogen increases NE turnover. Its effects on plasma levels of the NE metabolite 3-methoxy-4-hydroxyphenylglycol (MHPG) are mixed. Increased NE activity may be due to decreased NE reuptake and decreased NE metabolism due to inhibition of monoamine oxidase and decreased catechol-o-methyl

R. Paoletti et al. (eds.), Women's Health and Menopause, 131–136.

transferase (COMT) activity. Estrogen's effect on α_2-andrenoceptor binding is mixed, and it increases β-adrenoceptor binding [5].

Given estrogen's multiple effects on neural functions, it would appear that estrogen should act as a potent antidepressant, but this has not yet been proven. Probably estrogen does not act as an antidepressant on its own right but acts as a potent adjunct to antidepressants in cases of nonresponse. Even though estrogen alone does not improve mood in women with major depression, it does improve mood and feeling of well-being in healthy, nondepressed postmenopausal women [7-16]. In addition to epidemiological studies, there are numerous case studies, anecdotal clinical reports, and spontaneous patient reports of increased well-being, energy, libido, sexual performance, desire, and affection in elderly women who have received HRT for long periods of time [5].

Whether postmenopausal women are more vulnerable to the development of depression compared to women of reproductive age remains controversial. It was once widely believed that depression was more common after menopause, which led to the term "involutional depression." We still do not know whether a distinct subtype of "postmenopausal depression" actually exists. However, 65% of women seeking medical consultation for menopausal symptoms have varying degrees of depression [17].

Due to its accumulated action on several neurotransmitter systems, estrogen might act prophylactically in women who are vulnerable to affective disorders, even though this possibility has not yet been fully confirmed. The almost lack of estrogen during menopause might decrease the threshold for depression in women who are vulnerable to developing them. For these women, HRT might be a useful preventative measure.

Estrogen and Cognition

Performance on specific cognitive tasks, particularly those involving short- and long-term memory, have been reported to decrease in postmenopausal women, compared to women of reproductive age [18,19]. It has been suggested that the deterioration of specific cognitive functions may be linked to the lack of estrogen during menopause [20].

Estrogen replacement therapy (ERT) might enhance memory and other cognitive functions, and be protective against age-related decline in certain cognitive functions. Studies of surgically and naturally menopausal women have found that ERT selectively improves specific types of cognitive functions in postmenopausal women: short- and long-term memory and the capacity for learning new associations [19]. However, ERT has not been shown to improve general attention or visuospatial memory. Yet, it is significant that healthy 65-year-old women who have taken estrogen since menopause scored higher than nonusers on tests of delayed and immediate verbal memory tests [17]. Studies comparing 65- and 72-year-old estrogen-users and nonusers found that the nonusers complained more profoundly about memory loss [21].

Estrogen and Alzheimer's Dementia

Current research suggests a link between the decline in estrogen among some postmenopausal women and the onset of Alzheimer's dementia as well as a lower prevalence of Alzheimer's dementia among women who receive HRT.

The number of individuals with Alzheimer's disease dramatically increases every 5 years after 65 [22,23]. Up to three times more women than men are affected in each age group and the onset is earlier among women [22]. In addition to environment and genetics, some key risk factors have been shown to be more prevalent among women and to be influenced by estrogen. These include: female gender itself, hysterectomy, hip fractures, hypertension, myocardial infarction, diabetes, hypothyroidism, increased hematocrit, and depression [24].

Therefore it appears plausible that the lack of estrogen during menopause would be implicated in increasing the risk for Alzheimer's dementia and that ERT would be suggested as a preventive treatment modality.

Several epidemiological studies have reported that women who received ERT for an extended period of time had a significantly lower incidence of Alzheimer's dementia than their age-matched peers [25-28]. Recent clinical studies have reported that women who received higher doses of estrogen for longer times had lower relative risks for developing Alzheimer's dementia [20]. Patients with Alzheimer's who received HRT showed less cognitive impairment.

Estrogen's neurostructural effects suggest far-reaching impacts on the central nervous system (CNS) and human cognitive abilities. Estrogen stimulates expression of neurotrophic factors (NGF, BDNF) [29]. It maintains the viability of neurons and inhibits β-amyloid toxicity [30]. It also stimulates axonal regeneration and synaptogenesis [31]. It has been shown to prevent loss of learning and memory performance following ovariectomy [32]. Estrogen also increases cerebral blood flow and glucose transport [24,33]. Research on the effects of estrogen treatment on female rats suggests that estrogen induces dendritic spines, creating new synapses in the ventromedial hypothalamus. It also increases dendritic spinal density on pyramidal neurons in the hippocampus [6,34]. Thus, the CNS effect of estrogen might eventually be an additional major indication for HRT.

Progesterone and HRT Regimens

Although much research has focussed on the effects of estrogen on mood and cognition, less is known about the effects of progesterone [28]. Several regimen options for menopausal HRT exist. They include: cyclic estrogen and a progestin, cyclic estrogen and androgen, continuous estrogen and cyclic progestin, continuous combined estrogen and progestin, and continuous single multieffect compound.

Most HRT regimens employ sequential administration of estrogen and progesterone. For women with an intact uterus, progesterone is added to induce endometrial shedding and bleeding, and thus simulate the normal menstrual cycle and reduce the risk of estrogen-induced endometrial hyperplasia and endometrial cancer. Progesterone, however, counteracts some of the beneficial aspects of estrogen, particularly those involving mood and behavior. Some progestins might cause depression, anxiety, and other dysphoric symptoms even when not given in a sequential cyclic fashion. The cyclic addition of progesterone can produce PMS-like symptoms, especially in women who were vulnerable to PMS during their reproductive lives. Finally there remains the issue of whether progesterone counteracts the positive effects of estrogen on cognitive function [6,21,24].

The administration of estrogen with no progesterone is not common even though cyclic administration with short withdrawal periods every 30-60 days induces endometrial bleeding in almost all cases. An alternative HRT is the combination of estrogen and androgen, with no progesterone. Androgens have been shown to have a positive cognitive effect, they also increase libido, assertiveness, and energy. However, negative side effects can include masculinization and increased aggression and irritability. In an adequate dosage of an adequate androgen, the combination of estrogen and androgen approximates the normal hormonal secretion of the ovary. Its cyclic administration with no progesterone shows a promising potential [35].

Acknowledgements

Parts of this chapter were previously published in:
Halbreich U. Role of estrogen in postmenopausal depression. Neurology 1997; 48(Suppl.7): S16-S20.
Halbreich, U. Psychiatric Disorders in Women. In: Weller M, van Kammen D, editors. Progress in Clinical Psychiatry. London: Saunders Company, 1997: 96-115.

Some of the studies reported here were supported by NIMH grant RO7MH46901.

References

1. Belchetz P. Hormonal treatment of postmenopausal women. N Engl J Med 1994;330:1062-71.
2. Lindsay R. Why do oestrogens prevent bone loss? Baillieres Clin Obstet Gynaecol 1991;5: 837-52.
3. Stampfer M, Colditz G, Willet W, et al. Postmenopausal estrogen therapy and cardiovascular disease: Ten-year follow-up from the Nurses Health Study. N Engl J Med 1991;325:756-62.
4. American College of Obstetricians and Gynecologists. Hormonal replacement therapy. Technical Bulletin #166. Washington: American College of Obstetricians and Gynecologists, 1992.
5. Halbreich U. Role of estrogen in postmenopausal depression. Neurology 1997;48(Suppl.7): S16-S20.
6. McEwen BS, Alves SE, Bulloch K, et al. Ovarian steroids and the brain: Implications for cognition and aging. Neurology 1997;48(Suppl.7):S8-S15.
7. Schneider MA, Brotherton PL, Hailes J. The effect of exogenous oestrogens on depression in menopausal women. Med J Aust 1977;2:162-63.
8. Michael C, Kantor H, Shore H. Further psychometric evaluation of older women–the effect of estrogen administration. J Gerontol 1970;25:337-41.

9. Aylward M. Estrogens and plasma tryptophan levels in perimenopausal patients. In: Campbell S, editor. The management of the menopause and post-menopausal years. Baltimore, Maryland: University Park Press, 1976:135-47.
10. Furuhjelm M, Carlstrom K. Treatment of climacteric and postmenopausal women with 17-beta-oestradiol and norethisterone acetate. Acta Obstet Gynecol Scand 1977;56:351-61.
11. Furuhjelm M, Fedor-Freybergh P. The influence of estrogens on the psyche in climacteric and post-menopausal women. In: van Keep PA, Albeaux M, Greenblatt R, editors. Consensus on menopause research. Baltimore, Maryland: University Press, 1976:84-93.
12. Ditkoff EC, Crary WG, Cristo M, Lobo R. Estrogen improves psychological function in asymptomatic postmenopausal women. Obstet Gynecol 1991;78:991-95.
13. Best, NR, Rees MP, Barlow DH, Cowen PJ. Effect of estradiol implant on noradrenergic function and mood in menopausal subjects. Psychoneuroendocrinology 1992;17:87-93.
14. Palinkas LA, Barrett-Connor E. Estrogen use and depressive symptoms in postmenopausal women. Obstet Gynecol 1992;80:30-36.
15. Daly E, Gray A, Barlow D, McPherson K, Roche M, Vessey M. Measuring the impact of menopausal symptoms on quality of life. Br Med J 1993;307:836-40.
16. Limouzin-Lamothe M, Mairon N, Joyce CR, LeGal M. Quality of life after the menopause: Influence of hormonal replacement therapy. Am J Obstet Gynecol 1994;170:618-24.
17. Sherwin BB. Menopause, early aging, and elderly women. In: Jensvold MF, Halbreich U, Hamilton JA, editors. Psychopharmacology and women: Sex, gender, and hormones. Washington, D.C.: American University Press, 1996:225-37.
18. Halbreich U, Lumley LA, Palter S, Manning C, et al. Possible acceleration of age effects on cognition following menopause. J Psychiatr Res 1995;29:153-63.
19. Sherwin BB. Estrogen effects on cognition in menopausal women. Neurology 1997;48 (Suppl.7):S17-20.
20. Halbreich U. Psychotropic effects of estrogen replacement therapy in menopause and Alzheimer's disease. In: Brunello N, Langer SJ, Racagni G, editors. Mental disorders in the elderly: New therapeutic approaches. Int Acad Biomed Drug Res. Basel: Karger, 1998;13: 117-24.
21. Sherwin BB. Cognitive assessment for postmenopausal women and general assessment of their mental health. Psychopharm Bull 1998; in press.
22. Jorm AF, Korten AE, Henderson AS. The prevalence of dementia: A quantitative integration of the literature. Acta Psychiatr Scand 1987;76:465-79.
23. Herbert LE, Scherr PA, Beckett LA, Albert MS, et al. Age-specific incidence of Alzheimer's disease in a community population. JAMA 1995;273:1354-59.
24. Birge S. The role of estrogen in the treatment of Alzheimer's disease. Neurology 1997;48 (Suppl.7):S36-S41.
25. Birge SJ. The role of estrogen deficiency in the aging central nervous system. In: Lobo RA, editor. Treatment of the postmenopausal woman: Basic clinical aspects. New York: Raven Press, 1994:153-57.
26. Henderson VW, Paganini-Hill A, Emanuel CK, Dunn ME, et al. Estrogen replacement therapy in older women: Comparisons between Alzheimer's disease cases and non-demented control subjects. Arch Neurol 1994;51:896-900.
27. Mortel KF, Meyer JS. Lack of postmenopausal estrogen replacement therapy and risk of dementia. J Neuropsychiatry Clin Neurosci 1995;334-37.
28. Paganini-Hill A, Henderson VW. Estrogen replacement therapy and risk of Alzheimer's disease. Arch Intern Med 1996;156:2213-17.

29. Singh M, Meyer EM, Simpkins JW. The effect of ovariectomy and estradiol replacement on brain-derived neurotrophic factor messenger ribonucleic acid expression in cortical and hippocampal brain regions of female Sprague-Dawley rats. Endocrinology 1995;136:2320-24.
30. Simkins JW, Singh M, Bishop J. The potential role for estrogen replacement therapy in the treatment of the cognitive decline and neurodegeneration associated with Alzheimer's disease. Neurobiology of Aging 1994;15(Suppl.2):S195-S197.
31. Matsumoto A, Arai Y, Osanai M. Estrogen stimulates neuronal plasticity in the deafferented hypothalamic arcuate nucleus in aged female rats. Neuroscience Research 1985;2(5):412-18.
32. Singh M, Meyer EM, Simpkins JW. Ovarian steroid deprivation results in a reversible learning impairment and compromised cholinergic function in female Sprague-Dawley rats. Brain Res 1994;644:305-12.
33. Ohkura T, Teshima Y, Isse K. Estrogen increases cerebral and cerebellar blood flow in postmenopausal women. Menopause 1994;2:13-18.
34. McEwen BS, Woolley CS. Estradiol and progesterone regulate neuronal structure and synaptic connectivity in adult as well as developing brain. Exp Gerontol 1994;29:431-36.
35. Halbreich U. Psychiatric disorders in women. In: Weller M, van Kammen D, editors. Progress in clinical psychiatry. London: Saunders Company, 1997: 96-115.

THE BRAIN: TARGET AND SOURCE FOR SEX STEROID HORMONES

Francesca Bernardi and Andrea Riccardo Genazzani

Introduction

There is a large body of evidence indicating that in the brain the action of sex hormones is not limited to the regulation of endocrine functions and mating behavior. In fact, the brain is one of the specific target tissues for sex steroid hormones and the identification of estrogen, progestin, and androgen receptors in numerous regions of the central nervous system (CNS) suggests a role for sex hormones in modulating different brain functions. The mechanism of action of these steroids in the CNS is similar to that observed in the peripheral target organs, producing both genomic and nongenomic effects. In the classical genomic mechanism of steroids, estrogen induces relatively long-term actions on neurons by activating specific intracellular receptors that modulate gene transcription and protein synthesis. Thus, gonadal steroids modulate the synthesis, release, and metabolism of many neuropeptides and neuroactive transmitters and the expression of their receptors (Table 1). Among the neurotransmitters, noradrenaline, dopamine, γ-aminobutirric acid (GABA), acetylcholine, serotonin, and melatonin are modulated by sex steroid hormones. The neuropeptides, which are directly influenced by gonadal hormones, include the opioid peptides, gonadotropin-releasing hormone (GnRH), corticotropin-releasing factor (CRF), neuropeptide Y (NPY), and galanin [1]. Moreover, estrogen exerts very rapid effects in the brain that cannot be attributed to genomic mechanisms. These nongenomic effects of estrogen modulate electrical excitability, synaptic functioning, and morphological features, and are involved in many of the physiological functions and clinical effects of estrogen in the brain [2,3]. The interaction of genomic and nongenomic mechanisms allows for the wide range of estrogen actions in the regulation of cerebral functions.

Several data indicate that the brain is not only a target for sex steroid hormones but also a source of steroids, named neurosteroids, produced *de novo* or from blood-born precursors [4]. Neurosteroids exert nonclassical rapid actions as allosteric agonists of $GABA_A$ receptor [4] and also modulate classic neurotransmitters in the brain. Several studies have shown a relationship between the fluctuation of the synthesis and release of neurosteroids and psychological symptoms such as depression, anxiety, and irritability [4,5].

R. Paoletti et al. (eds.), Women's Health and Menopause, 137–143.

Table 1. Neurotransmitters and Neuropeptides Modulated by Sex Steroids

Neurotransmitters	Noradrenaline
	Dopamine
	Acetylcholine
	Serotonin
	γ-aminobutirric acid (GABA)
	Melatonin
Neuropeptides	Opioid peptides
	Neuropeptide Y (NPY)
	Galanin
	Corticotropin-releasing factor (CRF)

Brain as Target for Sex Steroid Hormones

Sex steroid hormones modulate the hypothalamic and extrahypothalamic noradrenergic and dopaminergic systems, controlling movement and behavior in both animals and humans. An impairment of catecholaminergic neurons with an increase in noradrenaline release and a decrease in dopamine has been demonstrated in castrated female rats and estrogen administration is able to decrease noradrenaline hypothalamic release and to increase dopaminergic neuronal activity and dopamine release in medio-basal hypothalamus [6]. The effect of estrogens in modulating adrenergic receptors appears to be bimodal by up-regulating α_1-adrenergic and down regulating β-adrenergic receptors activity. Few data are available regarding the effects of progesterone and progestins in experimental or clinical models on the catecholaminergic system. The association of progesterone to estrogens in ovariectomized female rats suppress estrogen effects on noradrenergic neurons in the pineal gland [1] while in another animal model estradiol and progesterone enhance noradrenaline release, leading to increased excitability of ventromedial hypothalamus neuronal activity and to the expression of lordosis behavior [6]. The serotoninergic system is also modulated by endogenous or exogenous sex steroids. A sex difference in the central serotoninergic system exists and ovarian hormone fluctuation has been considered responsible for changes in cerebral serotonin activity during rat estrous cycle [7]. Estrogens can modify the concentration and the availability of serotonin, by increasing the degradation of monoamino oxidase (MAO), by enhancing the transport of serotonin, and by displacing tryptophan from its binding sites to plasma albumin, helping the metabolization of tryptophan into serotonin [8,9].

Estrogen affects basal forebrain cholinergic neurons that project to the cerebral cortex and hippocampus in rat brain. Studies on steroid effects on the expression of cholinergic enzymes demonstrated that estrogens act as a cholinergic agonist by inducing synthesis and activation of choline acetyltransferase (ChAT), the rate-limiting enzyme for acetylcholine formation.

Among the neuropeptides regulated by gonadal steroids, endogenous opioids exert inhibitory or excitatory signals on GnRH hypothalamic neurons. A decrease in plasma β-EP levels has been shown in postmenopausal women after surgical or spontaneous menopause suggesting a role in the mechanisms of hot flushes and sweat episodes [10,11]. The decrease in plasma β-EP levels has also been related to the pathogenesis of mood, behavior, and nociceptive disturbances of the postmenopausal period. Findings from our recent studies show that in postmenopausal women the administration of transdermal estradiol, independent of the type of progestin associated, restores β-EP levels in the range of premenopausal values [12]. The lack of response to clonidine, an α_2-presynaptic receptor agonist, and to naloxone, an opiate receptor antagonist, before HRT suggests a postmenopausal impairment of adrenergic and opiatergic receptors in modulating β-EP release [13]. The restoration of the β-EP response to naloxone and clonidine tests in postmenopausal women undergoing HRT, represents a valid tool that expresses a restored neuroendocrine regulation [13].

Other neuropeptides such as neuropeptide Y (NPY) and galanin, produced by neurons modulated by gonadal steroids, influence central behaviors and neuroendocrine functions by stimulating the pulsatile release of GnRH and gonadotropins. In castrated female rats, gonadal steroid deficiency reduces neurosecretion of NPY-producing neurons [14] and estrogen administration increases NPY content in the median eminence and the synthesis of NPY in arcuate nucleus. Several interactions between NPY and β-EP neurons in the hypothalamus have been described and it has been hypothesized that estrogens and progestagens modulate NPY release, inducing β-EP secretion. Galanin is a neuropeptide isolated from anterior pituitary of rat and man, whose synthesis is under the control of estrogens. Galanin stimulates hypothalamic GnRH release through a PGE_2- and α_1-adrenoceptor-related mechanism and a common pulse-generating mechanism between galanin and NPY has been hypothesized [15].

Recent data indicate that estrogens and progestagens play a relevant role in neuronal plasticity promoting interactive communications among neurons by affecting synaptic contact formation, or synaptogenesis [16]. Estrogen treatment determines an increase in dendritic spines and new synapses in the ventromedial hypothalamus, as well as in the density of dendritic spines of the CA1 pyramidal hippocampal neurons [17]. In addition, dendritic spine density in female rats cyclically changes during the estrous cycle, suggesting that synapses rapidly are formed and broken under estradiol level control [17]. Estradiol seem to act also with a N-methyl-D-aspartate (NMDA) receptor-mediated mechanism, by increasing NMDA receptor binding sites that are able to induce spine formation on the synapses on the CA1 pyramidal neurons [17]. Therefore estrogen induction of the NMDA receptor may be an important event in synapsis formation [18].

Brain as Source for Sex Steroid Hormones

Recently, the attention focused on GABA-mediated modulation of hypothalamic-pituitary-gonadal axis has led to the identification of the brain as a source *de novo* of sex steroid hormones, named neurosteroids, that bind to $GABA_A$ receptors. The term neurosteroids has subsequently come to include progesterone, its metabolites, and other steroids produced in the brain from blood-born precursors. The neurosteroids have a 3α-hydroxy-Δ5 structure and the first step in the synthesis of steroids is the chain cleavage of cholesterol, involving cytochrome P450 activity. The glial cells either contain cytochrome P450 or are able to transform the classical steroid hormones to a variety of neuroactive compounds. Therefore, the glia may be described as a "neuroparacrine gland" that is able to produce steroids and directly communicate with neurons of specific brain areas. Neurosteroids, as well as estrogen, influence brain function via both genomic and nongenomic mechanisms; the first include the induction of progesterone receptors as demonstrated in cultured oligodendrocytes; the latter include the modulation of calcium channel currents and of chloride channel opening [19]. Electrophysiological and ligand experiments showed that allopregnanolone, pregnanolone, and allotetrahydroDOC act as $GABA_A$ agonists while progesterone, DHEA, and pregnenolone sulphate exert a noncompetitive antagonist action on the $GABA_A$ receptor. In addition, neurosteroids may modulate neural function by acting on glial cells, activating myelination. In these complex ways, neurosteroids appear to be involved in the modulation of behavioral mechanisms and psychophysiological phenomena, such the response to stressful stimuli, anxiety, seizure disorders, memory, depression, and sleep. Recently, the relationship between neurosteroids and the hypothalamus-pituitary-gonadal (HPG) axis has been investigated [20]. Brain allopregnanolone concentration in female rats change according to the estrous cycle: hypothalamic allopregnanolone concentration is lower in proestrus than in diestrus or in estrus, while opposite changes are observed in the hippocampus. In addition, the modifications of hippocampal allopregnanolone concentrations from prepuberty to adulthood suggests that neurosteroids modulate the neuroendocrine changes occurring in the puberal period. The evidence that the central injection of allopregnanolone has an inhibitory effect on ovulation and that the antiserum to allopregnanolone enhances ovulation and sexual behavior, support the role of allopregnanolone as a inhibitor of hypothalamus-pituitary control of the ovulatory process. These findings support a role of allopregnanolone as a neuromodulator in the central mechanisms related to the HPG axis. In addition, an age-related variation of allopregnanolone in serum and in cerebral areas has been described in rats [21]. These data and the increase of cerebral allopregnanolone contents under stressful conditions, suggest that the stress-induced increase and the age-related decrease of brain allopregnanolone may have important behavioral and/or neuroendocrine consequences [22]. Rat adrenal glands express a 5α-steroid-reductase and produce allopregnanolone, with a significant increase of allopregnanolone adrenal contents in aged rats. The synthesis of pregnane compounds by rat ovary has been clearly demonstrated and a novel highly specific radioimmunoassay has been able to detect allopregnanolone also in rat testis, showing a serum-parallel age-related increase [21].

The discussion is open on the possible target of brain allopregnanolone. The increased synthesis of allopregnanolone in rat brain following acute stress suggested that this steroid may play a role as endogenous stress-protective compound [23,24]. Pretreatment with high doses of allopregnanolone significantly decreased corticotrophin-releasing hormone (CRH)-induced behavioral manifestations of stress and anxiety, indicating an anxiolytic, sedative-hypnotic, and antiaggressive effect similar to that produced by benzodiazepines [24].

In humans, modifications in plasma allopregnanolone levels throughout the menstrual cycle have been reported [25-27]. High levels were observed in the luteal phase of the menstrual cycle, with controversial results on the possible involvement of allopregnanolone in the premenstrual syndrome [25,26]. Recent data obtained in our laboratory seem to confirm that women suffering from premenstrual syndrome have low levels of allopregnanolone (unpublished data). Modifications of allopregnanolone and tetrahydrodesossicorticosterone plasma levels in pregnancy and in the postpartum period have been also reported. An increase in serum allopregnanolone levels in women in the third trimester of pregnancy has been correlated with an increase in $GABA_A$ receptor sensitivity to GABA agonists and with mood alterations [28]. High levels of allopregnanolone may determine the drowsiness observed in pregnancy, while its decrease after delivery may lead to postpartum blues. Moreover, neurosteroids may be implicated in memory mechanisms modulating the acquisition or the loss of memory, thus suggesting that the reduced memory performance that occurs in humans with aging could be related to a modification of the steroidogenesis. Concerning this issue, while scanty information is available on allopregnanolone effect on cognitive functions, it is known that DHEA and its sulfated conjugated metabolite (DHEAS), neuroactive steroid $GABA_A$ antagonist, are able to improve physical and psychological well-being and cognitive performances in postmenopausal women. Preliminary results suggest that DHEA may act in the CNS both via a direct effect on neuronal cells and through its conversion into other steroids. These findings open several avenues in the study of various types of hormonal replacement therapy.

A recent study by our group aimed to identify the possible sources of allopregnanolone in humans [29] analyzing the response of the steroid to GnRH test, CRF test and ACTH test. While the GnRH test activates the ovarian function, the last two tests stimulate adrenal cortex function. The significant increase of serum allopregnanolone in response to CRF and ACTH tests and the reduction of plasma allopregnanolone after dexamethasone-mediated suppression indicate that the adrenal cortex is a site of production of circulating allopregnanolone. In addition, the increase of serum allopregnanolone in response to GnRH test, and the presence of an amount of serum allopregnanolone not suppressed by dexamethasone, indicate that in women the ovary significantly contributes to circulating allopregnanolone. Other studies are necessary to better clarify whether additional sites of production of neurosteroids exist. The elucidation of the exact role of neurosteroids in the regulation of reproductive function might help to understand the modifications of cerebral activity, mood, and sexual behavior associated with the fluctuations of sex steroid levels.

Conclusion

The role of the CNS as target of sex steroid hormones is well recognized and it is known that the effects of gonadal hormones on the CNS are mediated by both nongenomic and genomic actions. Gonadal hormones are of primary importance for the physiological brain function acting both on the development and on the maintenance of the female behavior, cognition, and reproductive function. The complex and fascinating mechanisms of brain function regulation are understood with the finding that the CNS is also a source of neuroactive steroids. In fact, neurosteroids, by influencing neurotransmission, are involved in the modulation of behavioral mechanisms and psychophysiological phenomena, such the response to stressful stimuli, anxiety, seizure disorders, memory, depression, and sleep. Starting with this information, new and broad perspectives will be opened for clinical research concerning brain function.

References

1. Speroff L, Glass RH, Kase NH. Clinical gynecological endocrinology and infertility. 5th edition. Baltimore, Maryland:Williams and Wilkins, 1995.
2. Alonso-Soleis R, Abreu P, Leopez-Coviella I, Hernandez G, Fajardo N. Gonadal steroid modulation of neuroendocrine transduction: A transynaptic view. Cell Mol Neurobiol 1996; 3:357-82.
3. Sherwin BB. Hormones, mood and cognitive functioning in postmenopausal women. Obstet Gynaecol 1996;87:20-26.
4. Mellon SH. Neurosteroids: Action and clinical relevance. J Clin Endocrinol Metab 1994;78: 1003-8.
5. Bohus B, Koolhaas JM, Korte SM. Psychological stress, anxiety and depression: Physiological and neuroendocrine correlates in animal model In: Genazzani AR, Nappi G, Petraglia F, Martignoni E, editors. Stress and related disorders from adaptation to dysfunction. Carnforth, UK: The Parthenon Publishing Group, 1991:28-31.
6. Etgen AM, Karkanias GB. Estrogen regulation of noradrenergic signaling in the hypothalamus. Psychoneuroendocrinology 1994;19:603-10.
7. Biegon A, Bercovitz H, Samuel D. Serotonin receptor concentration during the estrous cycle of the rat. Brain Res 1980;187:221-25.
8. Luine VN, McEwen BS. Effect of estradiol on turnover of Type A monoamine oxidase in the brain. J Neurochem 1977;28:1221-27.
9. Panay N, Sands RH, Studd JWW. Estrogen and behavior. In: Genazzani AR, Petraglia F, Purdy RH, editors.The brain: Source and target for sex steroid hormones. Carnforth, UK: The Parthenon Publishing Group, 1996:257-76.
10. Genazzani AR, Petraglia F, Facchinetti F, et al. Increase of proopiomelanocortin- related peptides during subjective menopausal flushes Am J Obstet Gynec 1984;149:775-79.
11. Linghtman SL, Jacobs HS, Maguire AK, et al. Climateric flushing: Clinical and endocrine response to infusion of naloxone. Br J Obstet Gynec 1981;88:919-24.
12. Stomati M, Bersi C, Rubino S, et al. Neuroendocrine effects of different oestradiol-progestin regimens in postmenopausal women. Maturitas 1997;28:127-35.
13. Mc Loughin L, Grossman A, Tomlin S, et al. CRF-41 stimulates the release of beta-lipotropin and beta-endorphin in normal human subjects. Neuroendocrinology 1984;38:

282-84.

14. Karla SP. Gonadal steroid hormones promote interactive communication. In: Genazzani A R, Petraglia F, Purdy RH, editors. The brain: Source and target for sex steroid hormones. Carnforth, UK: The Parthenon Publishing Group, 1996:257-76.
15. Kaplan LM, Gabriel SM, Koenig JL, et al. Galanin is an estrogen inducible secretory product of the rat anterior pituitary. Proc Natl Acad Sci USA 1988;85:7408-12.
16. Matsumoto A. Synaptogenic action of sex steroids in developing and adult neuroendocrine brain. Psychoneuroendocrinology 1991;16:25-40.
17. McEwen BS, Wooley CS. Estradiol and progesterone regulate neuronal structure and synaptic connectivity in adult as well as developing brain. Exp Gerontol 1994;29:431-36.
18. Wooley C, McEwen BS. Estradiol regulates hippocampal dendritic spine density via an N-methyl-D-aspartate receptor-dependent mechanism. J Neurosci 1994;14:7680-87.
19. Morrow AL, Pace JR, Purdy RH. Characterization of steroid interactions with gamma-aminobutirric acid receptor-gated chloride ion channels: Evidence for multiple steroid recognition sites. Mol Pharmacol 1990;37:263-70.
20. Palumbo MA, Salvestroni C, Gallo R, et al. Allopregnanolone concentration in hippocampus of prepubertal rats and female rats throughout estrous cycle. J Endocrinol Invest 1995;18:853-56.
21. Bernardi F, Salvestroni C, Casarosa E, et al. Aging is associated with changes in allopregnanolone concentrations in brain, endocrine glands and serum in male rats. Eur J Endocrinol 1998;138:316-21.
22. Paul SM, Purdy RH. Neuroactive steroids. FASEB J 1992;6:2311-22.
23. Barbaccia ML, Roscetti G, Trabucchi M, et al. Brain allopregnanolone concentrations and GABA A receptor function in stressed rats. Society for neuroscience, 1995. Abstract 408.
24. Patchev VK, Shoaib M, Holsboer F, Almeida OFX. The neurosteroid tetrahydro-progesterone counteracts corticotropin-releasing hormone-induced anxiety and alters the release and gene expression of corticotropin-releasing hormone in the rat hypothalamus. Neuroscience 1994;62:265.
25. Genazzani AR, Salvestroni C, Guo A-L, et al. Neurosteroids and regulation of neuroendocrine function. In: Genazzani A R, Petraglia F, Purdy RH, editors. The brain: Source and target for sex steroid hormones. Carnforth, UK: The Parthenon Publishing Group, 1996: 83-91.
26. Wang M, Seippel L, Purdy RH, Backstrom T. Relationship between symptom severity and steroid variation in women with premenstrual syndrome: Study on serum pregnenolone sulfate, 5α-pregnan-3,20-dione and 3α-hydroxy-5α-pregnan-20-one. J Clin Endocrinol Metab 1996; 81:1076-82.
27. Schmidt PJ, Purdy RH, Moore PH, Paul SM, Rubinow DR. Circulating levels of anxiolytic steroids in the luteal phase in women with premenstrual syndrome and in control subjects. J Clin Endocrinol Metab 1994;79:1256-60.
28. Majewska MD. Neurosteroids: Endogenous bimodal modulators of the GABA-A receptor. Mechanism of action and physiological significance. Progress in Neurobiology 1992;38: 379-95.
29. Genazzani AR, Petraglia F, Bernardi F, et al. Circulating levels of allopregnanolone in humans: Gender, age and endocrine influences. J Clin Endocrinol Metab 1998;83:2099-103.

THE ROLE OF ESTROGEN IN BRAIN AGING AND ALZHEIMER'S DISEASE

Stanley J. Birge

Introduction

There is a growing appreciation of the role of ovarian hormones as modulators of neuronal function within the central nervous system (CNS). Ovarian failure has long been known to result in reversible changes in mental function, affect, and behavior. More recent epidemiological investigations demonstrate that ovarian failure may result in changes in mental function that can be characterized by an acceleration of brain aging. The clinical expression of these changes are injurious falls, hip fracture, and automobile accidents which can be attributed to a slowing of brain processing of complex sensory information and the generation of a timely response. Estrogen deficiency is also associated with the earlier expression of Alzheimer's disease. These effects of estrogen can be attributed to the multiple mechanisms whereby estrogens modify brain and neuronal function.

Acute Effects of Estrogen on the CNS

For most women, the gradual decline in ovarian function associated with the menopause is not accompanied by significant changes in affect or behavior. However, in some, the changes are dramatic and incapacitating. Symptoms associated with the menopause that are responsive to hormone replacement include anxiety, depression, and memory impairment. In prospective controlled trials, estrogen administration improves mood as assessed by a reduction in the number of depressive symptoms [1-3] and symptoms of anxiety [4]. These effects of estrogen on affect may be related to the hormone's ability to enhance the neurotransmitters, serotonin and acetylcholine within the CNS.

In surgically postmenopausal women randomly assigned to estrogen replacement or placebo following surgery, Phillips and Sherwin [5] demonstrated improved performance on tests of short-term verbal memory compared to those women who received placebo. This effect could not be accounted for by estrogen's effect on mood or insomnia. The deterioration in memory performance was associated with the frequency of hot flushes. These observations are consistent with the concept that the hot flush is a consequence of CNS glucose deprivation. Glucose transport across the blood brain barrier is uniquely dependent on estrogen [6]. Until adaptive changes can be marshaled to compensate for impaired estrogen-dependent glucose delivery following the abrupt loss of estrogen with

R. Paoletti et al. (eds.), Women's Health and Menopause, 145–149.

oophorectomy or the menopause, the CNS may experience glucose deprivation. This triggers an autonomic discharge in the brain's effort to stimulate hepatic gluconeogenesis and to restore glucose levels [7]. This autonomic discharge causes the physiologic changes associated with the hot flush. Within the CNS, there is a highly select group of cells within the hippocampus that are sensitive to glucose deprivation as evidenced by autopsies of insulin-dependent diabetics. It is these cells within the hippocampus that subserve the function of verbal memory and are the primary target of the neurodegenerative process of Alzheimer's disease.

To test the hypothesis that the hot flush would result in the earlier expression of Alzheimer's disease, we studied women attending a geriatric assessment clinic. Recognizing that a history of hot flush frequency and even a history total versus simple hysterectomy would be unreliable in this elderly population, which included women with dementia of the Alzheimer's type, we used a history of hysterectomy as a proxy for frequent hot flushes. In this cohort, about 80% of women with a hysterectomy would have had a bilateral oophorectomy. The performance on measures of cognitive function used in the diagnosis of Alzheimer's disease, specifically, the Clinical Dementia Rating scale and the Short Blessed Test of orientation, concentration, and memory were compared in women with a hysterectomy (average age 44), in women without a hysterectomy (average age 50), and, to adjust for the longer duration of estrogen deficiency, women with an early natural menopause (average age 44). Women with a hysterectomy had more severe impairment of cognitive function than both women with a natural menopause and women with an early natural menopause ($P < .001$). This difference was equivalent to an additional 3 years of progression of the disease process. Thus, the hot flush may be associated with a limited destruction of neurons within the hippocampus resulting in a decrease in the neuronal reserve in this region subserving memory thereby predisposing the individual to the earlier expression of Alzheimer's disease.

Estrogen Deficiency and Alzheimer's Disease

Alzheimer's disease (AD) is a slowly progressive neurodegenerative process characterized by the loss of neurons and the deposition of β-amyloid protein plaques within specific regions of the brain. Loss of neurons and their synaptic connections is associated with a progressive dementia. It is postulated that the expression of the disease is a consequence of an imbalance between neuronal injury and repair in which the neuronal injury may be a consequence of age-related factors such as free radical generation, anoxia, nutritional deficiencies, or cytotoxic products of the repair (inflammatory) process including aberrantly processed or transcribed β-amyloid protein. The antioxidant effects of estrogen may attenuate age-related injury and facilitate neuronal repair through its effects on the expression of a variety of neurotrophic factors [8]. In cultured neurons, estrogen stimulates neurite outgrowth and synaptogenesis and protects these cultured neurons from toxicity of β-amyloid protein and oxidative stress [9].

The first clue as to a potential role of estrogen in the expression of AD are the observations that AD occurs 2-3 times more frequently in women than in men after adjusting for age [10]. The age-adjusted prevalence of AD increases exponentially after age 70 and after age 80, 50% of women will be affected. This earlier expression of AD in women may be attributed to the 2-3 times greater levels of circulating estradiol in men at age 70 than in women at age 70. More compelling evidence is that 5 recent studies have demonstrated that the long-term use of hormone replacement decreases a woman's risk of AD by 40-70% [11-15].

Estrogen in the Treatment of Alzheimer's Disease

The first study to examine the ability of estrogen to effect the decline in cognitive function we now associate with AD was conducted by Caldwell [16] in 30 residents of a nursing home. In this placebo-controlled trial of i.m. estradiol, she found an improvement in selected areas of mental function including memory over the first 12 months of treatment whereas women receiving the placebo demonstrated the expected decline in mental function over this time interval. The magnitude of the improvement was equivalent to about 1.5 years of progression of the disease. Discontinuation of the estrogen did not result in a loss of the treatment effect indicating the improvement reflected an effect of the estrogen on the underlying disease process and not just the symptoms of the disease. Similar results were obtained by Kantor and colleagues [17] in a placebo-controlled trial in 50 residents of a nursing home randomized to either Premarin, 0.625 mg daily, or a placebo for 3 years. Again, improvement in the outcome measure of self-care skills was observed only during the first year of treatment. Subsequent studies, although positive, have been small, largely uncontrolled, and confounded by effects on affect. In a *post hoc* analysis of the tacrine trial that led to its approval for the treatment of AD, Schneider and co-workers [18] found that there was no improvement in mental function in women on active drug unless they were concurrently receiving estrogen replacement therapy. These observations suggest that there may be a synergistic interaction between estrogen and the choline esterase inhibitor. Other drugs approved or to be approved for the treatment of AD are choline esterase inhibitors. Their efficacy in women may also depend upon the co-administration of estrogen.

Estrogen Deficiency and Brain Aging

Brain aging affects primarily frontal and subcortical brain regions and to lesser extent the hippocampus, the primary target of AD. Consequently, brain aging is manifested by more prominent changes in executive function than in memory. These changes involve the speed by which the brain processes complex sensory information. This decline in central processing speed plays a critical role in postural stability and an individual's risk of experiencing an injurious fall and hip fracture. It is postulated that because of the slowing of central processing speed, older adults are unable to extend their forearm in time to break their fall and, therefore, the full energy of the fall is more likely to be directed to the hip. Support of this hypothesis is the observation that both osteoporotic fractures, the wrist and hip, increase in parallel after the menopause but after age 70 hip fractures increase exponentially whereas wrist fractures plateau or decline [19]. This decline in central processing speed begins only after the menopause [20] and may account for the 2-3 times greater incidence of falls in elderly women compared to men of the same age. This deterioration in brain function appears to be delayed in women on estrogen replacement therapy [21].

Summary

Estrogen deficiency appears to play an important role in brain function and in the aging of the brain. Acute withdrawal may lead to hypoglycemic damage to neurons involved in memory and predispose the individual to the earlier expression of Alzheimer's disease. Chronic estrogen deficiency may contribute to an acceleration of brain aging and the earlier expression of Alzheimer's disease relative to men who maintain higher levels of estrogen after age 60 than women. Estrogen replacement promises to slow the progression of these neurodegenerative changes.

References

1. Sherwin BB. Affective changes with estrogen and androgen replacement therapy in surgically menopausal women. J Affect Disord 1988;14:177-87.
2. Sherwin BB. A prospective one year study of estrogen and progestin in postmenopausal women: effects on clinical symptoms and lipoprotein lipids. Obstet Gynecol 1989;73:759-66.
3. Ditkoff EC, Crary WG, Cristo M, Lobo RA. Estrogen improves psychological function in asymptomatic postmenopausal women. Obstet Gynecol 1991;78:991-95.
4. Gerdes LC, Sonnendecker EW, Polakow ES. Psychological changes effected by estrogen-progestogen and clonidine treatment in climacteric women. Am J Obstet Gynecol 1989;142: 98-104.
5. Phillips S. Sherwin BB. Effects of estrogen on memory function in surgically menopausal women. Psycho Neuroendocrinology 1991;17:485-95.
6. Bishop J, Simpkins JW. Role of estrogens in peripheral and cerebral glucose utilization in ovariectomized rats. Brain Res Bulletin 1992;36:315-20.
7. Simpkins JW, Katovich MJ. Hypoglycemia causes hot flushes in animal models. In: Flint M, Kronenberg F, Utian W, editors. Multidisciplinary perspectives on menopause. New York: Annals of the New York Academy of Sciences (vol 592), 1990:433-35.
8. Singh M, Meyer EM, Isaackson PJ, Simpkins JW. The effect of ovariectomy and estradiol-replacement on brain derived neurotrophic factor mRNA expression in brain regions of female Sprague-Dawley rats. Endocrinology 1995;136:2320-24.
9. Goodman Y, Bruce AJ, Cheng B, Mattson MP. Estrogens attenuate and cortisone exacerbates excitotoxicity, oxidative injury, and amyloid-β peptide toxicity in hippocampal neurons. J Neurochem 1996;66:1836-44.
10. Jorm AF, Korten AE, Henderson AS. The prevalence of dementia: A quantitative integration of the literature. 1987;76:465-79.
11. Paganini-Hill A, Henderson VW. Estrogen deficiency and risk of Alzheimer's disease in women. Am J Epidemiology 1994;140:256-61.
12. Brenner DE, Kukull WA, Storgachis A, et al. Postmenopausal estrogen replacement therapy

and the risk of Alzheimer's disease: A population-based case-controlled study. Am J Epidemiol 1994;140:262-67.

13. Kawas C, Resnick S, Morrison A, et al. A prospective study of estrogen replacement therapy and the risk of developing Alzheimer's disease. The Baltimore Longitudinal Study of Aging. Neurology 1997;48:1517-21.
14. Tang M-X, Jacobs D, Stern Y, et al. Effect of estrogen during menopause on risk and age at onset of Alzheimer's disease. Lancet 1996;348:429-32.
15. Lerner AJ, Koss E, Debanne SM, et al. Interactions of smoking history with estrogen replacement therapy as protective factors for Alzheimer's disease. Presentation, 26th Annual Meeting, Society of Neuroscience, Washington, DC, 1996.
16. Caldwell BM. An evaluation of psychological effects of sex hormone administration in aged women. J Gerontology 1954;9:168-74.
17. Kantor HI, Michael CM, Shore H. Estrogen for older women. Am J Obstet Gynecol 1973;116: 115-18.
18. Schneider LS, Farlow MR, Henderson VW, Pogoda JM. Effects of estrogen replacement therapy on response to tacrine in patients with Alzheimer's disease. Neurology 1996;46:1580-84.
19. Birge SJ. Osteoporosis and hip fracture. Geriatric Medicine Clinics 1993;9:69-86.
20. Halbreich U, Lumley LA, Palter S, et al. Possible acceleration of age effects on cognition following menopause. J Psychiat Res 1995;29:153-63.
21. Resnick SM, Metter EJ, Zonderman AB. Estrogen replacement therapy and longitudinal decline in visual memory. Amer Acad Neurology 1997;49:1491-97.

Role of Estrogens in Dementing Illnesses: Hypotheses on the Biological Rationale

Stefano Govoni, Daniela Solano, Bruno S. Solerte, Antonio Guaita, and Marco Racchi

Epidemiological data [1] indicate that postmenopausal estrogen replacement therapy (ERT) may be associated with a reduced risk of Alzheimer's disease (AD) and ERT may also improve the cognitive performance of AD affected women. Moreover epidemiological data [2] suggest a sex difference in the prevalence of AD. The decline in sex steroid levels during aging has been widely studied; the receptor distribution in brain mirrors the distribution of the neuropathological markers of AD, therefore it is possible that estrogen levels decay in postmenopause contributes to the exacerbation of molecular events taking place with aging, thus leading to the development of AD.

One of the most prominent neuropathological features of AD is the amyloid deposit of the brain. The major component of neuritic plaques is the β-amyloid protein (βA4), derived from a large precursor, the amyloid precursor protein (APP), through proteolysis. A delicate balance exists between the proteolytic action of α-secretase that releases soluble APP (sAPP) extracellularly and alternative pathways leading to the formation of amyloidogenic fragments. Since βA4 is neurotoxic, working pathogenetic hypotheses, according to which an excess of βA4 is a key event starting neurodegeneration, have been proposed [3].

Defects in the secretory mechanism of the amyloid precursor protein leading to excess amyloid can be demonstrated directly in tissues from AD donors. Fibroblasts from patients affected by sporadic AD show a defective mechanism of sAPP secretion with correlated increase in βA4 [reviewed in 3]. Such defect is correlated to a reduction of PKC activity [3] and it is susceptible to drug, oxidative stress, and hormonal regulation [4]. Within this context it has been demonstrated that chronic 17β-estradiol (17βE) treatment modulates APP, increasing the secretion of the soluble fragment without modifying the cellular content of the protein (see below). Estrogens (E) may act at various levels of the proposed chain of pathogenetic events exerting ameliorating actions on brain oxygen and nutrient supply, through a vascular effect, or they may increase energy balance efficiency through an action on insulin sensitivity. Finally E may protect against oxidative stress and βA4-induced neurodegeneration and may improve the response to mood elevating neurotransmitters such as catecholamines.

R. Paoletti et al. (eds.), Women's Health and Menopause, 151–156.

Neurotrophic and Neuroprotective Actions of Estrogens

Several studies *in vitro* have demonstrated that sex steroids are capable of promoting neuronal survival or neurite extension in dissociated cell cultures [5]. Estrogen receptors (ER) colocalize with nerve growth factor (NGF) receptors in cholinergic neurons of the basal forebrain [6] participating to the development and survival of the basal forebrain target neurons. In addition E, such as 17βE, have been shown to protect neuronal cells against oxidative stress-induced cell damage and death [7]. The mechanism through which E exert antioxidant effects are largely unknown and may involve both the interaction with their cytosolic receptors regulating gene expression as well as novel mechanisms. In turn oxidative conditions and energy shortage may unbalance normal APP metabolism as outlined below.

Estrogens and Amyloid Precursor Protein Metabolism (APP)

Jaffe et al. [8] reported the ability of 17βE to modulate the levels of sAPP released from human breast carcinoma cells. The levels of the nonamyloidogenic sAPP in the conditioned medium of these cells treated with nM concentrations of estradiol were significantly higher compared to untreated cells. It has been suggested that the hormone may act stimulating the amount or activity of APP processing enzyme(s). While these data were obtained in a cellular model overexpressing ER, other groups [9] replicated the observation in neuronal-like cells. In our hands, the effect of 17βE was observed using SH-SY5Y cells (Figure 1), while we failed to obtain a similar E effect on sAPP secretion working with COS cells and with primary cultures of fibroblasts derived from control and AD donors [4,10]. These results further evidence the need for detailed studies aimed at determining the physiological significance of E effect on APP metabolism. The antioxidant actions of E may be important not only in preventing neurodegeneration, but also in avoiding oxidative stress-driven unbalance of amyloid precursor metabolism. In fact we were able to show that fibroblasts from AD donors were more sensitive to energy and oxidative metabolism inhibition. In fact, the addition to the incubation medium of sodium azide, which interferes with cytochrome C oxidase and inhibits sAPP release from COS cells [11], significantly inhibited sAPP secretion from AD fibroblasts at concentrations barely affecting control cells (Figure 2). The data support the concept that interference with oxidative metabolism in AD tissues, if occurring in brain, may contribute to aberrant APP metabolism and to βA4 deposition.

Estrogens and Genes Associated with Increased Risk for AD

It is widely accepted that the e4 allele of ApoE is a risk factor for AD thus the interaction between E and lipoproteins might assume a significant role in AD studies. It is known that E have cardioprotective action through changes in the lipoprotein metabolism. However, the effect on plasma lipoprotein levels that implies a direct effect on the arterial wall, does not fully explain the cardioprotectant action of E observed in pre- and postmenopausal women receiving ERT, thus other vascular actions in the ameliorative effect of steroids is a possibility to be explored and would be important in mixed dementias.

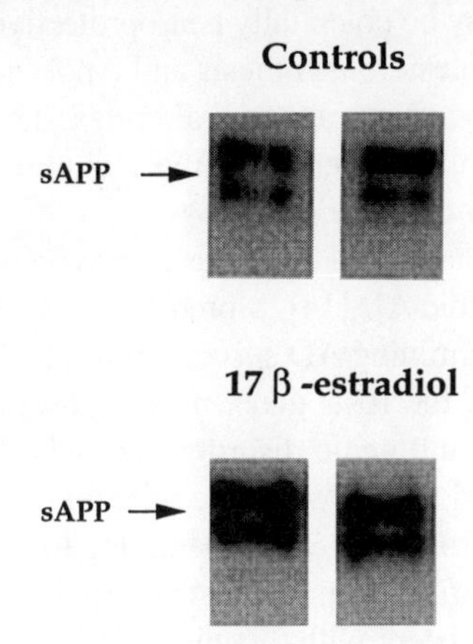

Figure 1. Effect of long term (9 days) 17-β-estradiol (2nM) treatment of SH-SY5Y human neuroblastoma cells on sAPP secretion. The figure shows a sample experiment on different cell plates. The response to treatment was rather variable. It should be noted that the other described effects on sAPP secretion (see text) can be observed after acute treatment (2 hrs) while estrogen acted only after long term exposure. Note also that in these cells sAPP appears as a doublet.

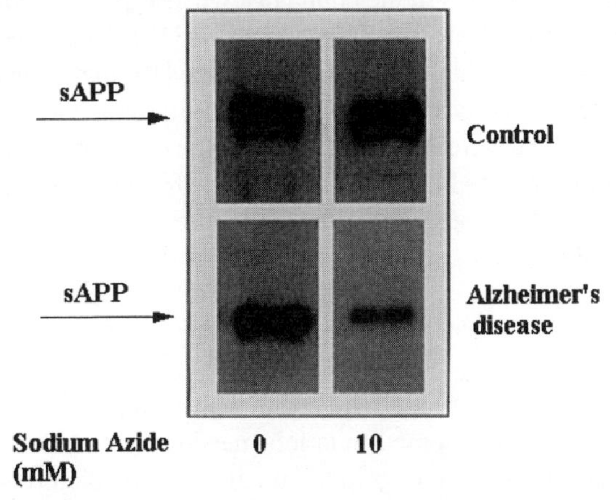

Figure 2. Effect of sodium azide on sAPP secretion from control and AD fibroblasts. The figure shows a representative experiment, the averaged data from 8 AD cases gave mean inhibition of 51%, $P < 0.01$ compared with controls.

On the other hand there is the possibility that E differentially modulate ApoE expression in various tissues: ApoE levels are significantly increased in the brain of a strain of mice treated with high doses of estradiol [12]. Since ApoE may be important in neuronal repair mechanism and neurite extension [13], the action exerted by E on brain ApoE, if proved in further experiments, may be potentially neuroprotective. On the other hand, E at high concentration may affect cholesterol synthesis and ApoE containing lipoproteins, as shown in primary neuronal cultures from rat cerebral cortex. This effect may influence in the long term, events sensitive to cholesterol loading including nonamyloidogenic APP metabolism which is inhibited by cholesterol [4].

More recently we found that the PPXX ERα genotype is associated with a higher risk of developing late onset sporadic AD [14]. Moreover, the ERα gene appears to interact with the ApoE-4 genotype in determining AD susceptibility. It should be stressed that the introduction of another variable, the ERα gene, may be useful in the identification of patients at risk of developing a multigenic disorder, like AD. Moreover since the same PPXX genotype is also associated with lower bone mineral density when in conjunction with a vitamin D receptor polymorphism [15], it is tempting to speculate that this genotype characterizes a phenotypic risk profile (slim, frail, osteoporotic). Some data on hip fractures in demented patients may be consistent with this view [16].

Estrogens and Glucose Utilization

Estrogens may have some positive effects on brain glucose metabolism increasing insulin sensitivity in the CNS and supporting the insulin-related improvement of cognitive function. In particular, the impairment of cholinergic pathways in AD has been associated with the decline of glucose turnover in the brain [17], whereas glucose administration may improve memory in moderately to severely demented [18]. The observation that AD patients may have hyperinsulinemia and insulin-resistance [19] may give a further support to introduce ERT in order to improve insulin sensitivity and glucose homeostasis in dementing illness, beyond the formal use during metabolic diseases of the postmenopausal period. Again it should be stressed that omitting glucose from the medium and impairing glucose utilization through the use of D-deoxyglucose in cultured cells impairs the nonamyloidogenic APP metabolism [4,11].

Concluding Remarks

While the above mentioned findings and hypotheses suggest a possible role of ERT in the prevention and treatment of AD, some limitations need to be emphasized. In particular, there is no substantial decline of cognitive functions immediately after menopause indicating the need for further investigations both at clinical and preclinical level to adequately determine whether the ERT has a specific add on value in relation to the prevention and treatment of dementia. On the other hand, since E may have a neuroprotective action, in particular with respect to the cholinergic system, they may be advantageous in combination with direct or indirect cholinomimetic drugs. In fact preliminary clinical data suggest that ERT can ameliorate the response to tacrine in AD female patients. Finally is open the question on the role of the newly discovered ERβ. The interest is increased by the observation that ERβ gene localization on chromosome 14 is in the same region involved in familial AD and by the fact that they are well represented at brain level.

References

1. Tang M, Jacobs D, Stern Y, et al. Effect of oestrogen during menopause on risk and age at onset of Alzheimer's disease. Lancet 1996;348:429-32.
2. Fratiglioni L, Viitanen M, Vonstrauss E, et al. Very old women at highest risk of dementia and AD - incidence data from the Kungsholmen project, Stockholm, Neurology 1997;48: 132-38.
3. Gasparini L, Racchi M, Binetti G, et al. Peripheral markers in testing pathophysiological hypotheses and diagnosing AD. FASEB J 1998;12:17-34.
4. Govoni S, Trabucchi M, Racchi M, Biological basis for the pharmacological modulation of APP metabolism. In: Brunello N, et al. (editors). Mental disorders in the elderly: New therapeutic approaches. Berlin: Karger, 1998: 84-90.
5. Sohrabji F, Miranda RC, Toran-Allerand CD. Estrogen differentially regulates estrogen and nerve growth factor receptor mRNAs in adult sensory neurons. J Neurosci 1994;14: 459-71.
6. Toran-Allerand CD, Miranda RC, Bentham WDL, et al. Estrogen receptors colocalize with low-affinity nerve growth factor receptors in cholinergic neurons of the basal forebrain. Proc Natl Acad Sci USA 1992;89:4668-72.
7. Keller JN, Germeyer A, Begley JG, Mattson MP. 17β-estradiol attenuates oxidative impairment of synaptic Na^+/K^+-ATPase activity, glucose transport, and glutamate transport induced by amyloid-peptide and iron. J Neurosci Res 1997;50:522-30.
8. Jaffe AB, Toran-Allerand CD, Greengard P, Gandy SE. Estrogen regulates metabolism of APP. J Biol Chem 1994;269:13065-68.
9. Xu H, Gouras GK, Greenfield JP, et al. Estrogen reduces neuronal generation of Alzheimer β-amyloid peptides. Nature Med 1998;4:447-51.
10. Ianna P, Racchi M, Cattaneo E, et al. Estrogen effect on brain biology and cognition. In: Paoletti R, et al (editors). Women's health and menopause. Dordrecht: Kluwer Academic Publisher, 1997: 191-98.
11. Gasparini L, Racchi M, Benussi L, et al. Effect of energy shortage and oxidative stress on APP metabolism in cos cells. Neurosci Lett 1997;23:113-17.
12. Srivastava RAK, Bhasin N, Srivastava N. ApoE gene expression in various tissues of mouse and regulation by estrogen. Biochem Mol Biol Intern 1996;38:91-101.
13. Bellosta S, Nathan BP, Orth M, et al. Stable expression and secretion of apo E3 and E4 in mouse neuroblastoma cells. J Biol Chem 1995;270:27063-61.
14. Brandi L, Becherini L, Gennari L, et al. Association of the estrogen receptor gene polymorphisms with sporadic AD. Neurobiol Aging 1998;19(Supp.1):124.
15. Gennari L, Becherini L, Masi L, et al. Vitamin D and estrogen receptor allelic variants in Italian postmenopausal women: Evidence of multiple gene contribution to bone mineral

density. J Clin Endocrinol Metab 1998;83:938-44.

16. Johansson C, Skogg L. A population based study on the association between dementia and hip fractures in 85 year olds. Aging Clin & Expt Res 1996;8:189-96.
17. Meier-Ruge W, Bertoni-Freddari C, Iwangoff P. Changes in brain glucose metabolism as a key to the pathogenesis of AD. Gerontology 1994;40:246-52.
18. Manning CA, Ragozzino ME, Gold PE. Glucose enhancement of memory in patients with probable SDAT. Neurobiol Aging 1993;14:523-28.
19. Solerte SB, Cerutti N, Fioravanti M, et al. Alterations of 24-h serum insulin profile in patients with SDAT: Potential link with the cognitive derangement. Aging Clin Exp Res 1997;9(Suppl.4):85.

WOMEN'S HEALTH AND MENOPAUSE EPIDEMIOLOGY: THE USA EXPERIENCE

Vivian W. Pinn, Mary T. Chunko, and Teri Manolio

CVD Morbidity and Mortality

For many years, cardiovascular disease (CVD) has been the leading cause of death for women of all races in the United States, as it is for men, and kills more women than men each year (Figure 1) [1,2]. In fact, CVD kills nearly twice as many women as cancer (the next leading cause) and is responsible for a slightly higher proportion of total deaths in Caucasian than African-American women [2].

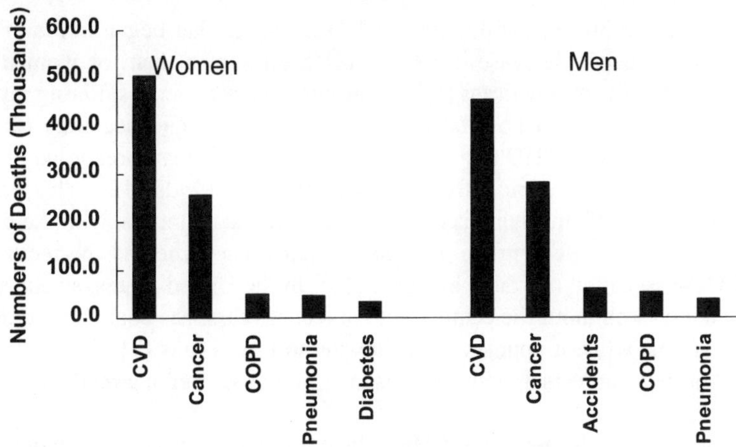

Figure 1. Leading causes of death of women and men, U.S. 1995.

Compared with women in 26 countries, women in the United States rank eleventh highest in mortality associated with CVD and coronary heart disease (CHD), while death rates from stroke place U.S. women twenty-second highest among women in 26 countries [3].

Importance of CVD at Advanced Ages

As a proportion of all causes of death, CVD increases almost linearly in women from age 25 on, but does not exceed the proportion of deaths due to CVD in men until age 75 [4].

R. Paoletti et al. (eds.), Women's Health and Menopause, 157–164.

If present trends continue, in the next century it is expected that the proportion of the United States population between the ages of 65 and 84, and over the age of 85, will increase dramatically, with women constituting a majority of the aged population [5]. Mortality rates from CHD increase exponentially with age and remain higher in men than in women throughout the life span, with women's rates comparable to those of men who are approximately 10 years younger [4]. While the reliability of data classifying cause of death decreases at very advanced ages, it is clear that the incidence of cardiovascular and heart disease rises with age in both sexes, with the incidence in women remaining one-half to two-thirds that of men, up until the age of 85 [6]. The reported prevalence of CHD increases with age, but is much more similar in women and men than are incidence or mortality rates [7].

Like CHD death rates, the death rate from stroke increases exponentially with age, but shows much less sex difference than CHD mortality figures, although there is a substantial difference in death rates from stroke for Caucasian and African-American women and men [2]. As is the case with CVD, the incidence of congestive heart failure (CHF) increases exponentially with age, but the male-female risk differential is much less than with CHD [6,8]. For women, younger age at menopause is associated with a significantly higher risk of CVD death at advanced ages [9,10].

In the United States, the decline in CHD mortality that began in the mid-1960s continues, with a steeper decrease in CHD mortality for men than for women and for Caucasians than for African Americans [1,11]. The survival of patients suffering myocardial infarction (MI) who reach a hospital continues to improve and accounts for a large proportion of the decline in CHD mortality, as well as for the greater recorded prevalence of CHD cases during the period from 1970 to the mid-1990s [12]. This pattern of substantial decline in CHD mortality is observed in nearly all countries, with the exception of Eastern Europe; of 26 countries, the United States has witnessed the tenth greatest decline in CHD-associated mortality in women [3]. In the United States, stroke mortality has been declining throughout the century and, in recent years, has continued to decline in African Americans, while it appears to be slowing in Caucasians [7]. Declines in stroke mortality appear near-universal, with declines in U.S. women ranking tenth greatest of 26 countries [3].

By contrast, death rates from CHF, which are higher in men than women and in African Americans than in Caucasians, increased during the period 1979 to 1989, but have declined since then [1]. From 1965 to 1994, hospitalizations due to CHF rose dramatically, with very little difference in men's and women's rates; much of the increase can be attributed to decreased fatality following myocardial infarction (MI) and increased survival after CHF onset [12].

Trends in CVD Risk Factors

Of the risk factors associated with CVD, one of the greatest is tobacco use. According to the National Health Interview Survey, in the United States the prevalence of smoking is distressingly similar in women and men, particularly among young adults, and declines slightly with age beyond middle age. Indeed, the prevalence of smoking declines fairly uniformly over time, but more precipitously in men than in women, with the result that one-time gender differences in smoking rates were much diminished by 1994 [13]. Hypertension, another important risk factor for CVD, has a prevalence which increases almost linearly with age but more steeply in women than in men, so that higher prevalence in men at young ages is crossed and exceeded by prevalence in women at older ages. Fortunately, inconsistent patterns of prevalence over time have given way to sharp declines in both sexes and all racial groups in the United States in the most recent national health survey [14].

The prevalence of high cholesterol, which also places one at increased risk for CVD, peaks in mid-life but earlier in men than in women, with prevalence remaining high in older women and being roughly double that found in men past the age of 65. Fairly substantial declines in high cholesterol among both women and men of all racial groups have occurred over time, although the decline has been greater among some groups than others [7]. Obesity is the CVD risk factor that is almost always more common in women than in men across the life span, with prevalence peaking for both sexes in mid-life. The prevalence of obesity has risen consistently in all racial groups and in women and men over time, but has shown a marked increase in the most recent survey [7].

Changes in CVD Risk Factors at Menopause

In view of the fact that the incidence of CVD increases in women after the menopause, it might be expected that the risk factors for CVD increase at the menopause. At least one study has found that, over a four-year period, blood pressure did not differ between women experiencing menopause and those remaining premenopausal, but that high density lipoprotein (HDL)-cholesterol declined more and low density lipoprotein (LDL)-cholesterol and triglycerides increased more in menopausal than premenopausal women [15]. Another study documented greater declines in resting metabolic rate and fat-free mass, and greater increases in fat mass, fasting insulin, and waist-hip ratio in women experiencing menopause than in those remaining premenopausal [16]. In a study by Pierre-Yves Scarabin and his colleagues, levels of Factor VII, the vitamin K-dependent protein that plays an important role in coagulation, were assessed by a variety of assays and found to be greater in menopausal than premenopausal women [17].

The Postmenopausal Estrogen/Progestin Interventions (PEPI) trial supported by the National Heart, Lung, and Blood Institute (NHLBI) was a three-year, randomized, double-blind, placebo-controlled clinical study conducted at multiple centers across the United States and designed to assess pairwise differences among placebo, unopposed estrogen, and each of three estrogen/progestin regimens on selected heart disease risk factors in healthy postmenopausal women. The PEPI study found that postmenopausal women randomized to receive estrogen and estrogen plus cyclic micronized progesterone had greater decreases in LDL-C and greater increases in triglyceride levels at follow-up than women receiving placebo. Treatment effects on total cholesterol levels were conflicting and no differences by treatment were noted for blood pressure levels (Figure 2) [18]. PEPI also documented that postmenopausal women randomized to receive estrogen and estrogen plus cyclic micronized progesterone had greater decreases in levels of fasting and postchallenge glucose and smaller increases in fibrinogen and body weight at follow-up than women receiving placebo [18].

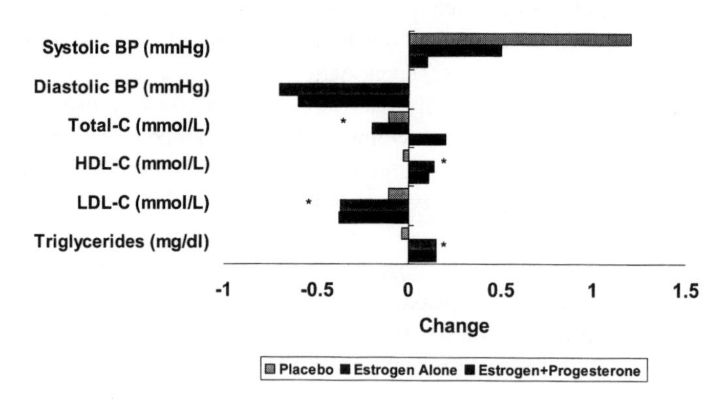

* Differences among treatment groups significant at p < 0.001.

Figure 2. Three-year changes in blood pressure and lipids in postmenopausal women by treatment group. From [18].

Studies have indicated that changes in lipid profile may account for a mere 25-50% of the CVD-protective effect of estrogen replacement therapy [19]. In an effort to understand the role of other risk factors for CVD, researchers have studied endothelin levels and fibrinolytic factor levels in postmenopausal women. In a study of endothelin levels in women who had received 30 days of treatment with oral or transdermal estrogen, it was found that endothelin levels declined 10-14% ($p < 0.02$) in postmenopausal women randomized to receive estrogen therapy compared to no significant change in women randomized to placebo [20]. An investigation of fibrinolytic factor levels indicated that concentrations of antithrombin III increased significantly and tissue plasminogen activator and protein S tended to increase, while plasminogen activator inhibitor-1 tended to decrease, after six months of oral estrogen therapy in healthy postmenopausal women [21].

Prospective, nonrandomized studies of such CVD risk factors as blood pressure, lipids, and clotting factors have suggested that LDL-cholesterol was significantly lower and HDL-cholesterol significantly higher in Native American women using postmenopausal hormone replacement therapy (HRT) than in nonusers, though blood pressure levels and other lipids did not differ [22]. Data derived from the Cardiovascular Health Study indicate that LDL-cholesterol and total cholesterol were significantly lower and HDL-cholesterol significantly higher in older women currently using HRT, compared with women who had used HRT in the past or who had never used HRT (Table 1). In addition, fibrinogen was lower and factor VII higher in current HRT users [23]. In the Cardiovascular Health Study, Manolio and her colleagues also found that older women using HRT had significantly thinner carotid walls after multiple adjustment, though much of the difference was attenuated after further adjustment for lipids, which may be in the potential causal pathway. Moreover, older women using HRT had significantly smaller ECG-defined left ventricular mass after multiple adjustment with little change after further adjustment for lipids [23].

Table 1. Age-Adjusted CVD Risk Factors by HRT Use: Cardiovascular Health Study. From [23].

	Estrogen Use Group		
	Current	Past	Never
HDL-C (mg/dl)	70	59	56
LDL-C (mg/dl)	119	139	141
Total-C (mg/dl)	218	226	225
Fibrinogen (mg/dl)	302	317	327
Factor VII (% act)	152	132	132
BMI (kg/m^2)	24.9	26.1	26.9

All differences among groups significant at $p > 0.0001$.

Examination of data from multiple studies suggests that HRT use is associated with 50-70% reduction in CHD and MI incidence, CVD death, total mortality, and angiographic stenosis, although no differences in stroke incidence have been discerned in large-scale prospective studies (Table 2) [24-27]. In a study of HRT use and survival following coronary artery bypass grafting (CABG), it was found that HRT use was associated with a 62% reduction in mortality following CABG in postmenopausal women who were followed for up to 10 years, even after adjustment for number of diseased vessels, left main coronary disease, and diabetes [26].

Conclusion

Cardiovascular disease is the leading cause of death for women in the United States. Studies have demonstrated that postmenopausal women are at increased risk for CVD. A number of clinical trials conducted to date have examined the effects of various interventions, including regimens of HRT, on CVD risk factors, incidence, and mortality. Such studies suggest that several interventions, including HRT, can ameliorate the rates of CVD risk, incidence, and mortality in postmenopausal women.

Table 2. Estimated CVD Risk Reduction Associated with Postmenopausal Estrogen Use

Author/Endpoint	Adjusted Relative Risk [95% CI]	P-Value
Stampfer, 1985		
Total CHD	0.30 [0.14,0.64]	0.002
Nonfatal MI	0.34 [0.14, 0.82]	0.02
Bush, 1987		
CVD death	0.37 [0.16,0.88]	0.03
Sullivan, 1988		
70% stenosis	0.44 [0.29, 0.67]	0.04
Henderson, 1991		
Mortality	0.80 [0.70, 0.87]	0.0001
Grodstein, 1996		
CHD	0.60[0.43-0.83]	N/A
Stroke	1.27 [0.95-1.69]	N/A

Acknowledgements

The authors wish to acknowledge the assistance of Mr. Thomas Thom of the National Heart, Lung, and Blood Institute and Ms. Clarissa LaSalle of the Office of Research on Women's Health in preparing this article for publication.

References

1. National Center for Health Statistics. Public-use mortality data tapes and unpublished data from the Division of Vital Statistics, 1968 to 1995.
2. National Office of Vital Statistics. Death rates by age, race, and sex: United States, 1900-1953: All causes. Vital Statistics Special Reports 1956;43(1):9.
3. World Health Organization. World Health Statistics Annual. Selected issues for years 1969 to 1994.
4. National Center for Health Statistics. Vital statistics of the United States, 1993. Vol. 2, pt. A. Washington, D.C.: U.S. Government Printing Office, 1993.
5. U.S. Bureau of the Census. Projections of the population of the United States by age, sex, race, and Hispanic origin: 1992 to 2050. Washington, D.C.: U.S. Government Printing Office, 1992 (current population reports: series p-25, no. 1092).
6. Kannel WB, Wilson PWF. Risk factors that attenuate the female coronary disease advantage.

Arch Intern Med 1995;155:57-61.

7. National Center for Health Statistics. Unpublished tabulations from the National Health and Nutrition Examination Survey, 1988-1994, furnished by Thomas Thom (National Heart, Lung, and Blood Institute).
8. Levy D, Larson MG, Vasan RS, Kannel WB, Ho KK. The progression from hypertension to congestive heart failure. JAMA 1996;275(20):15557-62.
9. van der Schouw YT, van der Graaf Y, Steyerberg EW, Eijkemans MJC, Banga JD. Age at menopause as a risk factor for cardiovascular mortality. Lancet 1996;347:714-18.
10. Gensini GF, Micheli S, Prisco D, Abbate R. Menopause and risk of cardiovascular disease. Thrombosis Res 1996;84(1):1-19.
11. National Center for Health Statistics. Chartbook of the conference on the decline in coronary heart disease mortality, 1978.
12. National Center for Health Statistics. National Hospital Discharge Survey. Vital and health statistics: Series 13 (issues from 1970 to 1995).
13. National Center for Health Statistics. Unpublished tabulations from the National Health Interview Survey, 1988-1990 and 1991-1993, furnished by Thomas Thom (National Heart, Lung, and Blood Institute).
14. Burt VL, Whelton P, Roccella EJ, et al. Trends in the prevalence, awareness, treatment, and control of hypertension in the adult US population. Data from the health examination surveys, 1960 to 1991. Hypertension 1995;26:60-69.
15. Wing RR, Marcus MD, Salata R, Epstein LH, Miaskiewicz S, Blair EH. Effects of a very-low-calorie diet on long-term glycemic control in obese type 2 diabetic subjects. Arch Intern Med 1991;151:1334-40.
16. Poehlman ET, Toth MJ, Gardner AW. Changes in energy balance and body composition at menopause: A controlled longitudinal study. Ann Intern Med 1995;123:673-75.
17. Scarabin P-Y, Vissac A-M, Kirzin J-M, et al. Population correlates of coagulation factor VII: Importance of age, sex, and menopausal status as determinants of activated factor VII. Thromb Vasc Biol 1996;16:1170-76.
18. The Writing Group for the PEPI Trial. Effects of estrogen or estrogen/progestin regimens on heart disease risk factors in postmenopausal women. JAMA 1995;273(3):199-208.
19. Bush TL, Barrett-Connor E, Cowan LD, Criqui MH, Wallace RB, Suchindran CM. Cardiovascular mortality and non-contraceptive estrogen use in women: Results from the Lipid-Research Clinics Program Follow-up Study. Circulation 1987;75:1002-9.
20. Wilcox JG, Hatch IE, Gentzschein BS, Stanczyk FZ, Lobo RA. Endothelin levels decrease after oral and nonoral estrogen in postmenopausal women with increased cardiovascular risk factors. Fertilit Sterility 1997;67(2):273-77.
21. Chen F-P, Lee N, Wang C-H, Cherng W-J, Soong Y-K. Effects of hormone replacement therapy on cardiovascular risk factors in postmenopausal women. Fertilit Sterility 1998;69(2): 267-73.
22. Cowan LD, Go OT, Howard BV, et al. Parity, postmenopausal estrogen use, and cardiovascular disease risk factors in American Indian women: The Strong Heart Study. J Women's Health 1997;6(4):441-49.
23. Manolio TA, Furberg CD, Shemanski L, et al. Associations of postmenopausal estrogen use with cardiovascular disease and its risk factors in older women. The CHS Collaborative Research Group. Circulation 1993;88(5, pt.1):2163-71.
24. Stampfer MJ, Willet WC, Colditz GA, Rosner B, Speizer FE, Hennekens CH. A prospective study of postmenopausal estrogen therapy and coronary heart disease. N Engl J Med 1985;

313(17):1044-49.

25. Sullivan JM, Vander Zwaag R, Lemp GF, et al. Postmenopausal estrogen use and coronary atherosclerosis. Ann Intern Med 1988;108:358-63.
26. Henderson BE, Paganini-Hill A, Ross, RK. Decreased mortality in users of estrogen replacement therapy. Arch Intern Med 1991;151:75-78.
27. Grodstein F, Stampfer MJ, Manson JE, et al. Postmenopausal estrogen and progestin use and the risk of cardiovascular disease. N Engl J Med 1996;335(7):453-61.

Doses, Duration, and Starting Age for HRT Treatment: The American Viewpoint

Wulf H. Utian

The United States must be recognized as both a melting pot and a mosaic. It would therefore be presumptuous to give an "American viewpoint" on anything. Nonetheless, it is possible to review the concept of postmenopausal hormone replacement therapy (HRT) with regard to dose, duration of therapy, and starting age, from several perspectives.

The Consumer Perspective

Discussions about menopause among women is now common behavior in the United States, and the broad availability of information, good and bad, has led to widespread interest. In general, the American culture is youth-oriented, and a major consideration in women's thinking regarding menopause management relates to issues of prevention of aging.

Scientific information published in weekly journals, such as *The New England Journal of Medicine*, are received by the media before the medical community. Presentation of this information, sometimes with appropriate review and sometimes with misinterpretation of data, is rapidly disseminated. The public is therefore often confused and the health provider not given the opportunity of adequate study of the scientific papers before bringing the information into line in therapeutic considerations. This can sometimes lead to miscommunication between the patient and doctor.

The North American Menopause Society (NAMS) has conducted Gallup Surveys in 1994 and 1997 on a broad range of questions, including women's perceptions of the menopause, use of hormone replacement, and reasons for discontinuation [1,2].

In the 1994 NAMS/Gallup Survey, almost 90% of women were aware of estrogen therapy and 20% were aware of androgens. The women aged 45-65 reported a current use of 34%, past use of 8%, with 58% having never taken any form of hormone replacement. Of the 34% current users, 59% were on estrogen, 37% on estrogen and progestin, and 2% on estrogen and androgen.

The main reasons for discontinuation were side effects reported by 34%, fear of cancer by 18%, and the rest believing that they did not need the medication or that the problem had been resolved.

In the 1997 Gallup Survey [2], a greater proportion of women reported that their physicians had spent considerable time with them discussing issues of hormone replacement

R. Paoletti et al. (eds.), Women's Health and Menopause. 165–167.

and in about 40% of instances had also discussed alternate therapies.

The overall lesson learned from this information is that women are far more aware of the issues concerning menopause and the potential treatments including hormone replacement, but that they maintain a serious concern regarding potential long-term side effects. Among these is certainly the fear of breast cancer, and despite the increasing information on this and on the risks of cardiovascular disease, concern for the problem persists.

Compliance

It is generally accepted that no more than 25% of women prescribed ERT or HRT will still be on therapy after 24 months. Moreover, the side effects induced by progestins result in many postmenopausal women never filling their prescription for this component, and either discontinuing therapy or remaining on estrogen only. Thus, continuance remains a major problem, and has lead to considerable rethinking amongst providers with regard to the dose and starting age for HRT therapy.

Dose and Starting Age for HRT

Traditionally, hormone replacement therapy was started after menopause or with the onset of specific symptoms such as hot flashes. The majority of HRT users are therefore in the age range 45-55. Poor continuance has led to far fewer users beyond the age of 55. The 45-55 year age range is also the time of maximal increase in the incidence of breast cancer. This probably accounts for part of the suspicion women have for long-term use of this therapy.

There appears to be a general consensus that the slight increase in the incidence of breast cancer in HRT users develops after 5-10 years of continuous use. The major increase in incidence of ischemic heart disease develops after the age of 70, as does the incidence of hip fractures. It must be recognized that current users of HRT show essential protection against coronary heart disease whereas previous users tend to lose this protective effect. Moreover, in relation to osteoporosis, women discontinuing therapy after 10 years of continuous use at age 60, would be at similar risk by age 70 for fracture as never-users.

The above factors have led to a trend to either recommend hormone therapy from menopause onwards without discontinuation, or a more serious consideration of starting HRT at an older age. The previous misconception that HRT was of minimal benefit if started beyond age 65 has now been replaced by a developing belief that it would be satisfactory and valuable to start therapy for the first time beyond age 65.

Clinical experience has suggested that side effects are higher on standard doses in older patients. There is therefore an increasing research focus on using lower doses of estrogen in older women. Good data with safety and efficacy results are still lacking, and this has become an area for increased research.

Conclusion

Use of HRT in the United States is undergoing a major evolution from both the provider and the consumer perspective. Women are questioning issues and seeking alternates. Health providers are considering alternates, but also open to utilizing lower doses started in older age. The effect of these trends on health outcomes such as the incidence of osteoporotic fractures, fatal coronary heart disease, and breast cancer are as yet unknown. There is an urgent need for more research on these therapeutic modalities and their outcomes in older populations.

References

1. Utian WH, Schiff I. NAMS/Gallup survey on women's knowledge, information sources, and attitudes to menopause and hormone replacement therapy. Menopause 1994;1:39-48.
2. 1997 NAMS/Gallup survey on menopause. Menopause In press.

Estrogen Complexes Contained in Conjugated Equine Estrogens (CEE): An Overview of Their Structure and Possible Effects on Target Tissue

C. Richard Lyttle and Michael Dey

Introduction

The basic concept of estrogen action was established many years ago by pioneering studies by Drs. Elwood Jensen and Jack Gorski [1,2]. In this presentation, this model will be reviewed and it will be suggested that several modifications to this model are required, as well as to the way clinicians and scientists think about the mechanism of steroid hormone action. First, data will be presented which suggest the need to rethink how estrogens work and how they interact with the receptor. Secondly, additional data on a new component found within the conjugated equine estrogens (CEE) family will be presented. Third, some speculation about the newer mechanisms that are being explored which can change the way estrogens work will be presented.

Original Mechanism of Action

Before beginning to discuss the need for changes in our concepts, we should examine the original models and discuss some of the thoughts and concepts regarding the mechanism of estrogen action. The original model as presented in Figure 1 described several key stages in the mechanism. The model in Figure 1, proposed in the late 1960s, was, in a general sense, correct, but our understanding of the details are changing and evolving. This initial concept was that an estrogen would interact with the receptor in the cytosol and then this estrogen receptor complex would change conformation and move into the nucleus [3]. Once the complex moves into the nucleus, it binds to an estrogen response element and modifies gene expression which then modulates the physiological response. The main view of the mechanism, however, was that any estrogen which translocated the estrogen receptor into the nucleus and induced binding to DNA functioned like any other estrogen. In fact, the concept implied that the stronger the estrogen bound to the receptor, the better an estrogen it was and thus, the single differentiating factor between estrogens was their binding affinity.

R. Paoletti et al. (eds.), Women's Health and Menopause, 169–181.

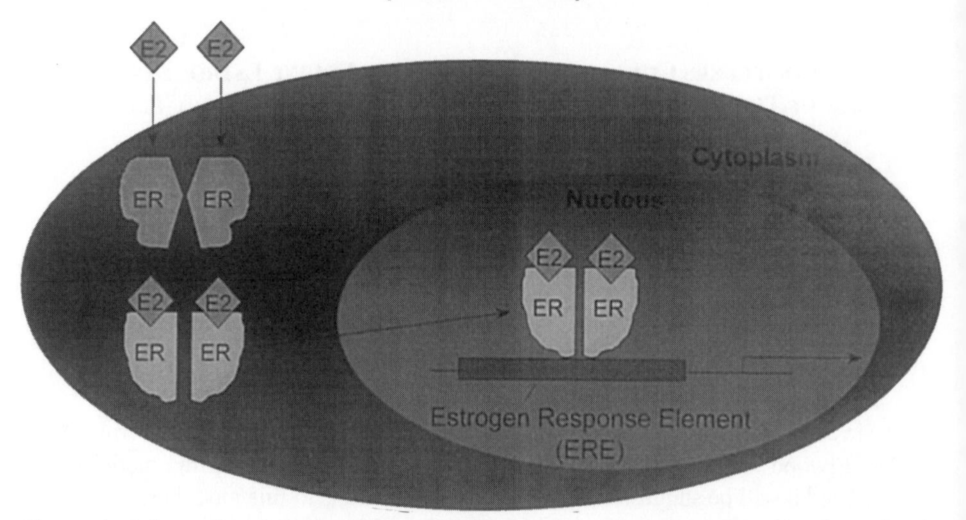

Figure 1. Schematic of original concepts of estrogen action demonstrating the activation and translocation of the estrogen receptor (ER). This model represents a composite of many experiments but is mainly based on early studies from Jensen and Gorski [1,2].

Recent data generated in a collaboration between Wyeth-Ayerst Research and Ligand Pharmaceuticals examined the correlation between hormone binding to the ER-α and gene activation. In Table 1, data are presented for a series of estrogens that are components found in conjugated equine estrogens (CEE). These data, displayed in rank order present the ability of the various equine estrogens simply to bind to the human estrogen receptor [4]. Estradiol-17β displayed the strongest binding affinity while equilenin was the weakest binder of the ten estrogens measured in this experiment. In the next column, we asked how this binding was related to their ability to modulate gene expression and thus, potentially modulate physiological responses [5]. We used a C3 promoter, that is the complement component 3 promoter which is a gene we discovered to be estrogen regulated in the epithelial cells of the rat uterus [6,7]. This gene is, in fact, a very good model for looking at gene activation in uterine epithelial cells. The data demonstrated that Δ8,9 dehydroestrone while being one of the weaker binders was active in this *in vitro* setting and in fact was one of the most potent in terms of activating this gene construct. What this tells us then is that the order of binding, i.e. binding affinity, does not necessarily reflect biological activity. These findings then suggest that it is necessary to rethink some of the basic concepts. This preclinical finding regarding the biological potency and activity of Δ8,9 dehydroestrone was examined in a clinical study. A clinical study was designed to establish this concept using 30 postmenopausal subjects. Women were treated either with

Table 1. Several Estrogens Contained in Premarin® Displayed Differences Between Binding Affinity and Gene Activation. Compounds were examined for their ability to compete for binding to the human estrogen receptor. These compounds were analyzed for C3 gene or ERE responses. The data indicate a lack of correlation between binding and gene activation.

Rank	Human ER Binding	Gene Activation C-3 Promoter	Gene Activation Consensus ERE
1.	17β-Estradiol	17β-Estradiol	17β-Estradiol
2.	17β-Dihydroequilin	Δ8,9-Dehydroestrone	Estrone
3.	17β-Dihydroequilenin	Estrone	Δ8,9-Dehydroestrone
4.	17α-Dihydroequilin	17β-Dihydroequilenin	Equilin
5.	17α-Estradiol	Equilenin	17β-Dihydroequilin
6.	Estrone	17β-Dihydroequilin	17β-Dihydroequilenin
7.	Equilin	Equilin	17α-Dihydroequilin
8.	17α-Dihydroequilenin	17α-Dihydroequilin	17α-Estradiol
9.	Δ8,9-Dehydroestrone	17α-Dihydroequilenin	Equilenin
10.	Equilenin	17β-Estradiol	17α-Dihydroequilenin

estrone sulfate (1.25 mg/day), Δ8,9 dehydroestrone sulfate (0.125 mg/day), or the two in combination. Within this open labeled, randomized study a series of parameters were measured at baseline and at 8 and 12 weeks after treatment. The data in Figure 2 are the results for one of the clinical endpoints evaluated in the study, the full results of which will be published shortly [8]. These data describe the number of hot flashes per day in patients either at 1) baseline, 2) treated with estrone sulfate, 3) Δ8,9 dehydroestrone sulfate, or 4) the two in combination. The results indicate that estrone sulfate is able to greatly reduce the number of hot flashes. Like estrone, Δ8,9 dehydroestrone sulfate, the compound under test, was also very effective in reducing the number of hot flashes. Again, we would like to point out that this is at 1/10th the dose of the estrone sulfate. The mixture was also effective in reducing the number of hot flashes. This study also examined a series of parameters including bone turnover markers and hormone levels; essentially most of the parameters demonstrated that Δ8,9 dehydroestrone sulfate was active as an estrogen in this study. These data thus support the findings from the *in vitro* binding and activity experiments which indicate that an estrogen's binding affinity is not necessarily a predictor of its biological activity. These clinical data further support the view some of the basic concepts of how an estrogen works must be reevaluated (Table 2).

EFFECTS OF ^{2}8,9 DEHYDROESTRONE SULFATE, ESTRONE SULFATE AND THE COMBINATION ON THE NUMBER OF HOT FLUSHES IN POSTMENOPAUSAL WOMEN

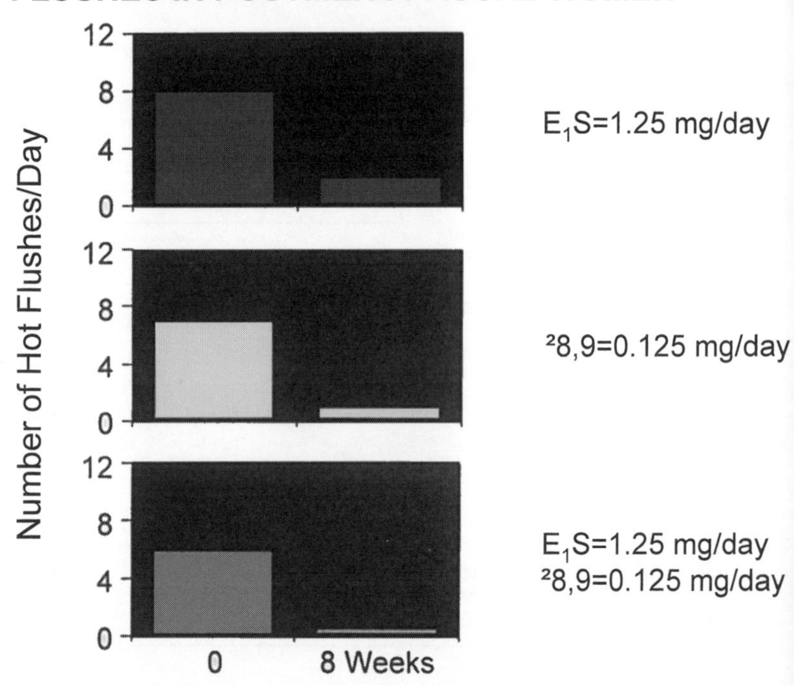

Figure 2. Δ8,9 dehydroestrone was administered to patients and the effect on number of hot flushes/day. Estrone sulfate (1.25 mg/day), Δ8,9 dehydroestrone (0.125 mg/day) and a combination of both estrone sulfate and Δ8,9 dehydroestrone sulfate. The data indicate that Δ8,9 dehydroestrone sulfate was effective in reducing the number of hot flushes/day.

Table 2. Summary of Outdated Concepts or Concepts Which Must Be Rethought as a Result of Many Studies from Data Such as Those Presented in Table 1

Outdated Concepts of Estrogen Action
• The biological activity of an estrogen receptor ligand is directly proportional to its binding affinity
• All estrogens are functinally the same and when corrected for affinity are indistinguishable
• The estrogen receptor works in isolation

As a result of these studies in conjunction with others [9-11], several concepts regarding the mechanism of estrogen action must be challenged. The first concept to be challenged is that the biological activity of an estrogen receptor ligand is directly proportional to its binding affinity. This is an important concept and one which must now changed. An associated or related concept is that estrogens are functionally the same when corrected for affinity and are, thus, essentially indistinguishable. Again, that is not correct based on the data above and, thus, must be re-evaluated. Therefore, just because an estrogen binds to the receptor and is able to turn on a gene, it does not follow that all estrogens are alike.

The third concept to be challenged is that the estrogen receptor works in isolation. Current data clearly suggest that other factors are very important in the underlying mechanism of estrogen action and thus, estrogen receptors do not work in isolation.

Conjugated Equine Estrogen

The data used to examine the relationship between estrogen binding and biological activity (Figure 1) were based upon the examination of a limited number of estrogens found in Premarin®. Using advanced analytical methods of analysis and separation, we have now been able to identify a large number of steroidal components in Premarin®.

One representative of some of these analyses is presented in Figure 3. The nature of many of these components is currently under investigation; however, either based on structural analysis or biological analysis many of these components have been determined to be either androstanes, estranes, or pregnanes.

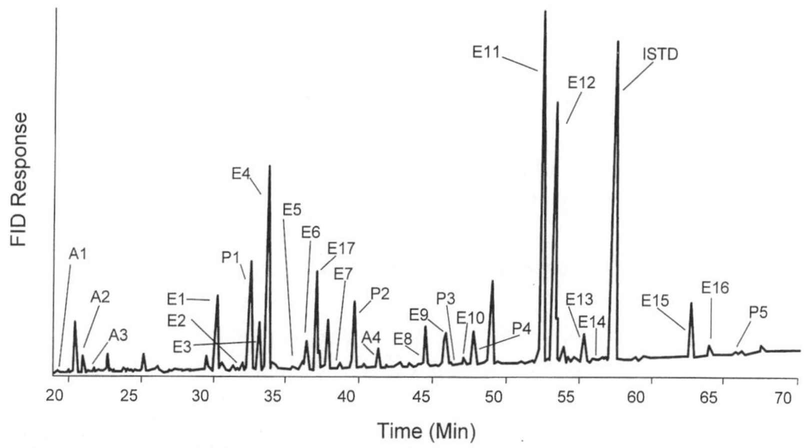

Figure 3. Chromatographic analysis of subfractions of extracted and analyzed. Peaks identified by E are estranes, P = pregnanes, and A = androstranes.

One example of further biological analysis is the activity of a component identified as estradiene sulfate. This compound has been shown to display affinity to the estrogen receptor and *in vitro* biological activity. *In vivo*, as demonstrated in Figure 4, estradiene sulfate has the ability to increase rat uterine wet weight, a good measure of classic estrogen activity in preclinical models [7]. The data indicate a dose-dependent increase in uterine weight. Therefore, this compound is biologically active and, in fact, has activity comparable to Δ8,9 dehydroestrone. In these ovariectomized rats, we also examined the changes in total serum cholesterol. While the rat is not an ideal model in terms of lipid modulation, it gives us a good indication of the ability of estradiene sulfate to modulate serum cholesterol concentrations (Figure 5). Taken together these data demonstrate a difference, in one case being estradiene sulfate is less potent and the other being more potent than Δ8,9 dehydroestrone, again, suggesting a possible tissue differential effect of some of these estrogens. Additional data demonstrate that estradiene is the fourth most abundant component that we see in the CEE having activity very similar to Δ8,9 dehydroestrone, but again, indicating some element of possible tissue selectivity.

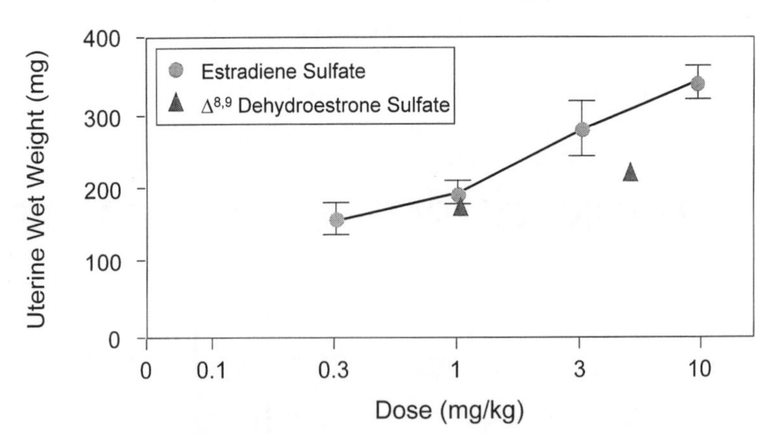

Figure 4. Estradiene sulfate administration to ovariectomized rats displayed a dose dependent increase in uterine wet weight.

Possible Mechanism of Estrogen Tissue Selectivity

As seen for the case of estradiene sulfate, some estrogens function differently in different tissues. Thus, overall, it is important to realize that an estrogen is not an estrogen, is not an estrogen. That is, estrogens can have different effects on different tissues as shown for estradiene, above; other examples exist in the literature [12,13]. For example, tamoxifen is a type of estrogen that has differential effects, as is raloxifene and other new compounds, which I call, tissue selective estrogens (TSE) [14-16].

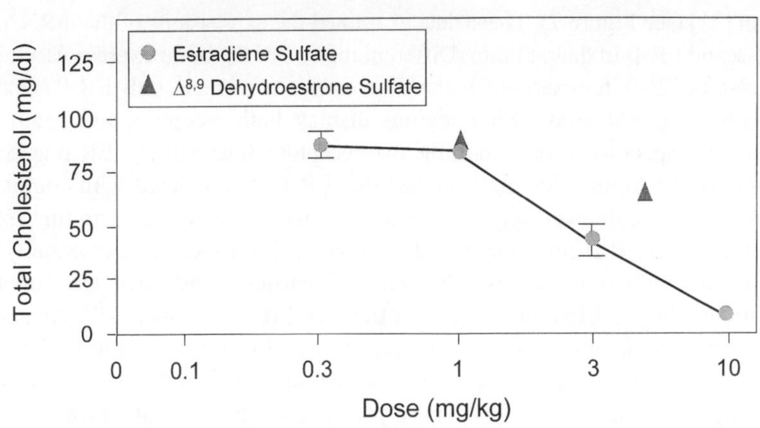

Figure 5. Estradiene sulfate administration effects on total cholesterol was examined in the study described in Figure 6. This estrogen was able to decrease total cholesterol with an activity comparable to Δ8,9 dehydroestrone sulfate.

The major question is what cellular or molecular mechanism makes this possible. What are the possible mechanisms which allow an estrogen to work one way in one tissue and another way in another or slightly better in one cell or differently on one gene or another. Listed in Table 3 is an overview of the various sites or mechanisms which can play key roles in facilitating tissue selective aspects to a variety of estrogens. The promoter context of a variety of genes and within various tissues can be very different [17,18]. The receptor conformation induced by the ligand is likely going to be a key component. The model in Figure 6 adapted from the studies of Dr. Donald McDonnell illustrates the receptor conformation concept [15]. The estrogen receptor has a particular shape with no hormone bound to it and no biological activity, per se. Should a hormone such as "Estrogen A" bind, it results in little change and may function on limited genes or tissues, in the extreme case where "Estrogen D" is 17β-estradiol, this results in a complete conformational change and functions as a full estrogen in all tissues. But other estrogens may have different effects on different tissues: Δ8,9 dehydroestrone, for example, may put the receptor in a different conformation and that resulting conformation can change depending on the estrogen which binds to the receptor. As a result of the changing conformation estrogens modulate different genes and have different effects on different tissues. Thus, the ligand-induced conformation of the receptor may result in gene or tissue specific regulation, thus a mixture of estrogens such as found in Premarin® can result in a variety of actions in multiple tissues. Another key mechanism we now know is the existence of a second receptor and you will read more about, the new ER-β in the paper by Dr. Gustafasson [18-19]. Very little data will be presented here since Dr. Gustafasson will provide in this volume a detailed exposé of his fine studies as recently reviewed [20]. Drs. Istvan Merchenthaler, Paul Shughrue, and colleagues of the Women's Health Research Institute at Wyeth-Ayerst Research have recently examined the distribution of ER-β within the rat brain [21] (see Figure 7). These data examined the expressions of the mRNA coding for the ER-α and ER-β in the rat brain. Different regions of the brain express varying levels of either ER-α or ER-β, for example, in the paraventricular nucleus, only ER-β is expressed while no ER-α is expressed. Other regions display both receptors and again we see differences in the specific expression of the two receptors. Interestingly, ER-β is expressed in the cortex of the brain. This may suggest that ER-β is associated with cognition and possibly Alzheimer's; interestingly, we do not see any ER-α expressed in this area [21]. Additionally, we see ER-β being expressed in the cerebellum and again, essentially no ER-α. This then is another mechanism by which a particular estrogen can function by preferentially binding to ER-α or ER-β and effecting different tissues [22]. An additional key mechanism which can result in tissue selectivity of estrogens is the find by several authors, of the presence of additional proteins which interact with the estrogen receptor in a ligand dependent manner [23-24]. These proteins which associate with the estrogen receptor and modulate the activity of the receptor. Finally, there are other modifications to the receptor, such as changing the state of phosphorylation of the receptor [25], tissue specific signaling systems, and other coactivators and corepressors [26-28]. Another key element in this selectivity process is the difference in the gene regulatory unit which is where the receptor binds thus regulating gene expression. This element and its surrounding region can vary from gene to gene.

Table 3. Tissue Selective Actions of Various Estrogens Can Be Due to a Variety of Molecular Mechanisms

Factors That Contribute to Estrogen Tissue Selectivity
• Receptor conformation
• ERα versus ERβ
• Receptor associated proteins
• Receptor phosphorylation
• Tissue specific signaling systems
• Gene regulatory region
• Tissue specific metabolism

The model presented in Figure 8 reflects some of the new concepts presented in this paper and described by many scientist in the current literature. The regulatory element responsible for the regulation of gene expression is the region where the liganded receptor will interact. Depending on the nature of the estrogen or ligand which binds, the receptor is placed into different conformations. This resultant conformation in turn is critical since it will allow for differential gene regulation, perhaps through interaction with co-activators or co-repressors. This in turn may result in a tissue specific expression.

REGULATION OF ESTROGEN RECEPTOR ACTION

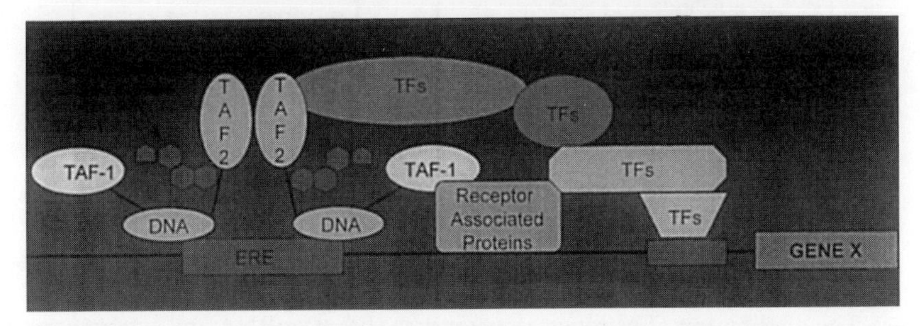

Cell Specific Expression of Co-Factors Can Regulate ER Activity

Figure 6. The interaction of estrogens with the estrogen receptor results in the potential to place the estrogen receptor into a specific conformation. This resultant conformation enables the ligand to display differential agonist or antagonist activity. This figure was adapted from the publication from Dr. McDonnell's laboratory [14,15].

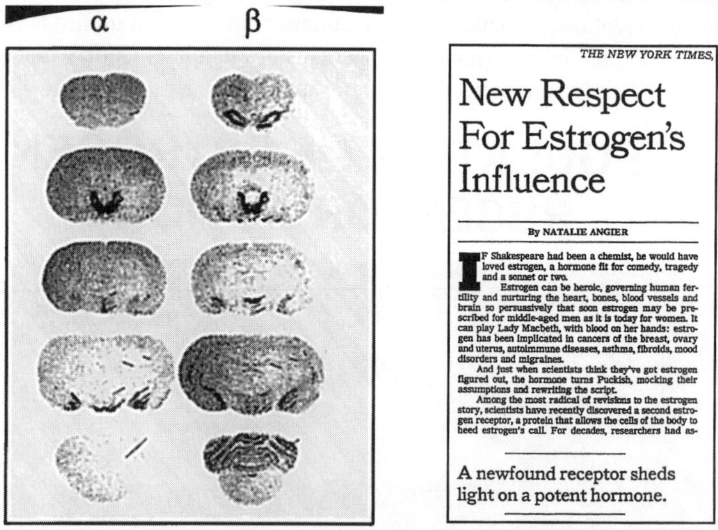

THE NEW YORK TIMES,

New Respect For Estrogen's Influence

By NATALIE ANGIER

IF Shakespeare had been a chemist, he would have loved estrogen, a hormone fit for comedy, tragedy and a sonnet or two.

Estrogen can be heroic, governing human fertility and nurturing the heart, bones, blood vessels and brain so persuasively that soon estrogen may be prescribed for middle-aged men as it is today for women. It can play Lady Macbeth, with blood on her hands: estrogen has been implicated in cancers of the breast, ovary and uterus, autoimmune diseases, asthma, fibroids, mood disorders and migraines.

And just when scientists think they've got estrogen figured out, the hormone turns Puckish, mocking their assumptions and rewriting the script.

Among the most radical of revisions to the estrogen story, scientists have recently discovered a second estrogen receptor, a protein that allows the cells of the body to heed estrogen's call. For decades, researchers had as-

A newfound receptor sheds light on a potent hormone.

Figure 7. The distribution of estrogen receptor beta (ER-β) mRNA and estrogen receptor alpha (ER-α) mRNA was determined using *in situ* hybridization was examined in the rodent brain. The details of these studies were published from studies of Dr. P. Shughrue and I. Merchenthaler [21].

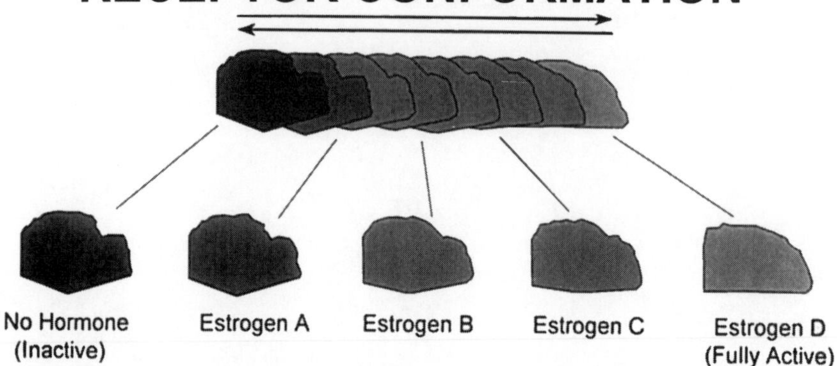

Figure 8. Schematic of estrogen model demonstrating current concepts and possible roles of transcription factors (TFs) and the role of co-activators or co-repressors (receptor associated proteins).

In conclusion, estrogens can function through a variety of mechanisms; thus, different estrogens can function differently in different tissues. As a result when there is a mixture, such as in Premarin®, each specific estrogen could function somewhat selectively or specifically. Since there are probably over 50 different steroids in Premarin® this could account for the large number of benefits resulting from selective and/or potentially different actions in a variety of tissues (Table 4). Finally, all the possible regulatory mechanisms thought to be associated with ER-α can now logically be applied to ER-β and thus, open new areas of research and clinical application for the steroids found in CEE.

Table 4. Based on Data Presented from Studies Presented in this Paper and Other Studies, Premarin® Derives its Activity from the Unique Composition of the Various Steroid Components Which in Turn Are Responsible for the Multitude of Benefits and Safety of this Mixture of Conjugated Equine Estrogens

Conclusions
• Premarin® is a complex mixture of multiple components
• Premarin's clinical activity is based on it's entire composition, not on a selected portion of its components
• The new data on Premarin's composition reinforces its uniqueness and may help explain its safety and efficacy

References

1. Jensen EV, Brecher PI, Numata M, Smith S, DeSombre ER. Estrogen interaction with target tissues; two-step transfer of receptor to the nucleus. Methods in Enzymology 1975; 36:267-75.
2. Toft D, Gorski J. A receptor molecule for estrogens; isolation from the rat uterus and preliminary characterization. Proc Nat Acad Sci USA 1966;55:1574-81.
3. Jensen EV, DeSombre ER. Estrogen-receptor interaction. Science 1973;182:126-34.
4. Reese JC, Katzenellenbogen BS. Differential DNA-binding abilities of estrogen receptor occupied with two classes of antiestrogens: Studies using human estrogen receptor overexpressed in mammalian cells. Nucleic Acid Res 1991;23:6595-6602.
5. Klein-Hitpass L, Tsai SY, Greene GL, Clark JH, Tsai M-J, O'Malley BW. Specific binding of estrogen receptor to the response element. Mol Cell Biol 1989;9:43-49.
6. Sundstrom SA, Komm BS, Xu Q, Boundy V, Lyttle CR. The stimulation of uterine complement component C3 gene expression by antiestrogens. Endo 1990;126:1449-56.
7. Sundstrom SA, Komm BS, Ponce-de-Leon H, Yi Z, Teuscher C, Lyttle CR. Estrogen regulation of tissue-specific expression of complement C3. J Biol Chem 1989;264:16941-47.
8. Baracat E, Haidar M, Lopez, FJ, Pickar, J, Dey M, Negro-Vilar A. Estrogen activity and

novel tissue selectivity of $\Delta^{8,9}$ dehydroestrone sulfate in postmenopausal women. JCEM Submitted.

9. Ward RL, Morgan G, Dalley D, Kelly PJ. Tamoxifen reduces bone turnover and prevents lumbar spine and proximal femoral bone loss in early postmenopausal women. Bone Miner 1993;22:87-94.
10. Colletta AA, et al. Anti-oestrogens induce the secretion of active transforming growth factor β from human fetal fibroblasts. British J Cancer 1990;62:405-9.
11. Turken S, Siris E, Seldin D, Flaster E, Hyman G, Lindsay R. Effects of tamoxifen on spinal bone density in women with breast cancer. Journal of the National Cancer Institute 1989;81(14);1086-89.
12. Fan JD, Wagner BL, McDonnell DP. Identification of the sequences within the human complement 3 promoter required for estrogen responsiveness provides insight into the mechanism of tamoxifen mixed agonist activity. Mol Endo 1996;10(12):1605-16.
13. Berry M, Metzger D, Chambon P. Role of the two activating domains of the oestrogen receptor in the cell-type and promoter-context dependent agonistic activity of the anti-oestrogen 4-hydroxytamoxifen. EMBO J 1990; 9:2811-18.
14. Tzukeman MT, Esty A, Santiso-Mere D, et al. Human estrogen receptor transactivational capacity is determined by both cellular and promoter context and mediated by two functionality distinct intermolecular regions. Mol Endo 1994;9:21-30.
15. McDonnell DP, Clemm DL, Hermann T, Goldman ME, Pike JW. Analysis of estrogen receptor function *in vitro* reveals three distinct classes of antiestrogens. Mol Endo 1995;9: 659-69.
16. Allan GF, Leng X, Tsai S-T, et al. Hormone and antihormone induce distinct conformational changes which are central to steroid receptor activation. J Biol Chem 1992;267:19513-20.
17. McDonnell DP, Vegeto E, O'Malley BW. Identification of a negative regulatory function for steroid receptors. Proc Natl Acad Sci USA 1992;89:10563-67.
18. Kuiper G, Enmark E, Pelto-Huikko M, Nilsson S, Gustafsson J-A. Cloning of a novel estrogen receptor expressed in rat prostate and ovary. Pro Natl Acad Sci USA 1996;93: 5925-30.
19. Katzenellenbogen BS, Korach KS. A new actor in the estrogen receptor drama- enter ERβ. Endo 1997;138:861-62.
20. Nilsson S, Kuiper G, Gustafsson JA. ER beta: A novel estrogen receptor offers the potential for new dug development. Trends Endo Metab 1998;9:387-95.
21. Kuiper GGJM, Shughrue PJ, Merchenthaler I, Gustaffsson JA. The estrogen receptor beta subtype—A novel mediator of estrogen action in neuroendocrine systems. Frontiers Neuroendo 1998;19:253-86.
22. Gustafsson JA. Therapeutic potential of selective estrogen receptor modulators. Current Opinion Chem Biol 1998;2:508-11.
23. Halachmi S, Marden E, Martin G, MacKay H, Abbondanza C, Brown M. Estrogen receptor-associated proteins: Possible mediators of hormone-induced transcription. Science 1994;264:1455-58.
24. Cavilles V, Dauvois S, Danielian PS, Parker MG. Interaction of proteins with transcriptionally active estrogen receptors. Proc Natl Acad Sci USA 1994;91:10000-13.
25. Arnold SF, Obourn JD, Jaffe H, Notides AC. Serine 167 is the major estradiol-induced phosphorylation site on the human estrogen receptor. Mol Endo 1994;8:1208-14.

26. Treuter E, Albrektsen T, Johansson L, Leers J, Gustafsson JA. A regulatory role for RIP140 in nuclear receptor activation. Mol Endo 1998;12:864-81.
27. Li H, Gomes PJ, Chen JD. RAC3, a steroid/nuclear receptor-associated coactivator that is related to SRC-1 and TIF2. Proc Natl Acad Sci USA 1997;94:8479-84.
28. Suen CS, Berrodin TJ, Mastroeni R, Cheskis BJ, Lyttle CR, Frail DE. A transcriptional coactivator, steroid receptor coactivator-3, selectively augments steroid receptor transcriptional activity. J Biol Chem 1998;273:27645-53.

Once a Week Transdermal Estrogen—Quality of Life Improvement in Long-Term Replacement Therapy: Newest Findings

Patrick Marquis, Ch. Lademacher, and H. Rozenbaum

Background

Menopause is characterized by the cessation of ovarian function, leading to permanent amenorrhoea and the absence of circulating estrogen in the systemic circulation [1]. Menopause has been viewed as a hormone deficiency or as a natural aging process in women. Nevertheless, it has been acknowledged to result in significant morbidity and mortality and associated reduction in quality of life [2]. All women who live long enough will experience menopause. The average age of menopause remains stable at around 50 years of age, and with increased life expectancy in the industrialized world, women may expect to live one third of their lifetime after menopause [3].

For some time around the cessation of menstruation, many women experience vasomotor symptoms such as hot flushes, sweating, and palpitations [4,5]. Other changes associated with the menopause include genitourinary atrophy, insomnia, and psychological changes such as depression and irritability and decreased sexual functioning. Treatment options have been aimed at the symptoms of menopause or at the prevention of long-term consequences. Thus hormone replacement therapy (HRT) has been used to provide immediate relief of such symptoms and has been associated with long-term prevention of osteoporosis and related fractures.

Several studies have shown that women on HRT experience improved quality of life in many domains in comparison to women on placebo [6,7]. A number of questionnaires have been developed and validated for use in the clinical assessment of women's health associated with changes around the menopause [8]. One of the most frequently used, the Women's Health Questionnaire (WHQ) [9], was developed to assess a wide range of physical and emotional symptoms or sensations experienced by women, specifically changes in heath and well-being associated with the climacteric symptoms of menopause.

Methods

The objectives of this study were to evaluate the effect of new 7-day transdermal matrix estradiol patch (Fem 7®, Merck KGaA) on the quality of life of postmenopausal women

R. Paoletti et al. (eds.), Women's Health and Menopause, 183–188.

outpatients with clinical symptoms of estrogen deficiency. The study was designed as a comparative, randomized, active, controlled, open, multicenter clinical trial. Patients were followed up at 5 visits over a period of 16 weeks. Two quality of life questionnaires, the WHQ [9] and the Sickness Impact Profile (SIP) [10-12], were completed by patients at baseline and after 4 cycles of treatment.

The WHQ comprises 36 items addressing 9 specific dimensions: depressed mood, somatic symptoms, memory/concentration, vasomotor symptoms, anxiety/fears, sexual behavior, sleep problems, menstrual symptoms, and attractiveness.

The Sickness Impact Profile (SIP) was developed to assess sickness-related dysfunction in 12 different areas of activity in order to provide a measure of change in perceived health status over a period of time. The total questionnaire comprises 136 items and for this study, four behavioral categories (home management, social interaction, recreation and pastimes, and work) were selected in order to assess the impact of transdermal estradiol on menopausal symptoms and resulting quality of life. Each of these scales provides a single score and has been extensively validated in a number of populations.

Both instruments have been translated into a number of languages and have been shown to be both conceptually and culturally equivalent thereby facilitating pooling of data from multinational studies. This is the first documented multinational study in hormone replacement therapy using cross-cultural instruments to measure quality of life.

Data from the trial were first analyzed with respect to the sociodemographic characteristics and the quality of completion of the quality of life questionnaires. In order to confirm the validity of the translated versions of the questionnaires, the homogeneity of the individual scales was assessed using the Cronbach alpha statistic as an indicator of internal consistency reliability. Finally, the distribution of quality of life scores at baseline and the changes over time were analyzed to assess the impact of treatment.

Results

A total of 244 postmenopausal women, 62% of whom were employed and with a mean age of 51.5 ± 4.8 years, were randomized to receive treatment. Of these, 125 received the 7-day transdermal matrix patch (50 μg/24h) and 119 received the active control, a 3.5-day matrix patch. Demographic characteristics differed between countries but were comparable between treatment groups (Table 1). Eighteen patients withdrew from the study such that 238 patients were analyzed on an intent-to-treat basis. Analyzed variables included quality of life and postmenopausal symptoms, including the frequency of vasomotor symptoms, hot flushes and night sweats.

From the analyzed population, 240 patients completed the WHQ and 242 completed the SIP, with the mean incidence of missing data of 0.8-3% at baseline and 7-9% after treatment. Where possible, missing data were interpolated using the half-scale method. The internal consistency reliability for each scale was similar across all countries, although some scales did not achieve the acceptable Cronbach alpha of 0.7 (Table 2), possibly indicative of some slight cultural differences, and because of low variability in those countries.

Table 1. Demographic Details of Study Population

	7-Day Transdermal Matrix Patch				3.5-Day Matrix Patch			
	Germany	Belgium	France	Spain	Germany	Belgium	France	Spain
n	35	15	42	31	35	15	35	34
Age	54.5±5.4	51.2±4.3	52.9±4.8	49.9±4.1	53.7±4.9	50.1±3.8	51.7±4.5	49.9±3.7
Weight (kg)	67.2±7.5	63.7±9.7	63.2±10.1	65.0±6.3	66.7±7.6	62.8±9.6	60.1±9.1	63.0±6.4
Smoker	14.3	49.7	11.9	16.1	20.0	49.7	5.7	14.7

Table 2. Internal Consistency Reliability of Individual Quality of Life Scales

	n	WHQ	SIP-HM	SIP-SI	SIP-RP	SIP-WO
Germany	69	0.85	0.49	0.71	0.34	0.62
Belgium	30	0.90	0.47	0.80	0.59	0.71
France	76	0.85	0.68	0.80	0.72	0.65
Spain	68	0.80	0.85	0.86	0.74	0.64
Total	240	0.85	0.74	0.81	0.64	0.65

The mean baseline scores were similar for each country (Figure 1) with no significant differences in any of the scales. Overall, all scales showed a significant improvement in quality of life over the period of treatment (Figure 2). A detailed analysis of the WHQ showed a significant improvement in all domains with the exception of menstrual symptoms. The greatest difference was seen with the vasomotor domain which was also associated with a clinical finding of a decrease in climacteric symptoms. Greatest improvement in quality of life was associated with high baseline levels (lowest quality of life scores) of night sweating, asthenia, irritability, and palpitations. The reliability of these findings suggest that the questionnaires could be used to predict outcome of, or compliance with, treatment.

Conclusion

Compliance with completing a quality of life questionnaire was extremely high in this study. Quality of life data from the four different countries based on both the WHQ and SIP scales were comparable at baseline and statistical analysis showed similar psychometric properties of the various language versions of the instruments. Overall, the 7-day transdermal matrix patch resulted in improved quality of life in major domains associated with a decrease in climacteric symptoms. These domains included daily activities, social life, well being, and perception of attractiveness, and were consistent across different cultures. Moreover, it was suggested that the measurement of quality of life in postmenopausal women on HRT might aid in assessing compliance and response to the prescribed treatment. Thus, the improvement of daily life by a 7-day patch may enhance compliance of symptomatic women undergoing long-term hormone replacement therapy.

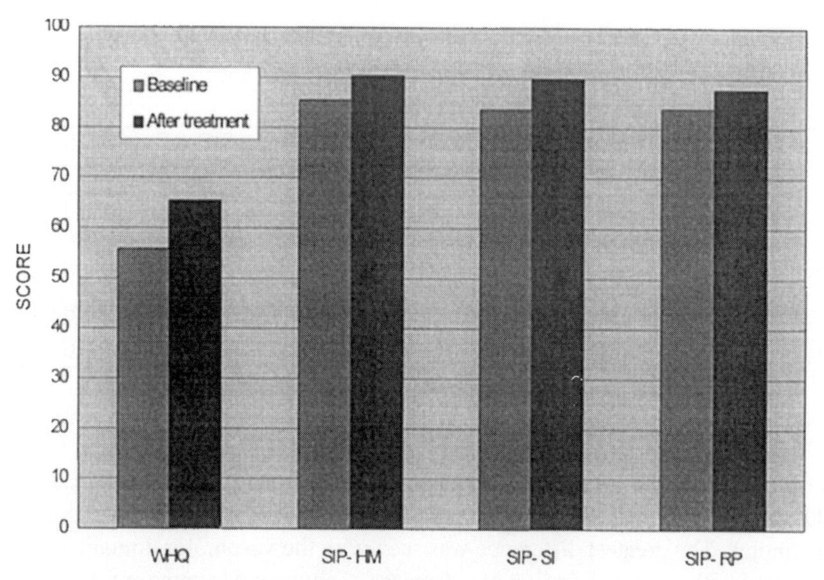

Figure 1. Baseline quality of life scores by country.

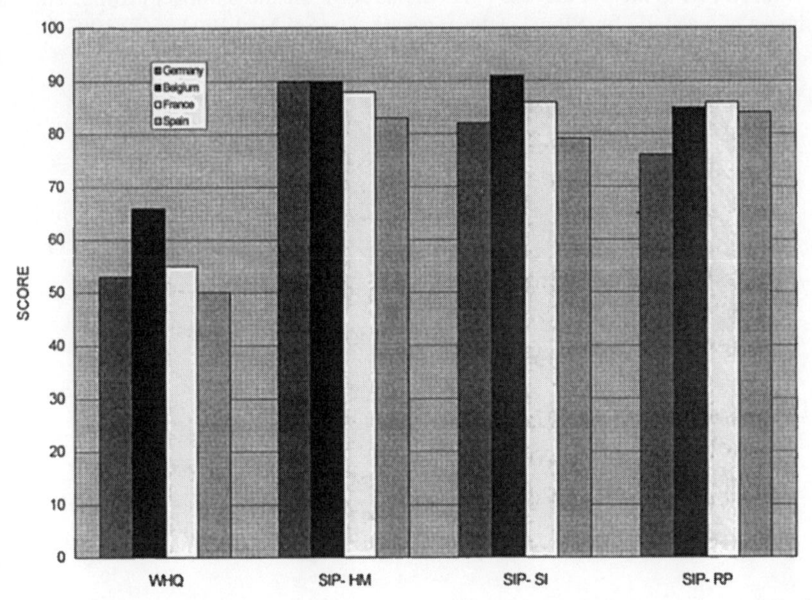

igure 2. Change in quality of life scores after 16 weeks' treatment with the 7-day ransdermal matrix patch.

eferences

1. Burger HG. The endocrinology of the menopause. Maturitas 1996;23:129-36.
2. Oldenhave A, Jaszmann LJB, Haspels AA, Everaerd TAM. Impact of the climacteric on well-being. Am J Obstet Gynecol 1993;168(3):772-80.
3. Hill K. The demography of menopause. Maturitas 1996;23:113-27.
4. Dennerstein L. Well-being, symptoms, and the menopausal transition. Maturitas 1996;23: 147-57.
5. Lomax P, Schönbaum S. Postmenopausal hot flushed and their management. Pharmac Therap 1993;57:347-58.
6. Wiklund I, et al. Quality of life of postmenopausal women on a regimen of transdermal estradiol therapy. Am J Obstet Gynecol 1993;168:824-30.
7. Nathorst-Böös J, et al. Is sexual life influenced by transdermal estrogen therapy? Acta Obstet Gynecol Scand 1993;72:656-60.
8. Limouzin-Lamothe MA, et al. Quality of life after the menopause: Influence of hormonal replacement therapy. Am J Obstet Gynecol 1994;170(2):618-24.
9. Hunter M. The Women's Health Questionnaire: A measure of mid-aged women's perceptions of their emotional and physical health. Psychology and Health 1992;7:45-55.
10. Bergner M, Bobbitt RA, Pollard WE, Martin DP, Gilson BS. The Sickness Impact Profile: Validation of a health status measure. Med Care 1976:14(1):57-67.
11. Bergner M, The Sickness Impact Profile: A brief summary of its purpose, uses, and

administration. Baltimore, Maryland: The Johns Hopkins University , 1977.

12. DeBruin AF, De Witte LP, Stevens F, Diederiks JPM. Sickness Impact Profile: The state of the art of a generic functional status measure. Soc Sci Med 1992;35(8):1003-14.

Estrogen-Androgen Hormone Replacement Therapy

Morrie M. Gelfand

Estrogen-Androgen Hormone Replacement Therapy

The term "menopause" is a key word for health professionals caring for the aging woman. It makes no difference whether you call it "change of life" or "climacteric" or "the change" or in fact "normal aging." It is a fact of life that there are significant changes in hormone production whether the ovaries are removed surgically or whether they atrophy normally. There are 3 major hormones produced by the ovary, namely, estrogen, progestin, and androgen. There is therefore a responsibility for the health professional to come to grips with the care of the aging woman. The hormone deficiency should be taken into account. During the next 10 years, there will be a dramatic increase in the number of women who are entering the menopause and we as obstetricians and gynecologists should be the prime health providers for the care of this population.

The objectives for the management of a menopausal patient are (1) to increase longevity, (2) to prevent age-related deterioration, (3) to maintain a quality of life, (4) to prevent induced risks, and (5) to maintain a physiologic hormone balance. The role of estrogen-androgen hormone replacement therapy will be measured in terms of all 5 parameters but with special emphasis on maintaining "a quality of life." Androgens in the form of androstendione, dehydroepiandrosterone, and testosterone are produced in the ovary [1,2]. Testosterone is the most potent androgen, followed in order of decreasing activity by androstendione and dehydroepiandrosterone; however, testosterone must be converted to dihydrotestosterone by 5 α reductase to be effective at the cell target level. We are able to measure the levels of free serum testosterone, dihydrotestosterone, and biologically available testosterone to determine efficacy and therapy in estrogen-androgen hormone replacement therapy (EA-HRT). In North America, estrogen-androgen replacement therapy was first employed in the early 1950s. Greenblatt et al. [3] in 1950 compared the use of estrogen, estrogen-androgen, androgen alone, and placebo in the treatment of symptoms and of the menopause. The results were encouraging in terms of increased libido and sexual functioning but the number of cases evaluated were very small and there was no extensive basic psychological evaluation or determination of blood levels of the hormones. Furthermore, the androgen preparations used high dose oral androgen combined with oral estrogen. In 1953 Grody et al. [4] reported the results of estrogen-androgen substitution therapy in menopausal females. Studd [5] and coworkers described

R. Paoletti et al. (eds.), Women's Health and Menopause, 189–194.

a study of 76 women who received pellet implants of estradiol in combination with estradiol and testosterone or a placebo. They showed that only on a combination therapy patients experienced decided increase in sexual response and frequency of coitus. Brincat et al. [6] also showed in a study that sexual status was enhanced by the use of combination estrogen-androgen subcutaneous pellets. In Canada, Sherwin and Gelfand [7-11] have done extensive research in this field and in 1998 the use of estrogen-androgen hormone replacement therapy has become increasingly widespread in North America, in Europe, and in Australia. The use of estrogen-androgen or estrogen-progestin-androgen hormone replacement therapy are noted in Think-O-Grams III and IV [12]. It is quite evident that the enhancement of quality of life is a prime requisite in adding androgen to the hormone replacement therapy regimen.

The status of the pelvis in menopausal women who are being considered for HRT can be the following: the uterus and ovaries are intact; the uterus has been removed and ovary/ovaries still present; and the uterus and ovaries are removed. HRT can be considered as a mode of replacement therapy in terms of whether the patient will need estrogen alone, estrogen and progestin, estrogen-androgen alone, or estrogen-progestin-androgen. The presence of the uterus necessitates the addition of the progestin to prevent endometrial pathology, particularly hyperplasia and cancer of the endometrium [13].

In the process of making a decision as to whether the patient needs HRT and what kind of HRT, the physician must take into account the following parameters: (1) vasomotor symptoms [7], (2) urogenital atrophic changes, (3) prevention of osteoporosis or treatment of osteoporosis [14-19], (4) cardiovascular factors [19-22], (5) the risk of endometrial cancer [13], (6) the risk of breast cancer [23], (7) senile dementia [24,25], and (8) the quality of life with emphasis on psychosexual issues [7-11,26].

Indications and Clinical Guidelines For the Use of EA-HRT

The indications for the use of EA-HRT are determined by the needs of the patient. Women who have undergone removal of the uterus and ovaries premenopausally can benefit from enhanced libido, feelings of sexuality, increased energy, and diminution of breast tenderness. In women with intact ovaries but no uterus, the combination hormone therapy is indicated if hormone levels of estrogen and androgen undergo a significant decline following the menopause. For women with intact ovaries and uterus, progestin should be added to the estrogen-androgen combination to prevent uterine hyperplastic changes.

All patients who are scheduled for EA-HRT should have the major menopausal work-up that applies to all prehormonal patients. Blood levels of estradiol, free testosterone, and biologically available testosterone, if available, should be determined and monitored during the use of therapy. Hormone levels should be kept within physiological levels. Careful surveillance as with all modes of therapy should be kept in mind.

In the United States, EA therapy is available in the form of estratest regular and half dose. This is a combination of methyl testosterone and estrogen. In Canada, we have been using an injectable combination of testosterone enanthate and estradiol. Andriol (an oral androgen) is available in Canada.

Andriol (testosterone undecanoate) has been used effectively in aging men for the past 15 years. It was the purpose of our project to use Andriol as an effective androgen for the androgen component of estrogen-androgen hormone replacement therapy in women. We were particularly interested in the fact that andriol is effectively absorbed via the lymphatic route. Drug targeting via the lymphatic system enables therapeutic amounts of andriol to virtually escape first-pass hepatic inactivation and toxicity. The metabolism of andriol includes 3 binding sites. Andriol is bound to three different receptors. It is reduced by 5 α reductase to dehydrotestosterone and bound to its receptors. It is also bound directly to testosterone receptors and a small portion is aromatized to estradiol and bound to estrogen receptors. A pilot project was prepared so that we would have a prospective randomized double-blind study to evaluate the effects of a 16-week treatment period with oral combination therapy of estrogens and androgen on libido, climacteric complaints, mood, well-being, and metabolism in menopausal women and to compare these with the effects of estrogen and placebo. The thrust of the study was to be certain that there was a positive effect on sexuality, energy, and well-being which would allow us to persist with the use of andriol for hormone replacement therapy. At the same time, the study evaluated the physical examination of the patient, metabolic changes, and endocrine parameters, as well as adverse experiences if they occurred. From a psychological rating scale determination, the following scales were used: the Green climacteric scale, the McCoy female sexuality questionnaire, psychological general well-being index, irritability subscale of the Buss-Durkee, hostility inventory, the Center for Epidemiologic Studies depression scale, and a sleep dysfunction scale. Because of our experience with the injectable testosterone enanthate, we equated a daily dose of 40 mg of andriol (the actual capsule available) to the amount of testosterone which provided sufficient symptom relief during 28 days with the use of 1 cc of testosterone enanthate. After going through a long list of inclusion parameters/criteria as well as exclusion evaluations, we interviewed 120 patients and screened 68 of those for the study. One of the inclusion factors was to make sure that the free testosterone level was less than 3 pmols/l. This was difficult to attain in many of the patients and after a great deal of discussion and persisting with our original premise of the exclusion factors, we were left with 23 patients who were randomized, 2 of whom were excluded because of mild side effects. This left us with 21 patients in the group, 12 patients on 0.625 mg premarin and 40 mg andriol per day as well as 9 patients on 0.625 mg premarin and placebo. The patients were matched for age, body weight, blood pressure, height, and at the end of the studies, there were no changes in either group. All the psychological evaluation scales showed enhancement and improvement in quality of life and well-being in patients who were on the estrogen-androgen as compared to the estrogen-placebo. In the Greene climacteric scale, the psychological scale, the anxiety subscale, the depression subscale, the somatic scale, vasomotor scale, and the probe for sexual dysfunction were all superior in the andriol group as compared to the placebo group. The same applied to the psychological general well-being index where the well-being subscale, anxiety subscale, depressed mood subscale, general health subscale, self-control subscale, and vitality subscale all showed improvement with the addition of the andriol to the estrogen as compared to the addition of placebo to estrogen. The McCoy female sexuality

Thinkogram III

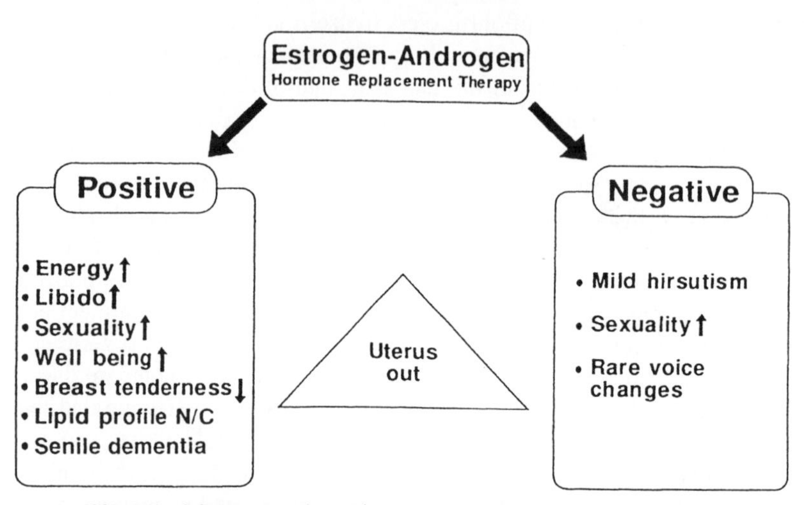
Estrogen-Androgen
Hormone Replacement Therapy
Positive
• Energy ↑
• Libido ↑
• Sexuality ↑
• Well being ↑
• Breast tenderness ↓
• Lipid profile N/C
• Senile dementia
Uterus out
Negative
• Mild hirsutism
• Sexuality ↑
• Rare voice changes
M.M. Gelfand. Estrogen-androgen hormone replacement therapy, chapter 10. In Hormone Replacement Therapy, Donald P. Swartz (ed). Williams & Wilkins, Baltimore, Maryland, September 1992

Thinkogram IV

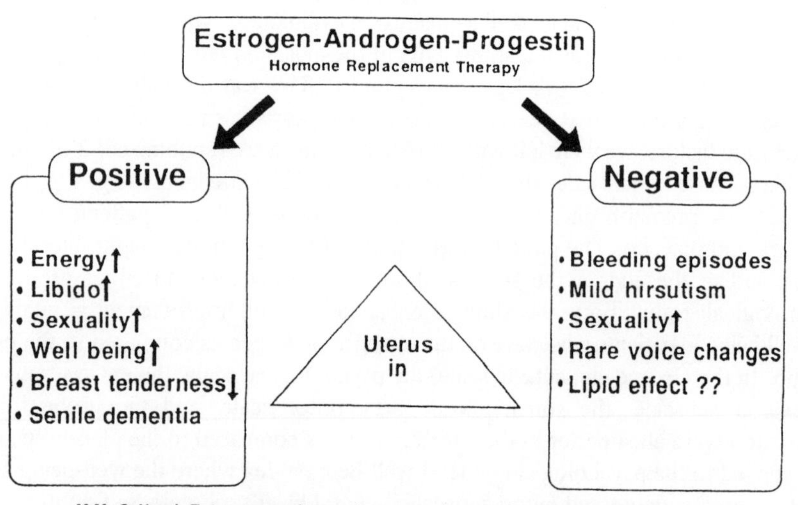
Estrogen-Androgen-Progestin
Hormone Replacement Therapy
Positive
• Energy ↑
• Libido ↑
• Sexuality ↑
• Well being ↑
• Breast tenderness ↓
• Senile dementia
Uterus in
Negative
• Bleeding episodes
• Mild hirsutism
• Sexuality ↑
• Rare voice changes
• Lipid effect ??
M.M. Gelfand. Estrogen-androgen hormone replacement therapy, chapter 10. In Hormone Replacement Therapy, Donald P. Swartz (ed). Williams & Wilkins, Baltimore, Maryland, September 1992

questionnaire showed a positive change. The depression scale was less for the patients on andriol than the placebo group. The hostility inventory was less in the andriol group as compared to the placebo group, and the sleep dysfunction scale showed an improved sleep pattern for patients on the andriol as compared to patients on the estrogen-placebo.

The above results are trends which are positive in every instance, and have been very encouraging so as to continue with a larger group evaluation and a continuing surveillance over the period of one year. At the same time, we are checking the metabolic and endocrine patterns and will be able to report on those in the near future.

Conclusion

It is clear that physicians need to address the full spectrum of menopause-related psychological effects, not just the physical symptoms. It is also essential to devote enough time with patients to discuss all the available options for hormone replacement therapy. The important role of EA-HRT in alleviating many of the symptoms of surgical and natural menopause can no longer be overlooked.

References

1. Longcope C. Adrenal and androgen secretions of normal women. Clin Endocrinol Metab 1986;15:213-28.
2. Judd HL, Judd GE, Lucas WE, et al. Endocrine function of the postmenopausal ovary--concentration of androgens and estrogens in ovarian and peripheral vein blood. J Clin Endocrinol Metab 1974;39:1020-24.
3. Greenblatt RB, Williams E, Barfield, et al. Evaluation of an estrogen-androgen combination and a placebo in the treatment of the menopause. J Clin Endocrinol Metab 1950;10:15,47-52.
4. Grody MH, Lampie EH. Estrogen-androgen substitution therapy in the aged female. Obstet Gynecol 1953;2:36-41.
5. Studd JWW, Collins WP. Estradiol and testosterone implants in the treatment of psychosexual problems in the postmenopausal woman. Br J Gynecol 1977;84:314-15.
6. Brincat M, Studd JWW, O'Dowd T, et al. Subcutaneous implants for control of climacteric symptoms. Lancet 1984;1:16-18.
7. Sherwin BB, Gelfand MM. Effects of parenteral administration of estrogen and androgen on plasma hormone levels and hot flashes in the surgical menopause. Am J Obstet Gynecol 1984;148:552-57.
8. Sherwin BB, Gelfand MM. Differential symptom response to parenteral estrogen and/or androgen administration in the surgical menopause. Am J Obstet Gynecol 1985;151:153-60.
9. Sherwin BB, Gelfand MM, Brender W. Androgen enhances sexual motivation in females: A cross-over study of sex steroid administration in the surgical menopause. Psychosom Med 1985;47:339-51.
10. Sherwin BB, Gelfand MM. Sex steroids and effect in the surgical menopause: A double-blind, cross-over study. Psychoneuroendocrinology 1985;10:325-35.
11. Sherwin BB, Gelfand MM. The role of androgen in the maintenance of sexual functioning in oophorectomized women. Psychosom Med 1987;49:397-409.

12. Gelfand MM. Estrogen-androgen hormone replacement. In: Swartz DP, editor. Hormone replacement therapy. London: Williams & Wilkins, 1992;221-34.
13. Gelfand MM, Ferenczy A, Bergeron C. Endometrial response to estrogen-androgen stimulation. In: Hammond CB, Haseltine FP, editors. Menopause: Evaluation, treatment and health concerns. New York: Alan R. Liss, 1989:29-40.
14. Ettinger B, Genant HK, Steiger P, et al. Low-dosage micronized 17-β estradiol prevents bone loss in postmenopausal women. Am J Obstet Gynecol 1992;166:479-88.
15. Notelovitz M. Osteoporosis: Screening, prevention, and management. Fertil Steril 1993;59: 707-25.
16. Longcope C, Baker RS, et al. Androgen and estrogen dynamics in women with vertebral crush fractures. Maturitas 1984;6:309.
17. Hassager C, Reis BJ, et al. Nandrolone decanoate treatment of postmenopausal osteoporosis for two years and effects of withdrawal. Maturitas 1989;11:305-17.
18. Passeri M, Pedrazzoni M, Pioli G, et al. Effects of nandrolone decanoate on bone mass in established osteoporosis. Maturitas 1993;17:211-19.
19. Watts NB, Notelovitz M, Timmons MC, et al. Comparison of oral estrogens and estrogens plus androgen on bone mineral density, menopausal symptoms, and lipid-lipoprotein profiles in surgically menopausal women. Obstet Gynecol 1995;85:529-37.
20. Sherwin BB, Gelfand MM. A prospective one-year study of estrogen and progestin in postmenopausal women: Effects on clinical symptoms and lipoprotein lipids. Obstet Gynecol 1989;73:759-66.
21. Friedl KE, Hannan Jr, CJ, Jones RE, et al. High density lipoprotein cholesterol is not decreased if an aromatizable androgen is administered. Metabolism 1990;39:69-74.
22. Sherwin BB, Gelfand MM, Schucher R, et al. Postmenopausal estrogen and androgen replacement and lipoprotein-lipid concentrations. Am J Obstet Gynecol 1987;4:414-19.
23. Lovell CW. Breast cancer incidence with parenteral estradiol and testosterone replacement therapy. Menopause. Abst North Am Menopause Soc 1994;1(3):150.
24. Henderson VW, Paganini-Hill A, Emanuel CK, Dunn ME, Buckwalter JG. Estrogen replacement therapy in older women. Arch Neurol 1994;51:896.
25. Paganini-Hill A, Henderson VW. Estrogen deficiency and risk of Alzheimer's disease in women. Am J Epidemiol 1994;140(3):256.
26. Davis SR, McCloud P, Strauss BJG, et al. Testosterone enhances estradiol's effects on postmenopausal bone density and sexuality. Maturitas 1995;21:227-36.

MANAGEMENT OF THE LATE MENOPAUSE: ULTRA LOW-DOSE ADJUSTIVE ESTROGEN THERAPY

Morris Notelovitz

Women over 65 years of age do benefit from hormone therapy, but compliance is often poor because of estrogen-induced side effects and fear of breast cancer. Ultra low-dose estrogen therapy (ULET), adjusted to the need of the individual is effective in meeting various therapeutic goals while minimizing unacceptable side effects. The four main principles of ULET include: individualizing therapy by patient examination and selective testing; prescription of the lowest effective dose of a natural estrogen; choosing the route of administration based on the patient's preference and medical need; adjusting the therapy over time by monitoring symptomatic and specific biologic end-point responses.

Urogenital aging affects 50% (or more) of elderly postmenopausal women. Two independent studies have confirmed that 40% of women on "adequate" systemic estrogen therapy, still have atrophic vaginitis as assessed by high lateral vaginal wall pH [1]. Rather than increase the systemic estrogen therapy, the addition of low dose local estrogen relieves the problem without significantly increasing the systemic absorption of estradiol. This can be conveniently achieved by 25 μg vaginal estradiol matrix tablets, inserted via a specially designed thin vaginal applicator thus enhancing patient comfort and compliance. In one study, the mean plasma estradiol value (after 12 weeks of therapy) was only 21 pg/dl. Follicle-stimulating hormone (FSH) values remained in the postmenopausal range. Improvement in vaginal integrity is mirrored by a similar response in the distal and proximal urethra, and an improvement in the chronic urinary urgency and frequency-related symptoms associated with the urethral syndrome.

The prevalence of osteoporosis-related fractures is reduced in elderly women, as long as they are compliant with therapy. Four clinically relevant factors govern the prescription of ULET for this indication: 1) the risk of vertebral and hip fractures is related to the endogenous production of estradiol in untreated women; this is 60% lower in women with values > 5 pg/dl [2]; 2) the individual's ability to absorb estrogen and its binding to sex hormone-binding globulin (SHBG); 3) the individual's biologic response to a given estradiol blood level. At the same plasma E_2 level, bone mineral density (BMD) increases vary [3]; 4) the degree of osteopenia. The dose needed to maintain the BMD is usually less than that required to increase (treat) significantly reduced BMD [3].

The dose of ULET selected for osteoporosis prevention, should also have

R. Paoletti et al. (eds.), Women's Health and Menopause, 195–198.

potential beneficial cardiovascular protection. Two years of ultra-low dose esterified estrogen (ESE)–0.3 mg, resulted in a significant increase in high density lipoprotein (HDL) cholesterol, and a significant decrease in low density lipoprotein (LDL) and total cholesterol. Triglycerides were unaffected [3]. The effect of this dose of ESE on coronary artery vascular reactivity and intimal physiology, is not known.

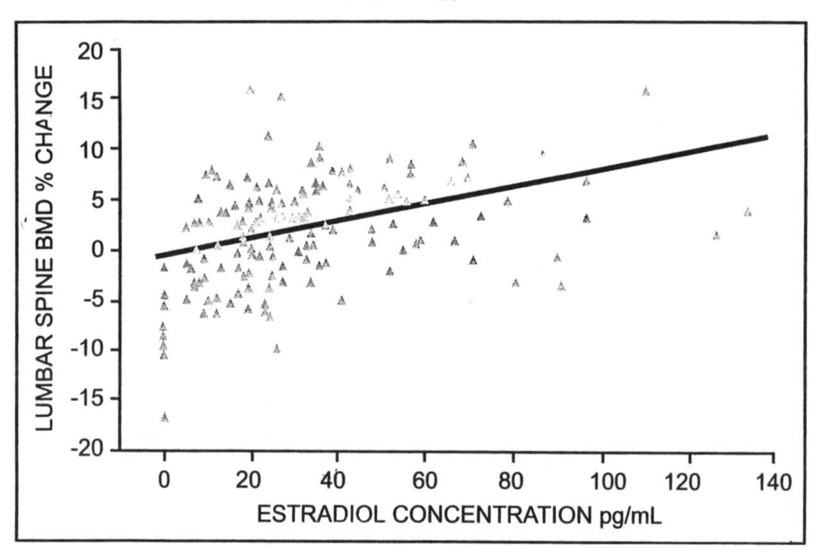

Figure 1. Individual response to esterified estrogen therapy after 24 months. BMD versus estradiol.

Table 1. Low Dose Esterified Estrogen Therapy Increases Lumbar Spine Bone Mineral Density

			ESE	Dose	Mg/d
Time (months)	Parameter	Placebo	0.3	0.625	1.25
12	E_2 pg/dl	22	29	43	64
	BMD	-1.6^+	1.2*	2.0*	3.7*
18	E_2 pg/dl	14	24	41	58
	BMD	-2.0^+	1.5*	3.0*	4.6*
24	E_2 pg/dl	16	26	40	60
	BMD	-2.9^+	1.9*	3.0*	5.0*

*$p < 0.05$ from baseline and placebo; $^+p < 0.05$ from baseline. Adapted from [3].

Ultra-low dose ET has a minimal endometrial stimulating effect. The latter allows for either ultra-low dose progestin (P) therapy (with reduced P-related side effects) or selectively, ULET alone therapy with annual endometrial monitoring. Whereas the 0.3 mg dose (after two years) resulted in a 1.7% prevalence of endometrial hyperplasia (equal to placebo-treated subjects), increasing the dose of ESE to 0.625 mg and 1.25 mg produced an endometrial hyperplasia incidence of 28.2% and 53.3%, respectively [4]. At the "higher" doses of estrogen therapy, additional P therapy is mandatory. However, depending upon the dose of estrogen and progesterone used, ultra-low doses can effectively induce amenorrhea. The addition of 0.5 mg of NETA (to 1 mg of 17β estradiol) resulted in a 94% incidence of amenorrhea at 3 months, which was maintained for one year. Only 10% of subjects dropped from the study [5].

With the caveat that not all low-dose estrogens have the same biologic potency and target tissue effect, the principle of adjustive hormone therapy is summarized in Figure 2.

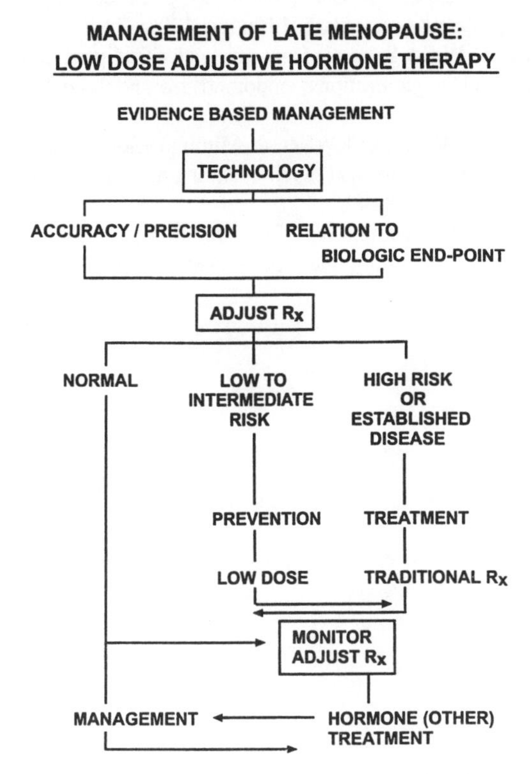

Figure 2. Management of late menopause: low dose adjustive hormone therapy.

Using evidence-based technology, the dose (and the route) of estrogen therapy is chosen: low dose to prevent (maintain) and a "traditional" dose to treat (correct) a given condition. By monitoring the patient on an annual basis, the treatment can be titrated (adjusted) upwards or downwards according to the patient's response and need. "Normal" patients can be managed without hormone therapy (for example with exercise, life style, and diet prescription), but given ULET if subsequent annual evaluation justifies the need. Finally, adding an exercise regimen to ultra-low dose estrogen therapy, may enhance the latter's efficacy.

References

1. Notelovitz M. Estrogen therapy in the management of problems associated with urogenital aging: A simple diagnostic test and the effect of the route of hormone administration. Maturitas 1995;22(Suppl.):31-33.
2. Bauer DC, Nevitt MC, Ettinger B, et al. Women with low serum estradiol have an increased risk of hip and vertebral fractures: A prospective study. In: Parapoulos SE, Lips P, Pols HAP et al., editors. Osteoporosis 1996. Proceedings of the 1996 World Congress on Osteoporosis. Amsterdam: Elsevier, 1996:271-75.
3. Genant HK, Lucas J, Weiss S, et al. Low-dose esterified estrogen therapy effects on bone, plasma estradiol concentrations, endometrium and lipid levels. Arch Int Med 1997;157:2609-15.
4. Notelovitz M, Varner RE, Rebar RW, et al. Minimal endometrial proliferation over a two-year period in postmenopausal women taking 0.3 mg of unopposed esterified estrogen. Menopause 1997;2:80-88.
5. Stadberg E, Mattson LA, Uvebrant M. 17β estradiol and norethindrone acetate in low doses as continuous combined hormone replacement therapy. Maturitas 1996;23:31-39.

Selective Estrogen Receptor Modulator (SERM) Drugs for the Prevention of Osteoporosis

Louis V. Avioli

It has been well established that estrogens in any form in doses equivalent to 0.3-1.25 mg esterified estrogens per day when administered by oral, percutaneous, or transdermal routes are effective in preventing postmenopausal bone loss [1-3]. The results of a variety of epidemiological studies designed to analyze the effects of hormone replacement therapy (HRT) on cardiovascular function also reveal a reduced relative risk of morbidity and mortality from cardiovascular diseases in woman taking HRT [4,5]. Despite these observations it is becoming increasingly obvious that a low prevalence of HRT use exists [6-8]. Factors which dispose to noncompliance and/or lack of acceptance of HRT treatment in the postmenopausal female include a dislike for withdrawal bleeding which results from cyclic estrogen/progestin therapy, breast engorgement and pain, poor compliance with progestins, and the fear of breast cancer [9,10]. In fact, many women either never have their prescriptions filled, follow treatment either intermittently or sporadically, or discontinue therapy completely after 21-24 months [7,8]. Although nonhormonal therapies for preventing bone loss in postmenopausal women such as the bisphosphonates have been recommended as effective substitutes for HRT [11-13], these pharmacological agents often require rigorous dosing schedules because of a variety of gastrointestinal complications [11]. SERM drugs which are also available to women who can not, will not, or should not utilize HRT characteristically react with estrogen receptors with resultant transcriptional pathways unlike those identified for estrogens [14-16].

SERM drugs currently available on the market such as tamoxifen and raloxifene have been proven effective in preventing bone loss in postmenopausal women [17-19]. Tamoxifen, in doses of 20 mg/day is partially effective in this regard in the axial and appendicular skeleton, although the vertebral effect is augmented when the dose is raised by 50% [20]. Although like HRT, tamoxifen produces changes in circulating lipoproteins which should minimize atherosclerotic disease [21,22], serious side effects such as endometrial carcinoma and possibly, after long-term treatment, recurrent breast carcinoma preclude the routine administration of tamoxifen for osteoporosis prevention [23-25]. Although unlike estrogen [25] raloxifene also shares tamoxifen's potential to prevent breast cancer, the drug possesses a clinical profile which is distinct from either tamoxifen or estrogen. Not only does this unique SERM drug decrease bone turnover [26] and prevent axial and appendicular bone loss [19], but like estrogen, raloxifene decreases circulating

R. Paoletti et al. (eds.), Women's Health and Menopause, 199–201.

cholesterol and low density lipoproteins, and decreases fibrinogen [19,27]. However, unlike estrogens which actually increase blood triglyceride levels, raloxifene has no effect on this blood lipid [19]. Finally, unlike tamoxifen and estrogen which have undesirable effects on uterine tissue, no significant uterine effects of raloxifene on endometrial tissue have been observed in patients treated with raloxifene [28]. In addition to raloxifene and tamoxifen, other drugs such as droloxifene, idoxifene, and toremifene have also been subjected to testing in animal models and limited trials in humans [29-31]. Unlike raloxifene which is a nonsteroidal benzothiophene, these SERM drugs are derivatives of tamoxifen [32] which is a triphenylethylene. Consequently we should anticipate that when these drugs are finally available for routine clinical use, their SERM activities will be comparable to those recorded for tamoxifen and not raloxifene.

References

1. Genant HK, Lucas J, Weiss S, et al. Low-dose esterified estrogen therapy. Arch Intern Med 1997;157:2609-15.
2. Lafferty FW, Fiske ME. Postmenopausal estrogen replacement: A long-term cohort study. Am J Med 1994;97:66-77.
3. Kanis JA. Estrogens, the menopause, and osteoporosis. Bone 1996;19:185S-190S.
4. Bush TL, Barrett-Connor E, Cowan LD, et al. Cardiovascular mortality in noncontraceptive use of estrogen in women: Results from the Lipid Research Clinics Program Follow-up Study. Circulation 1987;75:1102-9.
5. Stampfer MJ, Colditz GS, Willett WC, et al. Postmenopausal estrogen therapy and cardiovascular disease: Ten-year follow-up from the Nurses' Health Study. N Engl J Med 1991;325:756-62.
6. Utian WH, Schiff I. NAMS-Gallup survey on women's knowledge information sources, and attitudes to menopause and hormone replacement therapy. Menopause 1994;1:39-48.
7. Groeneveld FPMJ, Bareman FP, Barentsen R, et al. Determinants of first prescription of hormone replacement therapy. A follow-up study among 1689 women aged 45-60 years. Maturitas 1995;20:81-89.
8. Cano A. Compliance to hormone replacement therapy in menopausal women controlled in a third level academic centre. Maturitas 1995;20:91-99.
9. Gambrell RD Jr. Management of hormone replacement therapy side effects. Menopause 1994;1:67-72.
10. Smith RNJ, Holland EFN, Studd JWW. The symptomatology of progestogen intolerance. Maturitas 1994;18:87-91.
11. Hosking D, Chilvers CED, Christiansen C, et al. Prevention of bone loss with alendronate in postmenopausal women under 60 years of age. N Engl J Med 1998;338:485-92.
12. Thiebaud D, Burckhardt P, Kriegbaum H. et al. Three monthly intravenous injections of ibandronate in the treatment of postmenopausal osteoporosis. Am J Med 1997;103:298-307.
13. Herd RJM, Balena R, Blake GM, et al. The prevention of early postmenopausal bone loss by cyclical etidronate therapy: A 2-year, double-blind, placebo-controlled study. Am J Med 1997; 103:92-99.
14. McDonnell DP, Norris JD. Analysis of the molecular pharmacology of estrogen receptor

agonists and antagonists provides insights into the mechanism of action of estrogen in bone. Osteoporosis Int 1997;1:S29-S34.

15. Grese TA, Sluka JP, Bryant HU, et al. Molecular determinants of tissue selectivity in estrogen receptor modulators. Proc Natl Acad Sci 1997;94:14105-10.
16. Brzozowski AM, Pike ACW, Dauter Z, et al. Molecular basis of agonism and antagonism in the oestrogen receptor. Nature 1997;389:753-58.
17. Love RR, Barden HS, Mazess RB, et al. Effect of tamoxifen on lumbar spine bone mineral density in postmenopausal women after 5 years. Arch Intern Med 1994;154:2585-88.
18. Fornander T, Rutqvist LE, Sjoberg HE, et al. Long-term adjuvant tamoxifen in early breast cancer: Effect on bone mineral density in postmenopausal women. J Clin Onc 1990;8:1019-24.
19. Delmas PD, Bjarnason NH, Mitlak BH, et al. Effects of raloxifene on bone mineral density, serum cholesterol concentrations, and uterine endometrium in postmenopausal women. N Eng J Med 1997;337:1641-47.
20. Grey AB, Stapleton JP, Evans MC, et al. The effect of the antiestrogen tamoxifen on bone mineral density in normal late postmenopausal women. Am J Med 1995;99:636-41.
21. Rossner S, Wallgren A. Serum lipoproteins and proteins after breast cancer surgery and effects of tamoxifen. Atherosclerosis 1984;52:339-46.
22. Bagdade JD, Wolter J, Subbaiah PV, et al. Effects of tamoxifen treatment on plasma lipids and lipoprotein lipid composition. J Clin Endocrinol Metab 1990;70:1132-35.
23. Horwitz KD. Editorial: When tamoxifen turns bad. Endocrinol 1995;136:821-23.
24. Fornander T, Cedarmark B, Mattsson A, et al. Adjuvant tamoxifen in early breast cancer: Occurrence of new primary cancers. Lancet 1989;1(8630):117-20.
25. LaCroix AZ, Burke W. Commentary: Breast cancer and hormone replacement therapy. Lancet 1997;350:1042-43.
26. Heaney RP, Draper MW. Raloxifene and estrogen: Comparative bone-remodeling kinetics. J Clin Endocrinol Metab 1997;82:3425-29.
27. Kauffman RF, Bensch WR, Roudebush RE, et al. Hypocholesterolemic activity of raloxifene (LY139481): Pharmacological characterization as a selective estrogen receptor modulator. J Pharm Exp Ther 1997;280:146-53.
28. Boss SM, Huster WJ, Neild JA, et al. Effects of raloxifene hydrochloride on the endometrium of postmenopausal women. Am J Obstet Gynecol 1997;177:1458-64.
29. Ke HZ, Chen HK, Simmons HA, et al. Comparative effects of droloxifene, tamoxifen, and estrogen on bone, serum cholesterol, and uterine histology in the ovariectomized rat model. Bone 1997;20:31-39.
30. Chen HK, Ke HZ, Jee WSS, et al. Droloxifene prevents ovariectomy-induced bone loss in tibiae and femora of aged female rats: A dual-energy x-ray absorptiometric and histomorphometric study. J Bone Miner Res 1995;10:1256-62.
31. Marttunen MB, Hietanen P, Tiitinen A, et al. Comparison of effects of tamoxifen and toremifene on bone biochemistry and bone mineral density in postmenopausal breast cancer patients. J Clin Endocrinol Metabol 1998;83:1158-62.
32. Cos DA, Lakshmanan MC. Selective estrogen receptor modulators (SERMs): Current topics and future possibilities. Osteoporosis Japan 1997;5:24-34.

A Rationale for Hormone Replacement Therapy in Organ Transplant Patients

Marie L. Foegh

Introduction

Estrogens are steroid hormones that have important roles in the cardiovascular system and in maintenance of bone tissue. These effects are mediated by a ligand-activated transcription factor, the estrogen receptor (ER) [1]. Clinically the question arises as to whether estrogen replacement therapy (ERT) is indicated in postmenopausal recipients of organ transplants. Post-transplantation hypertension, hyperlipidemia, and insulin dependent diabetes develop in the immunosuppressed recipient. Renal transplant patients are already at high risk for cardiovascular complications of atherosclerosis and many patients have severe osteoporosis at the time they receive a renal allograft. The well-known female advantage in cardiovascular disease prior to menopause appears to be abrogated in transplant recipients. The female origin of the allograft appears to be a risk factor. The reason for the difference in risk of atherosclerosis/arteriosclerosis in the female organ in its natural host compared to in the transplant host is most likely of immune origin and is most spectacularly seen in the coronary arteries of the cardiac allograft. This graft arteriosclerosis presents as a characteristic vasculopathy consisting of a continous concentric myointimal thickening which is universal in all solid organ allografts and limits long-term graft and patient survival.

Gender Differences in Non-Immune Vascular Response to Injury

Myointimal thickening occurs also after non-immune injuries. We therefore investigated in animal models whether there are gender differences in the non-immune vascular response using the same strain of rats as donors and recipients. In such a model no immunological response occurs and the graft is only exposed to ischemic, surgical, and reperfusion injuries [2].

Syngeneic aorta grafts were transplanted into the same or the opposite sex as illustrated in Figure 1. When harvested 30 days later grafts of female gender transplanted into either sex responded with much less myointimal thickening than vessels from males. The sex of the host had no significant effect on myointimal thickening [3].

R. Paoletti et al. (eds.), Women's Health and Menopause, 203–209.

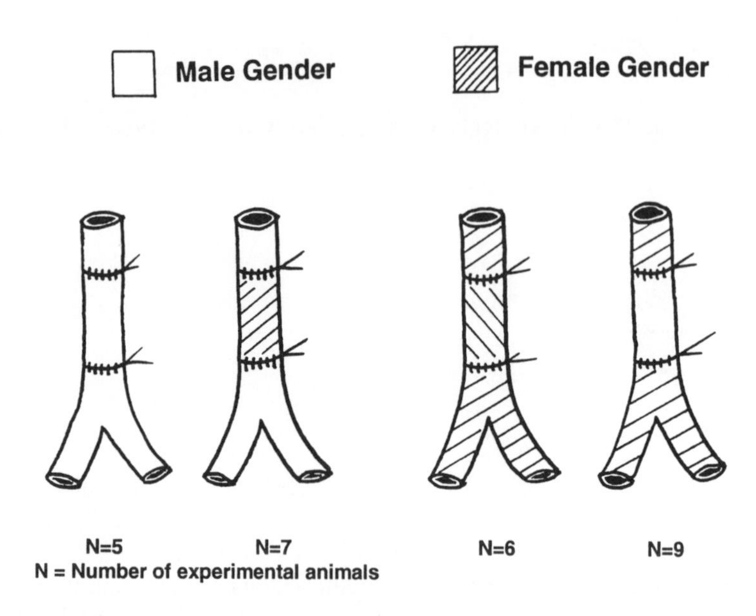

% Intimal Thickening			
% Intimal Thickening= $\frac{\text{Area of Myointimal Hyperplasia}}{\text{Total Vessel Area}}$ X 100%			
F→F	F→M	M→F	M→M
1.4 ± 0.9	2.2 ± 0.4	6.9 ± 1.8	7.3 ± 2.0

$P < 0.005$ M→F versus F→F
$P < 0.01$ M→M versus F→M

Figure 1. Effect of gender of aorta on myointimal hyperplasia in a male or female environment.

Estrogen Effect on Non-Immune Vascular Response to Injury

The protective effect of female graft gender was also observed by other investigators following balloon injury of the rat carotid artery [4]. They found that oophorectomy removed the female advantage and that exogenous estrogen administration restored the protection to that of the intact sexual mature female. Thus in this model the protective effect of female gender was linked to the estrogen environment rather than female gender. We found similar protective effects of estrogen in male rabbits undergoing balloon injury of the iliac arteries [5] which supports a vascular protective effect of estrogen independent of gender.

These effects of exogenous estrogen on the vascular wall do not need the classical estrogen receptor (ERα) but may be mediated through the recently discovered estrogen receptor (ERβ) as suggested in the vascular injury model in mice where the classical estrogen receptor (ERα) gene was disrupted [6,7].

Estrogen and Adhesion Molecules

An early feature of transplant rejection and experimental transplant arteriosclerosis is invasion of the vascular wall by monocytes/macrophages. This requires upregulation of adhesion molecules like E-selectin, ICAM, and VCAM. Recently it was shown that endothelial cells from human umbilical veins only express the newly discovered estrogen receptor (ERβ) [8]. Interestingly, 17β-estradiol but not its stereo-isomer 17α-estradiol suppresses IL-1-induced upregulation of these adhesion molecules on human umbilical vein endothelial cells *in vitro*. The estrogen suppression of IL-1-induced upregulation of these adhesion molecules may be interpreted as being elicited through an ERβ mediated mechanism since the binding affinity is poor for 17α-estradiol to the ERβ compared to the ERα. Thus the investigators may have used too low a concentration of the 17α-estradiol to elicit even an ERα mediated response. Interestingly ER is upregulated in the allograft [9].

Estrogen and the Immune System

It has long been known that estrogens have important effects in the immune system. The estrogen receptors together with other steroid receptors like glucocorticosteroids, belong to the same family of nuclear transcription receptors. Immune-related cells such as macrophages, monocytes, CD8 lymphocytes, and CD4 lymphocytes of subtype CD29 and CD34RO contain estrogen receptors (ER). In addition stromal cells in the bone marrow contains ER while B lymphocytes are described as being ER negative. However the recent discovery of a second estrogen receptor named ERβ to distinguish it from the old ER now renamed ERα, may change these findings, since the antibodies and oligonucleotide primers used previously for identification of ER will only identify ERα. The recent findings of transcripts of ERβ, but not ERα in the rat thymus and spleen supports the suggestion that the cells of the immune system may be ERβ positive [8] and that some of the immuno-modulatory effects of estrogen is mediated by the estrogen receptor-β (ERβ).

Gender Differences in Autoimmune Diseases

Female gender does not protect against the ravage of autoimmune diseases which are much more prevalent in women than in men (Table 1). Interestingly, although estrogen is known to augment many components of both the humoral- and cell-mediated immune response (Figure 2), ERT in women with rheumatoid arthritis reduces the symptoms. Furthermore high blood estrogen levels or estrogen treatment are associated with improved asthmatic disease. It is difficult to reconcile these reports with the current knowledge of estrogen's stimulatory effects on the different immune parameters, particularly its effect on the humoral-mediated immune response [10].

Table 1. Sex Ratio in Autoimmune Diseases

Autoimmune Disease	Female:Male Ratio
Scleroderma	15:1
Systemic lupus	10:1
Takayasu's arteritis	10:1
Rheumatoid Arthritis	3:1
Polymyalgia rheumatica	2:1

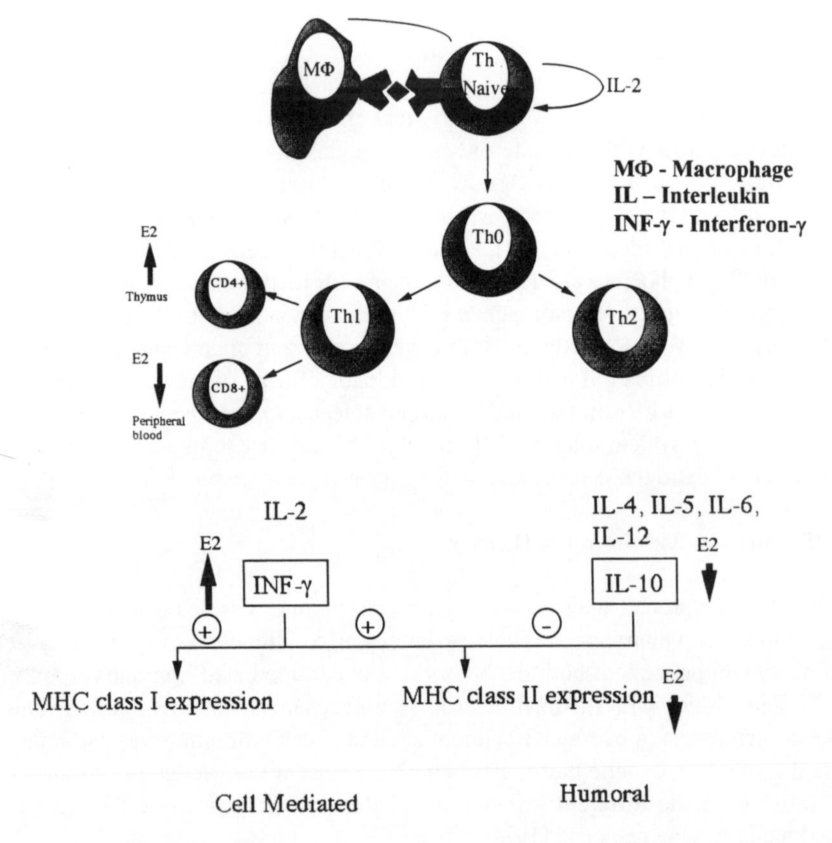

Figure 2. Effect of estrogen on cell-mediated and humoral immune response.

Estrogen Inhibition of Experimental Transplant Arteriosclerosis

ESTROGEN PROTECTION OF THE ENDOTHELIUM

In cardiac and aorta transplant models, using male rabbits as donors and recipients, we find estradiol inhibits accelerated transplant arteriosclerosis [11-15]. Four different doses of estradiol preserved the normal ultrastructure of the endothelium, up to the time of sacrifice. In contrast, the endothelium in the allograft arteries of the placebo-treated group underwent severe degenerative changes. Whether the preservation of the normal ultrastructure of the endothelium is synonymous with lack of activation and decreased release of cytokines and growth factors, remains to be determined. The elucidation of the effects of estrogen on the development of transplant arteriosclerosis in all graft organs may have a bearing on the role of estrogen in autoimmune diseases and immune responses in women in general.

ESTRADIOL-17ß AND SUPPRESSION OF MHC CLASS II UPREGULATION

Experimentally, we have repeatedly shown in different transplant arteriosclerosis models that estradiol-17β substantially attenuated the disease [11-18]. We believe that the most important mechanism is prevention of upregulation of inducible and constitutive MHC class II expression [13]. Secondarily, the inhibition of monocyte/macrophage infiltration of the graft contributes substantially to the beneficial effect of estradiol-17β [12]. The mechanism is probably through suppression of upregulation of the adhesion molecules.

Gender Effect in Transplant Patients

ESTRADIOL AND IMMUNE RESPONSE

The immuno-stimulatory effect of estrogen is seen in postmenopausal women receiving ERT where the peripheral blood lymphocyte CD4:CD8 ratio increases due to a decrease in CD8 cells. In rodents female thymocytes consists of more T-lymphocytes of the CD4 subgroup than their male counterpart. This suggests a superior response of the female immune system compared to the male. Whether this bears any relation to the higher risk of rejection of female organs is speculative.

In human transplantation more recent data show that the gender of the donor and not the gender of the recipient is a determent of survival with the exception of the kidney. Previously it was thought that a female kidney transplanted into a larger male put the survival of the transplanted kidney at risk due to hyperfiltration. Further investigations revealed that the most frequents cause of brain death in female donors are cerebrovascular accidents which are known to lessen the quality of the harvested kidney and thus is the culprit of the sex difference and not the female gender of the kidney graft as such. However in other transplanted organs the female origin of the graft is a risk factor [19-21]. The reason for the female sex of donors being a risk factor is not known. One may speculate that female gender may be associated with a greater expression of damaging cytokines from the allograft in response to immune injury.

EFFECT OF AGE

The only gender-related risk factor in renal transplantation is being a female recipient of > 50 years of age. We suggest that this is due to lack of estrogen and that ERT may eradicate this risk factor.

Summary

The immune-related cells express estrogen receptors and their ligands affect numerous mechanisms of both the humoral- and cell-mediated immune response. In transplant models we find estradiol-17β substantially attenuates transplant arteriosclerosis/chronic rejection and abrogates upregulation of MHC class II expression. The latter is an important mechanism of the allo-immune response. Other mechanisms involving suppression of adhesion molecule expression contribute to the beneficial effect of estrogen. These experimental data speaks to a beneficial effect of ERT in women transplant recipients at menopause.

References

1. Foegh ML, Ramwell PW. Cardiovascular effects of estrogen - Implications of the discovery of the estrogen receptor subtype ERβ. Current Opinion in Nephrol Hypertension 1998;7:83-9.
2. Foegh ML, Ramwell PW. Pharmacologic control of smooth muscle cells in allografts. Transplant Immunol 1997;5:267-75.
3. Foegh M, Rego A, Lou H, Katz N, Ramwell P. Gender effects on graft myointimal hyperplasia. Transpl Proc 1995;27:2070-73.
4. White CR, Shelton J, Chen S-J, et al. Estrogen restores endothelial cell function in an experimental model of vascular injury. Circulation 1997;96:1624-30.
5. Foegh ML, Asotra S, Howell MH, Ramwell PW. Estradiol inhibition of arterial neointimal hyperplasia after balloon injury. J Vasc Surg 1994;19:722-26.
6. Kuiper G, Enmark E, Pelto-Huiko M, Nielsson S, Gustafsson J-Å. Cloning of a novel estrogen receptor expressed in rat prostate and ovary. Proc Natl Acad Sci, USA 1996; 93:5925-30.
7. Iafrati MD, Karas RH, Aronovitz M, et al. Estrogen inhibits the vascular injury response in estrogen receptor α-deficient mice. Nature Med 1997;3:545-48.
8. Enmark E, Pelto-Huiko M, Grandien K, et al. Human estrogen receptor β-gene structure, chromosomal localization and expression pattern. J Clin Endocrinol Metab 1997;82: 4258-65.
9. Lou H, Martin MB, Stoica A, Ramwell PW, Foegh ML. Upregulation of estrogen receptor-α expression in cardiac allograft. Circ Res In press.
10. Olsen NJ, Kovacs WJ. Gonadal steroids and immunity. Endocrine Rev 1996;17:369-84.

11. Foegh ML, Khirabadi BS, Nakanishi T, Vargas R, Ramwell PW. Estradiol protects against experimental cardiac transplant atherosclerosis. Transpl Proc 1987;19:90-94.
12. Cheng LP, Kuwahara M, Jacobsson J, Foegh ML. Inhibition of myointimal hyperplasia and macrophage infiltration by estradiol in aorta allografts. Transplantation 1991;52:967-72
13. Lou H, Kodama T, Maurice P, Wang YN, Katz NM, Foegh ML. Inhibition of transplant coronary arteriosclerosis in rabbits by chronic estradiol treatment is associated with abolishing MHC class II antigen expression. Circulation 1996;94:3355-61.
14. Saito S. Foegh ML, Motomura N, Lou H, Kent K, Ramwell PW. Estradiol inhibits allograft inducible major histocompatibility MHC-class II antigen expression and transplant arteriosclerosis in the absence of immuno-suppression. Transplantation In press.
15. Foegh ML, Ramwell PW. The protective effect of estradiol in transplant arteriosclerosis. In: Rabuanyi G, Kauffman R, editors. Estrogens and the vessel wall. Amsterdam: Harwood Academic Publishers, 1998:213-24.
16. Motomura N, Lou H, Hong M, Tsutsumi Y, Mayumi T, Foegh ML. Local administration of estrogen inhibits transplant arteriosclerosis in rat aorta accelerated by topical exposure to IGF-I. Transpl Proc 1997;29:1118-20.
17. Lou H, Zhao Y, Delafontaine P, et al. Estrogen effects on insulin-like growth factor-I (IGF-I) induced cell proliferation and IGF-I expression in native and allograft vessels. Circulation. 1997;96:927-33.
18. Saito S, Motomura N, Lou H, Ramwell PW, Foegh ML. Specific effects of estrogen on growth factor and MHC class II expression in rat aorta allograft. J Thorac Cardiovasc Surg 1997;114:803-10.
19. Marino IR, Doyle HR, Aldringhetti L, et al. Effect of donor age and sex on the outcome of liver transplantation. Hepatology 1995;22:1754-62.
20. Velosa JA, Offord KP, Schroeder DR. Effect of age, sex and glomerular filtration on renal function outcome of living kidney donors. Transplantation 1995;60:1618-21.
21. Boucek MM, Novick RJ, Bennett LF, Fiol B, Keck BM, Hosenpud JD. The registry of the International Society of Heart and Lung Transplantation: First official pediatric report-1997. J Heart Lung Transplant 1997;16:1189-206.

Management of Ambivalence Towards Hormone Replacement Therapy

Alessandra Graziottin

Introduction

Indications for hormonal replacement therapy (HRT) after the menopause are twofold: 1) relief of postmenopausal symptoms secondary to estrogen deficiency and 2) risk reduction of diseases associated with estrogen deficiency [1]. Compliance is easier in symptomatic patients, when the efficacy in relieving disturbances is a major factor in satisfaction of use. Major compliance problems arise with the second indication, when, in absence of symptoms, the decision will rely not on the real benefits to be derived or the risks but on the perception of those benefits and risks [1]. Perception is a key mental step, as it involves subjective decision making and the many conflicting emotional and cognitive factors that may distort the decision-making process itself.

From the psychodynamic point of view, the conflicting attitude in the decision-making process and human behavior, i.e. the coexistence of conflicting fears and desires, is called "ambivalence" [2]. It has a particular role in preventive medicine, when one is expected to use a prophylactic pharmacologic treatment for years, in order to avoid some unwanted and/or potentially life threatening problems. It is a critical variable in accepting HRT after the menopause. Ambivalence is one of the most powerful and yet poorly recognized psychodynamic factors that may affect the decision-making process regarding HRT as well as compliance.

Because HRT only provides supplementation or replacement of deficient ovarian hormones, the most important medical benefits of HRT may require continued administration for an extended, indefinitely long period of time and may require lifelong administration. Motivation to continue HRT over the years, and compliance to it, must therefore be addressed in a serious way, if we are to have any hope of realizing the full impact of HRT [1]. Ambivalence is one of the critical factors to be understood and properly managed, if this goal is to be pursued.

Ambivalence Towards HRT

Ambivalence may involve intellect (when one expresses an idea and its opposite at the same time), will (when one is willing and not willing to do something at the same time), and affection (when one feels conflicting emotional states, such as love and hate, or sympathy

R. Paoletti et al. (eds.), Women's Health and Menopause, 211–215.

and aversion at the same time) [2]. All these aspects may be present in the request or avoidance, motivation, and compliance dynamics towards HRT.

Ambivalence towards HRT may be "spontaneous" in the woman, when it is an expression of a "natural" attitude towards life's seasons and indicates the reluctance or frank aversion to use "drugs" to modify the natural process of aging. It may be "mediagenic," affected by the pros and cons of HRT as reported in newspapers and magazines and contains a strong emotional component more influenced by ideologic backgrounds than scientific data. This source of ambivalence is critical as the majority of women get their information on HRT from the media [3]. In mediagenic ambivalence, fears focused on breast cancer are usually emphasized and dismiss many of the benefits derived from a well-prescribed HRT program. Additionally, the media has also encouraged unrealistic expectations from HRT, as when HRT is presented as a way to "regain youth" or to feel new again [3]. Ambivalence may also be "iatrogenic." This is one of the strongest and most insidious causes of ambivalence and poor compliance. The iatrogenic ambivalence has two main sources. It may be triggered by the individual doctor, who makes the prescription with verbal, and/or nonverbal, comments that may worry the woman. This "ambivalent" prescription could be partially responsible for the noncompliance to the prescription itself [4]. In one large survey, 20% to 30% of patients given a prescription for HRT never had the prescription filled; and of those who did, 10% took their medications sporadically and 20% stopped therapy within 9 months of starting [5]. The second iatrogenic cause of ambivalence is the reported conflicting attitude of different physicians towards HRT, reported by 47% of women who stopped the treatment [6]. When the attitude of doctors is clear and positive, adequacy of counselling becomes a major factor in compliance [1,3,7].

Different sources of ambivalence usually overlap in the single woman, making both the decision to start and the motivation to stay on treatment difficult even when faced with side effects. Ambivalence towards HRT arises from two main conflicting areas: the fears and desires with which sex hormones are associated. The most frequent fears involve: cancer [1,3-6], weight gain [1,3,6], being unnatural [6] or manipulated, having menses back [1,3,4,6], getting headaches, or premenstrual syndrome (PMS) [6]. These fears are usually so potent that they may explain the reduced number of women who start HRT and the very few who stay on treatment for more than one year [1,3-7].

Women who perceive these potent drawbacks are now confronted with the growing knowledge that women can positively affect the quality of aging and that HRT can improve that quality significantly. The increasing number of women requesting HRT in recent years are positively motivated by the desire for better health, improved well-being, a better self-image, empowerment towards aging, improved memory, and improved sexual satisfaction [3-7].

The final decision with regard to HRT depends on the balance of these two groups of factors. When the conflict is strong because both fears and desires are emotionally potent, the psychodynamic end-result may be a paralyzing effect on the decision-making process. The woman will not ask for HRT or will not comply with the program prescribed.

Management of Patient Ambivalence Towards HRT

Because of the perceptive basis and the inherent conflicting nature, ambivalence needs to be managed with two main objectives in mind:

1) Communications: Quality of communications (and counselling) is critical in medicine and in all human experiences where fear is high [1,5-8]. To reduce fears and the "infectious anxiety" they may convey, critical steps are:
• listen carefully to fears and expectations, paying attention to verbal and nonverbal language (the latter has the highest impact on the emotional perception of the counselling);
• speak simply, as cognitive difficulties reinforce the feeling of being manipulated, misunderstood, or not "listened to" [6];
• address fears properly. Precise answers have the highest reassuring effect and help the patient to perceive herself as an active participant in the decision-making process;
• transform expectations or desires into motivation to HRT. This is an easy and yet poorly used communicative medical strategy. Doctors should be flexible and not insist on osteoporosis, for example, if the major concern of the patient is the loss of memory or the worsening of skin. The more the patient understands that her expectations will be met, the more she will be compliant and the other benefits of HRT will come as "fringe benefits" of the treatment for her major concern;
• meet expectations and respect motivation to improve compliance. Again, a treatment focused on patient needs has the highest probability of being followed over time;
• last, but not least, give the patient time to think about HRT. Ambivalence, because of its conflicting nature, requires time to get through the decision-making process. The first counselling session clarifies the hottest issues; a second one, when the patient comes with the exams, may be essential in reinforcing motivation and alleviating fears.

2) Focus on the content and substance of fears: To be effective, reassurance should be convincing in terms of the facts and figures quoted on every single concern. Critical steps include:
• never dismiss fears as banal or trivial: they are central to the patient and this is the first good reason why they should be addressed properly. They also clarify the patient's priorities and help the understanding doctor to tailor both motivational techniques and treatment to the individual needs;
• address fears directly and specifically;
• counterbalance medical fears against medical advantages;
• distinguish emotional concerns from medical ones. These obvious recommendations are usually overlooked in clinical practice, where the counselling fluctuates from one mental domain to another, with substantial dismissal of the emotional issues critical to the final decision-making process.

Management of Iatrogenic Ambivalence

Attitudes of doctors towards any treatment are crucial in determining the success or failure of the specific treatment. When ambivalence towards HRT is strong in the physician, it will pervade the consultation causing a distortion in the information and affect the quality and outcome of the counselling. Guidelines for the physician to avoid unconscious negative dynamics are the following:

• listen to your inner attitude towards HRT and menopausal patients. The first point is critical from the scientific point of view, as the core of ambivalence centers on the uncertainty with regard to the potential increase in cancer, particularly breast cancer, due to HRT. The second is human, and relies on the physician's personal attitude towards menopausal women and the interpersonal skills requisite for successful doctor-patient relationships. Some doctors may be excellent surgeons or obstetricians, but do not like menopausal women with their complaints and need to discuss everything, and do not have the patience required to tailor treatment and answer many phone calls for "minor" side effects. Should these negative attitudes exist, a different selection of patients and/or different area of clinical practice should be considered.

• distinguish objective management of HRT side effects from subjective difficulties. Sometimes breakthrough bleeding or headaches are difficult to manage; it is important to accept that even HRT should be adjusted and the proper answer should not be an irritated "give it up" at the very first problem or side effect;

• cope with your personal ambivalence towards HRT. To get through them is a major step in helping the patient with her decisions.

• Fears are infectious and are often unconsciously shared between patient and physician. To understand this silent collusion is another major step in a confident and successful doctor-patient relationship.

Management of Mediagenic Ambivalence Towards HRT

To clarify the scenario, urgent, clear information on HRT is needed. Critical attention should be used in referring to the concept of relative risk (RR) that too often is reported in terms of absolute figures. This mistake causes a dramatic increase in fear towards HRT, particularly when breast cancer is the issue.

Clear data, avoidance of sensationalism, and a rigorous ethical review of doctors' interactions with the media are essential if the media are to become an informative and educative ally and not an enemy of HRT.

Conclusion

In spite of the many advantages HRT guarantees to postmenopausal women in the short and long term, only a minority of postmenopausal women comply with treatment for a significant length of time. Ambivalence, that is coexistence of fears and desires towards HRT, is the major multifaceted psychodynamic factor that may deeply affect the decision-making process towards hormonal treatment. Understanding the dimensions of ambivalence, recognizing the different sources, and actively working to get through them are critical steps if we really are willing to give women a concrete chance of a better life in older age.

References

1. Stumpf PG, Trolice MP. Compliance problems with hormone replacement therapy Obstet Gynecol Clinics North Am 1994;21(2):219-29
2. Galimberti U. Dizionario di psicologia. Torino: UTET, 1992:41-42.
3. Okon MA, Lee S, Li TC. A study to examine women's knowledge, perception and acceptability of hormone replacement therapy. European Menopausal Journal 1996;3(2): 47-52.
4 Hammond CB. Women's concerns with hormone replacement therapy. Fertil Steril 1994;62 (Suppl.2)6:157S-160S.
5 Ryan PJ, Harrison R, Blake GM. Compliance with hormone replacement therapy (HRT) after screening for postmenopausal osteoporosis. Br J Obstet Gynaecol 1992;99:325-28.
6. Graziottin A. HRT: From motivation to compliance. In: Paoletti R, Crosignani PG, Kenemans P, Samsioe G, Soma M, Jackson AS, editors. Women's health and menopause. Dordrecht, The Netherlands: Kluwer Academic Publishers, 1997: 263-73.
7 Coope J, Marsh J. Can we improve compliance with long-term HRT? Maturitas 1992;15: 151-58.
8. Gabbard GO. Psychodynamic psychiatry in clinical practice. The DSM IV edition. Washington, D.C.: American Psychiatric Press, 1994.

Breast and Female Genital Tract Neoplasms in Overweight Women

Carlo La Vecchia

Introduction: Excess Weight and Cancer

Overweight and obesity are consistently related to the risk of three cancer sites: gallbladder, endometrium, and breast in postmenopausal women [1]. The relative risks are of the order of five for obese individuals as compared to leaner ones for gallbladder and endometrium, and about 1.5 for breast cancer. Still, since mortality rates from breast cancer are substantially higher than from gallbladder and endometrial neoplasms, the public health implications of elevated breast cancer risks in overweight individuals are probably greater than for any other neoplasm [2].

Obesity is a major cause of gallstones [3], and this is likely to be at the basis of the elevated risk of gallbladder cancer, too, while for breast and endometrial cancer the underlying biological mechanism is likely related to increased estrogen levels in postmenopausal obese women [4].

Further, the American Cancer Society One Million Cohort Study [1], the cohort of 50,000 American Alumni [5] and some [6] case-control studies found a direct association between overweight and renal cell adenocarcinoma, mostly in women [7]. This may have hormonal correlates, too. Prostate cancer has been also related to overweight [1], although the evidence remains open to discussion.

It is more difficult, particularly using a case-control approach, to study the relationship between measures of body weight and digestive tract cancers, since early symptoms of these neoplasms induce modifications of dietary patterns and hence body weight changes [4]. It is thus possible that overweight may have some influence on colorectal cancer, too, as well as on other cancers linked to neoplastic cachexia, such as pancreas and ovary. All these sites, in fact, were linked to overweight in the American Cancer Society One Million Cohort Study [1] (Table 1).

It was estimated that approximately 2% of all cancer deaths in the United States were attributable to overweight [8]. Although in Europe, and specifically in Italy, this proportion may be somewhat lower since the prevalence of obesity is smaller [9], such a figure is not only relevant from a public health viewpoint, but has also important and immediate implications for prevention. Overweight, in fact is the single aspect of nutrition and diet to be so well defined in epidemiological terms as to open immediate perspectives for intervention and prevention.

R. Paoletti et al. (eds.), Women's Health and Menopause, 217–223.

Overweight and obesity are also related to diabetes and other metabolic disorders, digestive tract diseases, and cardiovascular diseases, in men as well as in women [10], thus being one of the major preventable causes of disease and death. In the present paper, however, the attention will be focused on cancer, and specifically on breast and endometrial cancers in postmenopausal women.

Table 1. Mortality Ratios for Selected Sites of Cancer According to Weight Index. The American Cancer Society One Million Study[1]

Site	Weight Index						
	< 80	80-90	90-109[2]	110-119	120-129	130-139	≥ 140
Males							
Stomach	134	61	100	122	97	73	188
Colorectum	90	86	100	126	123	153	173
Pancreas	120	82	100	91	88	76	162
Kidney	106	96	100	163	139	191	—
Prostate	102	92	100	90	137	133	129
Females							
Stomach	74	95	100	107	128	126	103
Colorectum	93	84	100	96	110	130	122
Breast	82	86	100	119	116	122	153
Cervix	76	77	100	124	151	142	239
Endometrium	89	109	100	136	185	230	542
Ovary	86	98	100	115	99	88	163
Kidney	112	70	100	109	130	182	203

[1]From Lew and Garfinkel, 1979 [1], simplified
[2]Reference category.

Breast Cancer

Overweight and obesity have been consistently associated to increased breast cancer risk in postmenopausal women, while in premenopausal ones overweight has, if anything, been inversely related to breast cancer risk, probably on account of obesity-related anovulation [11]. This excess breast cancer risk in postmenopausal women is attributed to elevated levels or availability of circulating estrogens in obese postmenopausal women, due to the conversion of androgens to estrogens in peripheral adipose tissue, and to lower levels of sex hormone-binding globulin (SHBG) in overweight women [11-13]

A direct trend in risk has been systematically observed between postmenopausal breast cancer and measures of body mass index (BMI) [11-14]. It is conceivable, moreover, that the association increases with long exposure to overweight-related high estrogen levels, i.e. with increased time after menopause [15].

The relationship between body mass index (BMI, Quetelet's index, kg/m^2) and postmenopausal breast cancer risk was considered in age-specific strata of a pooled analysis of three Italian case-control studies, including a total of 3,108 postmenopausal breast cancer cases aged 50 or over and 2,664 controls [16]. Overall, there was a significant association between BMI and postmenopausal breast cancer: the odds ratios (OR) were around 1.3 for the three intermediate quintiles as compared to the lowest one, and 1.4 for the highest one. The association was moderate among women aged 50 to 59 and 60 to 69, with ORs around 1.3 for the highest BMI quintiles, but stronger among elderly women, with ORs of 1.6 for the fourth and 2.1 for the fifth quintile. An 8-unit increase in BMI involved an OR of 1.18 at age 50 to 59, of 1.14 of age 60 to 69, and of 1.59 above age 70 (Table 2).

This pattern of risk is consistent with a duration-risk relationship between overweight and postmenopausal breast cancer, and has important implications for prevention, since the incidence of breast cancer increases with age [11,17]. Consequently, the absolute excess risk of breast cancer related to overweight is even larger for the elderly than indicated by the relative risk estimates.

This age-related pattern of risk is also similar to that observed for HRT in postmenopause, showing higher relative risk with advancing age [18-29]: for instance, the American Nurses' Health Study [18] found a relative risk of 1.4 for long-term ($\geq$ 5 years) current users of HRT below age 55, and of 1.7 at age 60 to 64. Besides confirming the similarities, for hormone-dependent carcinogenesis, of exogenous and endogenous estrogens [20,21], this may also reflect an increasing differential in estrogen levels between normal and overweight women with advancing age. There is, therefore, convincing biological support to the epidemiological observation of a stronger association between BMI and breast cancer risk with advancing age.

It has also been suggested that weight gain during adult life is a relevant factor for determining breast cancer risk [22,23]. In the Nurses' Health Study, the relative risk was 2.0 for nonestrogen users who had gained more than 20 kilograms after the age of 18.

In terms of population attributable risk [24,25], 20% of all postmenopausal breast cancer cases and 27% of those above age 70 were attributable to overweight and obesity in the Italian population [16,26]. These findings have therefore major preventive implications, since, to reduce breast cancer risk late in life, it is critically important to avoid weight gain during adult life, and control overweight in elderly women.

Table 2. Multivariate Relative Risks (95% Confidence Intervals) of Postmenopausal Breast Cancer in 3,108 Cases and 2,664 Controls According to Body Mass Index (BMI) in Separate Strata of Age[1]

BMI (quintile)	Age			Total
	50-59	60-69	≥ 70	
1 (lowest)	1[2]	1[2]	1[2]	1[2]
2	1.26 (1.0-1.6)	1.33 (1.0-1.7)	1.35 (0.9-2.0)	1.30 (1.1-1.5)
3	1.34 (1.0-1.8)	1.17 (0.9-1.5)	1.05 (0.7-1.6)	1.21 (1.0-1.5)
4	1.38 (1.1-1.8)	1.25 (1.0-1.6)	1.60 (1.1-2.4)	1.36 (1.2-1.6)
5 (highest)	1.30 (1.0-1.7)	1.24 (1.0-1.6)	2.14 (1.4-3.2)	1.40 (1.2-1.7)
8-unit increase (continuous)	1.18 (1.0-1.4)	1.14 (1.0-1.3)	1.59 (1.3-2.0)	1.23 (1.1-1.4)

[1]From La Vecchia et al., 1997 [16], simplified.
[2]Reference category.

Endometrial Cancer

Overweight and obesity are well-recognized risk factors for endometrial cancer [27]. The association is observed both in pre- and in postmenopausal women, although the underlying biological mechanisms are likely to differ at different ages. In fact, increased aromatization of androgens to estrogens and reduced SHBG are the relevant factors in postmenopause, while the influence of obesity on frequency of anovulation (and hence on relative progesterone deficiency) is the likely underlying mechanism in premenopause [26,28].

The relative risk for overweight tends to be smaller in HRT users than in nonusers, and the combined effect of exogenous and endogenous estrogens on the relative risk of the disease is additive rather than multiplicative, thus indicating that exogenous estrogens and obesity act through similar biological mechanisms [29,30]. This leads to a significant negative interaction on a multiplicative scale of estrogens and obesity, suggesting either an upper risk threshold and/or the existence of some limiting factor (e.g. sex hormone receptors) which impedes the estrogen-raising effect of obesity and exogenous hormone use to accumulate beyond a certain level [31].

Whatever the biological mechanism(s), and the interaction with other factors, overweight could explain one fourth of cases of the disease in North America [32] and as many as one third in Europe [24,25,33], where estrogen replacement therapy use (i.e. the other major recognized cause of the disease) is less widespread. In contrast, a case-control study from Hawaii [34] suggested that the role of a high fat diet on endometrial carcinogenesis is similar and, if anything, could be greater for overweight women.

A cooperative European study [35] was able to examine various time factors of the relationship between overweight and endometrial cancer risk. The risk of endometrial cancer increased with increasing body mass index (BMI) at various ages, although the risk estimates tended to be substantially higher at older ages, i.e. in proximity to cancer diagnosis: compared with women whose BMI (kg/m^2)was < 20, the ORs were 1.8 for BMI ≥ 25 at age 20 to 29, 2.7 for BMI ≥ 30 at age 40 to 49, and 3.8 at age 60 to 69 [35] (Table 3).

Table 3. Multivariate Relative Risks (and 95% Confidence Intervals) of Endometrial Cancer in Relation to Body Mass Index at Different Ages[1]

BMI (kg/m^2)	Age (years)		
	3rd decade (20 to 29)	5th decade (40 to 49)	7th decade (60 to 69)
< 20	1[2]	1[2]	1[2]
20-24	1.3 (0.9-1.8)	1.0 (0.6-1.7)	2.4 (0.7-8.3)
25-29	1.8 (1.1-3.0)	1.4 (0.8-2.5)	2.6 (0.8-9.1)
≥ 30	—	2.7 (1.3-5.6)	3.8 (1.0-14.5)

[1]From Levi et al., 1992 [35], simplified.
[2]Reference category.

When data were examined in separate strata of current BMI among women of normal body mass at diagnosis no significant effect of past overweight was observed. In contrast, among subjects overweight at diagnosis, there were significant direct relationships with BMI at younger ages.

Thus, overweight and obesity appear to have a late-stage effect in the process of endometrial carcinogenesis [17], providing relevant indications for prevention. As noted for breast cancer, also to reduce endometrial cancer risk, therefore, it is important essentially to avoid obesity in late middle age and older age, and the benefit could be greater for women who were overweight at a younger age. Consequently, weight loss, even late, is still worthwhile for the prevention of breast and endometrial cancers.

Acknowledgements

This work was conducted with the support of the Italian Association for Cancer Research,

Milan. The author thanks Mrs. M.P. Bonifacino for her editorial assistance.

References

1. Lew EA, Garfinkel L. Variations in mortality by weight among 750,000 men and women. J Chron Dis 1979;32:563-76.
2. La Vecchia C, Negri E. Public education on diet and cancer: Calories, weight and exercise. In: Public education on diet and cancer. Dordrecht: Kluwer, 1992: 91-9.
3. La Vecchia C, Negri E, D'Avanzo B, Boyle P. Risk factors for gallstone disease requiring surgery. Int J Epidemiol 1991;20:209-15.
4. La Vecchia C, Decarli A, Negri E, Parazzini F. Epidemiological aspects of diet and cancer: A summary review of case-control studies from northern Italy. Oncology 1988;45:364-70.
5. Whittemore AS, Paffenbarger RS Jr, Anderson K, Lee JF. Early precursors of site-specific cancers in college men and women. J Natl Cancer Inst 1985;74:43-51.
6. Goodman MT, Morgenstern H, Wynder EL. A case-control study of factors affecting the development of renal cell cancer. Am J Epidemiol 1986;124:926-41.
7. Tavani A, La Vecchia C. Epidemiology of renal-cell carcinoma. J Nephrol 1997;10:93-106.
8. Doll R, Peto R. The causes of cancer: Quantitative estimates of avoidable risks of cancer in United States today. J Natl Cancer Inst 1981;66:1191-308.
9. Pagano R, La Vecchia C, Decarli A, Negri E, Franceschi S. Trends in overweight and obesity among Italian adults, 1983 through 1994. Am J Publ Health 1997;87:1869-70.
10. Tavani A, Negri E, D'Avanzo B, La Vecchia C. Body weight and risk of nonfatal acute myocardial infarction among women: A case-control study from northern Italy. Prev Med 1997;26:550-55.
11. Pike MC, Krailo MD, Henderson BE, Casagrande JT, Hoel DG. "Hormonal" risk factors, "breast tissue age" and the age-incidence of breast cancer. Nature 1983;303:767-70.
12. Hunter DJ, Willett WC. Diet, body size, and breast cancer. Epidemiol Rev 1993;15:110-32.
13. Stoll BA. Breast cancer: The obesity connection. Br J Cancer 1994;69:799-801.
14. Hsieh C-C, Trichopoulos D, Katsouyanni K, Yuasa S. Age at menarche, age at menopause, height and obesity as risk factors for breast cancer: Associations and interactions in an international case-control study. Int J Cancer 1990;46:796-800.
15. Franceschi S, Favero A, La Vecchia C, et al. Body size indices and breast cancer risk before and after menopause. Int J Cancer 1996;67:181-86.
16. La Vecchia C, Negri E, Franceschi S, et al. Body mass index and post-menopausal breast cancer: An age-specific analysis. Br J Cancer 1997;75:441-44.
17. Day NE, Brown CC. Multistage models and primary prevention of cancer. J Natl Cancer Inst 1980;64:977-89.
18. Colditz GA, Hankinson SE, Hunter DJ, et al. The use of estrogens and progestins and the risk of breast cancer in postmenopausal women. N Engl J Med 1995;332:1589-93.
19. La Vecchia C, Negri E, Franceschi S, et al. Hormone replacement treatment and breast cancer risk: A cooperative Italian study. Br J Cancer 1995;72:244-48.
20. Tavani A, Braga C, La Vecchia C, Negri E, Franceschi S. Hormone replacement treatment and breast cancer risk: An age-specific analysis. Cancer Epidemiol Biomarkers Prev

1997;6:11-14.

21. Henderson BE. Hormones as a cause of human cancer. In: Fortner JG, Rhoads JE, editors. Accomplishments in Cancer Research 1984 Prize Year. Philadelphia: Lippincott, 1985: 152-68.
22. Huang Z, Hankinson SE, Colditz GA, et al. Dual effects of weight and weight gain on breast cancer risk. JAMA 1997;278:1407-11.
23. Kelsey JL, Baron J. Weight and risk for breast cancer. JAMA 1997;278:1448-49.
24. Bruzzi P, Green SB, Byar DP, Brinton LA, Schairer C. Estimating the population attributable risk for multiple risk factors using case-control data. Am J Epidemiol 1985;122:904-14.
25. Mezzetti M, Ferraroni M, Decarli A, La Vecchia C, Benichou J. Software for attributable risk and confidence interval estimation in case-control studies. Comput Biomed Res 1996;29:63-75.
26. Parazzini F, La Vecchia C, Bocciolone L, Franceschi S. The epidemiology of endometrial cancer. Gynecol Oncol 1991;41:1-16.
27. La Vecchia C, Negri E, Franceschi S, Levi F. An epidemiological study of endometrial cancer, nutrition and health. Eur J Cancer Prev 1997;6:171-74.
28. Key TJA, Pike MC. The dose-effect relationship between unopposed estrogens and endometrial mitotic rate: Its central role in explaining and predicting endometrial-cancer risk. Br J Cancer 1988;57:205-12.
29. La Vecchia C, Franceschi S, Gallus G, et al. Estrogens and obesity as risk factors for endometrial cancer in Italy. Int J Epidemiol 1982;11:120-26.
30. La Vecchia C, Franceschi S, Decarli A, Gallus G, Tognoni G. Risk factors for endometrial cancer at different ages. J Natl Cancer Inst 1984;73:667-71.
31. La Vecchia C. HRT and the risk of neoplasms other than in the breast. Eur Menopause J 1996;3:232-36.
32. McDonald TW, Annegers JF, O'Fallon WM, Dockerty MB, Malkasian GD Jr, Kurland LT. Exogenous estrogen and endometrial carcinoma: Case-control and incidence study. Am J Obstet Gynecol 1977;127: 572-80.
33. Parazzini F, Negri E, La Vecchia C, Bruzzi P, Decarli A. Population attributable risk for endometrial cancer in northern Italy. Eur J Cancer 1989;25:1451-56.
34. Goodman MT, Hankin JH, Wilkens LR, et al. Diet, body size, physical activity, and the risk of endometrial cancer. Cancer Res 1997;57:5077-85.
35. Levi F, La Vecchia C, Negri E, Parazzini F, Franceschi S. Body mass at different ages and subsequent endometrial cancer risk. Int J Cancer 1992;50:567-71.

Postmenopausal Hormone Replacement Therapy and Gynecological Malignancies

Peter Kenemans, Curt W. Burger, and Fedde Scheele

Introduction

Postmenopausal use of sex hormones (HRT) carries benefits as well as risks [1]. Both for unopposed estrogen use (ERT) in hysterectomized women as well as for combined estrogen-progestagen use (EPRT), the risk-benefit equation points to an overall benefit, with a calculable gain in postmenopausal years [2] and an actually demonstrated reduced mortality among women taking hormones [3-6].

Many of the beneficial actions of estrogens are mediated via estrogen receptors (ERs) present in various organs such as bone, brain, liver, and blood vessels. Most of our knowledge relates to the alpha-type of estrogen receptor and its genomic influence via the estrogen response elements. These ERs are also found in abundance in the female breast and lower genital tract. Here, estrogens are known to have a stimulative receptor-mediated action, leading to mitotic activity in normal cells (resulting in tissue proliferation) and in tumor cells (resulting in tumor promotion).

Classically, estrogens are considered to have only mitotic action and can therefore become tumor promoters. Estrogens are not considered to be mutagens and therefore do not initiate or induce cancer. However, in our view, estrogens should not be excluded from playing a role too in the initiation of epithelial cancer. Modern thinking is that carcinogenesis is a multi-step process, generally involving five to seven alterations in important regulatory genes. During DNA replication at the time of mitosis, DNA damage is likely to occur. Commonly, but not always, this is followed by correction via DNA repair mechanisms. An increased number of mitosis, thus implies an increased number of occasions during which DNA damage may occur. Moreover, incessant estrogen stimulation, leading to incessant mitotic activity of the ER-containing epithelial cells, could lead to impairment of the natural DNA repairing mechanisms (such as mis-match repair) or final destruction mechanisms (such as apoptosis).

In instants where the cell is driven constantly into the mitotic cycle without a moment of rest, the result could be an accumulation of unrepaired DNA damage with a subsequent inactivation of important regulatory genes (such as tumor suppressor genes), resulting ultimately in a malignant geno- and phenotype of the cell. The development of adenocarcinoma of the endometrium under constant unopposed estrogen stimulation should

R. Paoletti et al. (eds.), Women's Health and Menopause, 225–231.

be considered in such a way, leading to a ten-times increased incidence of endometrial carcinoma after ten years of ERT use [7].

In this paper, the question addressed is whether or not HRT increases the risk of cancer of the organs of the lower female genital tract by reviewing the available epidemiological data from case control and cohort studies.

Endometrial Cancer

Clinical conditions with long-lasting raised estrogen serum levels (without the cyclic influence of progestogens) like obesity, polycystic ovary syndrome and granulosa theca cell tumors, are accompanied by an increased risk of endometrial cancer. Similarly, unopposed exogenous estrogens increase endometrial cancer risk substantially among users [7]. This was first reported in 1975 [8,9] and confirmed in over 30 studies including recent ones [10-12].

The magnitude of risk is related to the duration of use. An increased risk is also found with lower, now commonly used doses of estrogen [7]. A unique, large cohort study [13] reports a RR of 20 (95% CI: 7.2-54) with unopposed estrogen use for fifteen years and more. Also, the risks of abnormal endometrial bleeding (RR 7.8), D&C (RR 4.9), and hysterectomy (RR 6.6) are significantly increased in ERT users [14].

Past users maintain an increased risk for many years after stopping ERT [7,15]. Even after a discontinuation of 10 years of more, the risk is still increased (RR 1.63; 95% CI: 1.04-2.53) [15].

During the 1980s, unopposed estrogens were prescribed less frequently, and prescriptions for progestagens increased dramatically, followed by a dramatic decline in the incidence of endometrial cancer [16].

The question, whether or not addition of a progestagen to estrogen totally eliminates the risk increase found with unopposed estrogens cannot be answered unequivocally. The data available are limited and conflicting. The meta-analysis of Grady [7], finds a reduced risk (RR 0.4; 95% CI: 0.2-0.6) with EPRT for cohort studies and an increased risk (RR 1.8; 95% CI: 1.1-3.1) for case control studies. A large European cohort study reported in 1996 [12], finds a RR of 1.0 (95% CI: 0.7-1.4) with EPRT, where unopposed estrogen raised the risk significantly (RR 2.8) within the same population.

The addition of progestagen for too short a time (fewer than 10 days per cycle) is found to be associated with an increased risk for endometrial carcinoma [11,15,17,18]. However, with adequate progesterone addition (that is for 10 days or more each artificial cycle) the RR is reported to decrease and to return to around unity in some studies [15,17], but not in others [11,18].

Beresford [11] finds that with progestagens added for 10 to 21 days per cycle, EPRT does not increase endometrial cancer risk during the first five years of hormone use, neither in ever users, nor in current users. However, use for five or more years was associated with a RR of 2.5 (95% CI: 1.1-5.5) for ever users and of 2.7 (95% CI: 1.2-6.0) for current users. These results are based on only 25 cases (with 64 controls) of which only

12 cases (16 controls) extended use over five years.

In contrast, Pike's study [15] based on 79 cases and 88 controls, finds essentially no increased risk, viz. an adjusted OR of 1.07 per five years of use (95% CI: 0.82-1.41). The adjusted OR for over five years of use (23 cases, 33 controls) is 1.09. Based on Pike's study, it can be concluded that sequentially EPRT will abolish the increased risk of endometrial carcinoma as seen with ERT when in each cycle a progestagen is added in an adequate dose for 10 days or more to a normal, daily dose of estrogens, at least during the first five years. Whether or not the risk stays at the level of a woman who never used HRT after five years of use, needs further study.

Reports on the effect of continuous combined HRT on endometrial cancer risk are sparse. Beresford [11] excluded women on continuous combined regimens (6 cases, 17 controls) from her analysis. Comerci et al. [19] reported on 8 cases found on low dose (2.5 mg medroxy-progesterone acetate) combined EPRT in a study biased by prior use of ERT and poor patient compliance. The most reliable data come from Pike [15]. In his study with 94 cases and 81 controls, continuous combined HRT was associated with essentially no increased risk (OR = 1.07 per 5 yrs of use; 95% CI: 0.80-1.43). The endometrial cancer found in ERT and EPRT users, often represents early and less aggressive disease, which may simply be an effect of increased surveillance in users [15].

Ovarian Carcinoma

The risk of cancer of the ovary is reduced by the use of oral contraceptives [20], but with HRT use there is no clear effect. Where oral contraceptives protect by ovulation suppression, HRT would theoretically protect by lowering the high serum gonadotrophin levels found in postmenopausal women, levels that have been implied as being a factor in ovarian carcinogenesis.

The use of unopposed estrogens was not associated with a significantly increased or decreased risk for ovarian cancer in various case control studies [21-24]. Some studies found an increase ranging from a nonsignificant 50% [22,23] to a near significant 77% (CI 95%: 0.98-3.20) [24]. With long-term use, the only cohort study reported thus far [25] found a significantly increased RR of 1.71 (95% CI: 1.06-2.77) for ERT, when used for eleven years or more. There is a lack of studies as to the relation between EPRT and ovarian cancer risk. Kaufmann reported a RR of 0.9 (95% CI: 0.4-2.3), when a progestagen or an androgen was added to estrogen [22].

A recent case control study, using 491 cases with epithelial ovarian cancer of which 100 had used HRT (predominantly (70%) ERT), reported an OR of 0.85 (95% CI: 0.62-1.2), but did not specify the OR found with regard to unopposed estrogen or combined use [26]. Also prolonged HRT use (10 years or more) did not influence risk (OR 0.6; 95% CI: 0.3-1.4). A significant association between the use of HRT and any specific histologic subtype was not demonstrable [26].

Cervical Cancer

The persistent presence of certain genotypes of the human papilloma virus (HPV) has been identified as the most important risk factor for cervical cancer cancerogenesis [27]. Two recent case control studies [28,29] did not find that HRT influenced the frequency of HPV infection in postmenopausal women. In epidemiological studies sex steroids do not seem to play an important role in cervical cancer, not even in adenocarcinoma of the cervix, although sex steroid receptors are present in the endocervix epithelium, even after menopause.

Two European cohort studies, looking at cancer incidence and mortality in women receiving HRT [12,30] reported a reduction of cervical neoplasia among women using HRT, which could probably be explained as a result of selection bias.

Other Gynecological Malignancies

Squamous cell carcinoma of the vulva and the vagina are rare, predominantly postmenopausal diseases. A relationship between these cancers and the use of estrogens [12,31-33] or EPRT [12,33] has not been demonstrated.

Uterine sarcomas, leiomyosarcoma, and particularly endometrial stromal sarcoma, commonly contain functional estrogen and progestagen receptors. Reports on the possible relationship between uterine sarcomas and HRT are anecdotal in nature [34].

Although there are no data as to the possible relationship between HRT use and other gynecological malignancies, such as ovarian germ cell tumors and sex cord stromal tumors or female tract melanomas, such a relationship seems unlikely.

Conclusion

Postmenopausal women using unopposed estrogens or sequentially combined hormone replacement therapy do not seem to have an increased risk for malignancies of organs that do not belong to the female tract. On the other hand, it has been found that colon carcinoma incidence and mortality is decreased in HRT users [35,36]. A possible causal relationship between HRT-use and breast cancer was not discussed in the present paper but it should be noted that this topic is still a highly controversial one [37-40].

With regard to gynecological malignancies, the following picture emerges. Cancer risk is not increased for squamous cell carcinomas of the female genital tract, such as vulva, vagina, and cervical carcinomas, and probably not increased for adenocarcinomas of the ovary, the cervix, and the vagina.

Cancer risk is increased for endometrial cancer with unopposed estrogens (relative risk 2.3 for ever use; relative risk 9.5 for 10 or more years of use), but probably not when using combined HRT with a progestagen added for 10 days or more every cycle. Data on cancer risk with continuous combined HRT are almost completely lacking.

Women who use unopposed estrogens or sequentially combined hormone replacement therapy and have been diagnosed as having endometrial cancer, have an improved prognosis and a reduced mortality compared to women of identical age and stage, not on HRT.

References

1. Crosignani PG, Kenemans P, Paoletti R, Soma MR, Woodford FP. Hormone replacement and the menopause: A European position paper. Eur J Obstet Gynecol Reprod Biol 1997; 74:67-72.
2. Grady D, Rubin SM, Petitti DB, et al. Hormone therapy to prevent disease and prolong life in postmenopausal women. Ann Intern Med 1992;117:1016-37.
3. Bush TL, Cowan LD, Barrett-Connor E, et al. Estrogen use and all-cause mortality: Preliminary results from the Lipid Research Clinics Program Follow-Up Study. JAMA 1983;294:903-6.
4. Henderson BE, Paganini-Hill A, Ross RK. Decreased mortality in users of estrogen replacement therapy. Arch Intern Med 1991;151:75-78.
5. Ettinger B, Friedman GD, Bush T, Quesenberry Jr CP. Reduced mortality associated with long-term postmenopausal estrogen therapy. Obstet Gynecol 1996;87:6-12.
6. Grodstein F, Stampfer M, Colditz GA, et al. Postmenopausal hormone replacement therapy and mortality. New Engl J Med 1997;336.25:1769-75,
7. Grady D, Gebretsadik T, Kerlikowke K, Ernster V, Petitti D. Hormone replacement therapy and endometrial cancer risk: A meta-analysis. Obstet Gynecol 1995;85:304-13.
8.. Ziel HK, Finkle WD. Increased risk of endometrial carcinoma among users of conjugated estrogens. N Engl J Med 1975;293:1167-70.
9. Smith DC, Prentice R, Thompson DJ, Herrmann WL. Association of exogenous estrogen and endometrial carcinoma. N Engl J Med 1975;293:1164-67.
10. Green PK, Weiss NS, McKnight B, Voigt LF, Beresford SAA. Risk of endometrial cancer following cessation of menopausal hormone use (Washington, United States). Cancer Causes and Control 1996;7:575-80.
11. Beresford SAA, Weiss NS, Voigt LF, B McKnight B. Risk of endometrial cancer in relation to estrogen combined with cyclic progestogen therapy in postmenopausal women. Lancet 1997;349:458-61.
12. Persson I, J Yuen J, Bergkvist L, Schairer C. Cancer incidence and mortality in women receiving estrogen and estrogen-progestin replacement therapy - long-term follow-up of a Swedish cohort. Int J Cancer 1996;67:327-32.
13. Paganini-Hill A, Ross RK, Henderson BE. Endometrial cancer and patterns of use of oestrogen replacement therapy: A cohort study. Br J Cancer 1989;59:445-47.
14. Ettinger B, Golditch IM, Friedman G. Gynecologic consequences of long-term, unopposed estrogen replacement therapy. Maturitas 1988;10:271-80.
15. Pike MC, Peters RK, Cozen W, Probst-Hensch NM, Felix JC, Wan PC, Mack TM. Estrogen-progestin replacement therapy and endometrial cancer. J Natl Cancer Inst 1997; 89:1110-16.
16. Ziel HK, Finkle WD, Greenland S. Decline in incidence of endometrial cancer following increase in prescriptions for opposed conjugated estrogens in a prepaid health plan. Gynecol Oncol 1998;68:253-55.
17. Voigt LF, Weiss NS, Chu J, Daling JR, McKnight B, Belle G van. Progestogen

supplementation of exogenous oestrogens and risk of endometrial cancer. Lancet 1991; 338:274-77.
18. Brinton LA, Hoover RN. Estrogen replacement therapy and endometrial cancer risk: Unresolved Issues. Obstet Gynecol 1993;81:265-71.
19. Comerci JT, Goldberg GL, Runowicz CD, Fields AL. Continuous low-dose combined hormone replacement therapy and the risk of endometrial cancer. Gynecol Oncol 1997;64: 425-30.
20. Burger CW, Bouwma AE, Stellingwerff GC, Kenemans P. "De pil" en kanker van de vrouwelijke geslachtsorganen en de mamma. Ned T Geneeskd 1994;138:16-21.
21. Booth M, Beral V, Smith P. Risk factors for ovarian cancer: A case-control study. Br J Cancer 1989;60:592-98.
22. Kaufman DW, Kelly JP, Welch WR, et al. Noncontraceptive estrogen use of epithelial ovarian cancer. Am J Epidemiol 1989:13-0:1142-51.
23. Whittemore AS, Harris R, Itnyre J, the Collaborative Ovarian Cancer Group. Characteristics relating to ovarian cancer risk: Collaborative analysis of 12 US case-control studies. II. Invasive epithelial ovarian cancers in white women. Am J Epidemiol 1992;136:1184-1203.
24. Risch HA. Estrogen replacement therapy and risk of epithelial ovarian cancer. Gynecol Oncol 1996;63:254-57.
25. Rodriguez C, Calle EE, Coates RJ, Miracle-McMahill HL, Thun MJ, Heath CW. Estrogen replacement therapy and fatal ovarian cancer. Am J Epidemiol 1995;141:828-35.
26. Hempling RE, Wong C, Piver MS, Natarajan N, Mettlin CJ. Hormone replace therapy as a risk factor for epithelial ovarian cancer: Results of a case-control study. Obstet Gynecol 1997;89:1012-16.
27. Remmink AJ, Walboomers JJM, Helmerhorst ThJM, et al. The presence of persistent high-risk HPV genotypes in dysplastic cervical lesions is associated with progressive disease: Natural history up to 36 months. Int J Cancer 1995;61:306-11.
28. Smith EM, Turek LP, Haugen TH, Fedderson D, Mendoza M, Figuerres EJ, Johnson SR. The frequency of human papillomavirus detection in postmenopausal women on hormone replacement therapy. Gynecol Oncol 1997;65:441-46.
29. Ferenczy A, Mansour N, Franco E, Gelfand MM. Human papillomavirus infection in postmenopausal women with and without hormone therapy. Obstet Gynecol 1997;90:7-11.
30. Hunt K, Vessey M, McPherson K, Coleman M. Long-term surveillance of mortality and cancer incidence in women receiving hormone replacement therapy. Br J Obstet Gynaecol 1987;94:620-35.
31. Newcomb PA, Weiss NS, Daling JR. Incidence of vulvar carcinoma in relation to menstrual, reproductive and medical factors. JNCI 1984;73:391-96.
32. Brinton LA, Nasca PC, Mallin K, et al. Case-control study of cancer of the vulva. Obstet Gynecol 1990; 75:859-66.
33. Sherman KJ, Daling JR, McKnight B, Chu J. Hormonal factors in vulvar cancer. A case-control study. J Reprod Med 1994;39:857-61.
34. Alteras MM, Jaffe R, Cohen I, et al. Role of prolonged excessive estrogen stimulation in the pathogenesis of endometrial sarcomas: Two cases and a review of the literature. Gynecol Oncol 1990;38:273-77.
35. Calle EE, Miracle-McMahill HL, Thun MJ, Heath Jr CW. Estrogen replacement therapy and risk of fatal colon cancer in a prospective cohort of postmenopausal women. J Natl

Cancer Inst 1995;87:517-23.
36. Newcomb PA, Storer BE. Postmenopausal hormone use and risk of large-bowel cancer. J Natl Cancer Inst 1995;87:1067-71.
37. Kenemans P, Stampfer M. Hormone replacement therapy and breast cancer. Eur J Obstet Gynecol Reprod Biol 1996;67:1-4.
38. Kenemans P, Scheele F, Burger CW. Hormone replacement therapy and breast cancer morbidity, mortality and recurrence. Eur J Obstet Gynecol Reprod Biol 1997;71:199-203.
39. Sismondi P, Biblia N, Giai M, Ponzone R, Campagnoli C. Hormone replacement therapy and breast and gynecologic cancers. In: Paoletti R, Crosignani P, Kenemans P, Samsioe G, Soma MR, Jackson AS, editors. Women's health and menopause: Risk reduction strategies. Dordrecht, The Netherlands: Kluwer Academic Press, 1997:303-13.
40. Burger CW, Kenemans P. Postmenopausal hormone replacement therapy and cancer of the female genital tract and the breast. Curr Opin Obstet Gynecol 1998;10:41-45.

Hormone Replacement Therapy and Breast Cancer Risk in High-Risk Subgroups

P. Sismondi, N. Biglia, M. Giai, R. Roagna, R. Ponzone, L.G. Sgro, and M. Cozzarella

Introduction

Epidemiologic studies have attempted to identify whether certain subgroups of postmenopausal women using hormone replacement therapy (HRT) are more susceptible to the effects of steroids in relation to breast cancer risk. Subgroups of interest include women with a family history of breast cancer, women with a history of benign breast disease (BBD), obese individuals, older hormone users, past users of oral contraceptives (OC), and women who consume alcohol.

The results of epidemiological studies are difficult to interpret, primarily because the numbers of users within subgroups are limited and no complete consensus has been reached on some aspects of HRT. In this paper the available data will be scrutinized with the objective of providing useful indications for the treatment of women at high risk of developing breast cancer who request HRT.

Family History of Breast Cancer

The data regarding a possible increase of breast cancer risk due to HRT in women with a family history of breast cancer are inconsistent. A meta-analysis of five studies suggested that the risk associated with HRT was greater among women with a family history of breast cancer (relative risk or RR 3.4; 95% confidence interval or CI 2.0-6.0) compared to women without a family history (RR 1.5; 95% CI 1.2-1.7) [1]. Conversely, in another meta-analysis combining data from ten studies the RR of breast cancer in women who had ever used HRT with (RR 1.1; 95% CI 0.73-1.56) and without (RR 1.11; 95%CI 0.94-1.31) a family history of breast cancer was not significantly different [2].

Brinton et al. [3] did not observe a greater risk among the subset of HRT users with a family history of breast cancer and several other studies do not show any influence of family history on breast cancer risk in women using HRT. Nevertheless, these studies may lack statistical power to detect such interaction as they usually include small numbers of cases [4-6].

More recently, two larger studies have been published which were not included in

R. Paoletti et al. (eds.), Women's Health and Menopause, 233–241.

the metanalysis [7,8]. One is the latest report of the Nurses' Health Study, where no difference of breast cancer risk was detected according to family history of breast cancer [7]. The other is a prospective cohort study on 41,837 women, suggesting that HRT use in women with a family history of breast cancer is associated with a nonsignificant increased incidence of breast cancer and with a significantly reduced total mortality rate [8]. Among women with a family history of breast cancer, those who currently used HRT and had done so for at least 5 years developed breast cancer at an annual rate of 61 cases per 10,000 persons-year (95% CI 28-94). This rate was not significantly higher than either the rate in current HRT users without family history (41 cases per 10,000 person-years) or in women with family history who had never used HRT (46 cases per 10,000 persons-year). The multivariate RR of death for current users of HRT with family history of breast cancer was 0.24 (95% CI 0.06-0.97) among women who had used hormones for less than 5 years and 0.55 (95% CI 0.28-1.07) for longer duration of use. Use of HRT among these women was significantly associated with a decreased rate of death from coronary heart disease, stroke, and all cancers combined; however, there was a nonsignificant increased rate of death from breast cancer (RR 1.9; 95% CI 0.6-5.7), although it was based on only 84 cases [8]. Similar mortality rates among women with family history of breast cancer have been shown in the Nurses' Health Study [9].

One possible explanation for the slightly increased risk associated with HRT use among women with a family history of breast cancer is the increased surveillance provided by screening mammography. In support of this view, it has been reported that women with a family history of breast cancer are more likely to have ever had mammography (75% versus 63%; $p < 0.001$) and to have had mammography in the 2 years before the survey ($p < 0.001$) than women without a family history [8].

The recent discovery of the BRCA genes has introduced a new dilemma for those who will have to take care of women who carry a mutated gene. Carriers develop breast and ovarian cancers at a very young age, with a penetrance approaching 90% in selected families. The substantial inefficacy of early detection tools such as mammography, pelvic sonography, and serum markers for detecting these cancers at a preclinical phase, makes the option of preventive measures even more appealing. Nevertheless, little is known on the efficacy of chemopreventive drugs such as tamoxifen or retinoids in BRCA-positive patients. Consequently, many women will decide to undergo prophylactic surgery involving the removal of both breast and ovaries.

According to the latest recommendations for BRCA carriers, bilateral prophylactic oophorectomy should be offered to BRCA positive patients as soon as they complete their family planning or before 35 years of age. Clearly, for these young women surgical menopause is a serious problem and HRT has to be considered. Current guidelines state that there are insufficient data to make recommendations concerning use or avoidance of HRT in BRCA carriers, so that many factors will have to be taken into account in making individual decisions.

Some authors suggest that as BRCA-related breast cancers are frequently of high grade and estrogen receptor negative, estrogen therapy should not increase the risk of breast cancer of these patients. Furthermore, there is experimental evidence suggesting that

BRCA1 and BRCA2 genes are not directly regulated by estrogens. As no prospective or even retrospective data are available on this issue, no recommendation can be made with confidence and only future studies will clarify the issue.

Benign Breast Disease (BBD)

One of the most debated aspects is the opportunity to offer HRT to women affected by benign breast diseases (BBD). Most of the studies conducted to assess the interactive effects of estrogens and benign breast disease were not designed to classify different types of BBD at the histological level. In the one study that was able to examine estrogen use in relation to specific types of BBD, the RR of breast cancer among patients with BBD was reduced from 1.8 for women who did not take estrogens to 0.98 for patients who did use estrogen replacement therapies [10]. Estrogens reduced breast cancer risk in each of the categories of BBD: the magnitude of this reduction was larger for patients with proliferative disease without atypia (RR 0.92 versus 1.9) or atypical hyperplasia (RR 3.0 versus 4.5) than in women without proliferative disease (RR 0.69 versus 0.91).

Of interest, women who took estrogens before the year 1956 had 2.3 times the risk of other estrogens users; this finding, also reported by Hoover [11], may presumably reflect the fact that these women had been treated with estrogens at high doses. Indeed, an increase of breast cancer risk associated with estrogens use in BBD patients may only be found in older studies, when estrogens doses were at least twice as high as those currently prescribed (0.625 mg/day of conjugated estrogens). For instance, Thomas et al. found that estrogen therapy started after a biopsy for benign breast disease could eliminate the protective effect of artificial menopause and act synergistically with epithelial hyperplasia and papillomatosis in increasing breast cancer risk (RR 1.8) [12]. Also Brinton et al. observed a threefold increase of breast cancer risk in women who took ERT for more than 10 years and who started taking hormones after a biopsy for BBD (95% CI 1.6-5.5). However, it must be noted that the conclusions reported by these authors are based on subgroup analyses of small numbers of patients [3].

The meta-analyses performed to evaluate the risk of breast cancer in women affected by BBD do not support a role of HRT in influencing such risk. Steinberg et al. found that the effects of estrogen use were similar among women with or without BBD [1]. In the meta-analysis of Dupont et al. [13], four out of five studies showed RRs around unit y either in women with or without BBD, while the only study reporting a positive correlation between HRT and increased breast cancer risk was based on only 14 patients with BBD. A third meta-analysis shows that ever-use of HRT is associated with a modest nonsignificant increase in risk of breast cancer among women with BBD (RR 1.11; 95% CI 0.86-1.43) [2]. Also more recent studies, which were not included in the meta-analysis, do not show any significant increase of breast cancer risk in women who underwent biopsies for BBD and used estrogens either alone or combined with progestins [4,14,15].

Speculations on the effects of estrogens on mammary carcinogenesis may be derived from experimental animal studies. A recent study in which 28 female monkeys with surgical menopause were treated with estradiol for 6 months, has shown that estradiol can induce mammary gland proliferation and hyperplasia but not atypical hyperplasia [16].

In the light of the available epidemiological data there are no reasons why HRT should not be recommended to BBD patients. Nevertheless, it must be underlined that the information available on this aspect were mainly gathered by women who are treated with estrogens alone, while the effects of long-term combined treatments are essentially unknown..

Age at Diagnosis

There is little information about the use of HRT beginning long after the menopause or about HRT use at older ages [17]. In the Collaborative Group of Hormonal Factors in Breast Cancer (CGHFBC) study 96% of hormone users started HRT before age 60, 92% of users stopped HRT before age 65, and 97% were aged under 70 at the time of cancer diagnosis [18]. Only a few studies have evaluated the effect of ever-use of HRT by age groups, and limited evidence suggests that the effects of HRT on breast cancer risk may increase with age [3,15,19,20]. In the Nurses' Health Study the elevated risk of breast cancer associated with current estrogen use was most evident among women aged 55 years of age or older; the RR associated with 5 or more years of HRT was 1.46 (95% CI 0.91-2.33) for women 50-54 years old and 1.71 (95% CI 1.34-2.18) for women of 60-64 [20]. The lack of association between current estrogen therapy and breast cancer among younger women may be due, at least in part, to the short time since menopause.

Data from two Italian case-control studies have confirmed this finding, as the excess risk of breast cancer associated with ever-use of HRT was not observed in the youngest age group (odds ratio or OR 0.9) and the risk increased with age at diagnosis to 1.2 (95% CI 0.9-1.5) for women 55-64 years old and 1.6 (95% CI 1.2-2.3) for those 65-74 years old at diagnosis [21]. A significant trend of increasing risk with duration of use was observed only in the oldest group (65-75 years old) with ORs of 1.6 (95% CI 1.1-2.3) and 2.2 (95% CI 1.1-4.7) for less than 60 and 60-or-more months of use respectively. In the oldest group the OR rose with age at which use was commenced: the OR was 2.0 for women of 50 years or more and 1.3 for those younger than 50 years.

Several years after the menopause, when the incidence of breast cancer is high, a greater excess risk in terms of OR means an even greater excess in term of absolute risk. The prolonged use of HRT for long periods beyond the time needed for the relief of perimenopausal symptoms, might considerably increase the risk of breast cancer. On the other hand, HRT should be taken for long periods beyond menopause to maintain its protective effect against cardiovascular diseases and osteoporosis. Thus, if these results are confirmed, the risk-benefit ratio of HRT in elderly women must be carefully and individually assessed.

Obesity

Several epidemiologic studies have shown that obesity is a risk factor for breast cancer [22]; women with body mass index (BMI) $\geq$ 25 kg/m^2 have a 35-40% risk increase compared to women with BMI < 25 [18]; in particular, abdominal obesity appears to be associated with a higher risk. Only a few investigators have examined the effects of estrogens according to body weight, finding a somewhat stronger effect of estrogen use in thin women [4, 14,23-26]. In the CGHFBC study, the increased risk associated with long-term (> 5 yrs) HRT in current and recent users is not apparent in the subgroup of overweight women [18]. The RR associated with a long duration of current or recent use decreased progressively with increasing weight, from 1.65 for weight < 60 kg, to 1.32 for 60-69 kg, to 1.05 for women ≥ 70 Kg ($p < 0.004$). The RR of breast cancer in long-term, recent hormone users is also related to BMI, being 1.52 for BMI < 25 kg/m2 and 1.02 for BMI ≥ 25 kg/m^2. This finding has two possible explanations: 1) women with abdominal obesity have a higher estrogenic activity which already causes maximal estrogen stimulation on the breast, independently on the use of exogenous estrogens; 2) oral estrogens, through their metabolic and hepatocellular effects, contrast other biological peculiarities of obesity potentially involved in breast cancer risk, so balancing the increased estrogen stimulation. In postmenopausal women we have demonstrated that oral estrogens cause an increase in sex hormone-binding globulin (SHBG) level [27], determining a reduction of free, bioavailable estrogens and androgens and a decrease of circulating insulin-like growth factor I (IGF-I) [28], a potent mitogen for breast cancer cells. These metabolic actions are particularly evident in obese women who show high basal IGF-I levels and low SHBG levels. [29].

Notably, the administration of nonoral estradiol, that has a limited hepatocellular action, does not produce such metabolic effects [28,29]. As it seems unwise to assume that information on compounds with different metabolic pathways may be extrapolated by those gathered on oral estrogens, further data regarding the effect of transdermal estradiol on breast cancer risk, both in normal weight and in obese women, are needed.

Alcohol Intake

There is a consensus in the literature that alcohol consumption is a risk factor for breast cancer. A meta-analysis of epidemiologic studies indicates both a modest positive association between alcohol and breast cancer (an approximately 25% increase in risk with daily intake of the equivalent of two drinks) and a dose-response relationship [30]. In a recent analysis including data from 6 prospective studies, women drinking a total of 30-60 grams of alcohol/day (about 2-5 drinks) had a 1.41 (95% CI 1.18-1.69) RR of developing breast cancer compared to nondrinkers [31].

In a study evaluating the causes of death associated with drinking, the mortality from breast cancer was 30% higher among women reporting at least one drink daily than among nondrinker (RR 1.3; 95% CI 1.1-1.6) [32]. Several factors including age, weight, family history, and postmenopausal estrogen use have been shown to modify this relation. Two large prospective American studies have shown that HRT effects could be enhanced by alcohol consumption. In the Iowa Women's Health Study, estrogen administration was associated with an increased risk of breast cancer only in women drinking at least 5 grams of alcohol per day; those who were drinking less than that or nondrinkers had the same risk as the general population [33]. According to the Nurses' Health Study, the daily intake of alcohol has no influence on breast cancer risk in current HRT users [34].

All the data available on this issue have been collected by the CGHFBC: in this study the RR of breast cancer for long-term HRT was greater in women drinking more than 50 grams/week (RR 1.64) compared to less than 50 grams/week of alcohol (RR 1.40), but the difference did not reach statistical significance [18]. Several pathophysiological mechanisms by which alcohol might enhance breast carcinogenesis have been suggested, including hepatic metabolism of carcinogens, immunologic surveillance, and production of cytotoxic protein products. The most convincing evidence of a link between alcohol and breast cancer derives from the observation that in a postmenopausal woman on HRT, alcohol intake produces a sudden rise of her estradiol levels by about 300% [35].

To explain these findings the following theory has been proposed: usual HRT doses produce relatively low estradiol levels in the blood which are equivalent to those of the follicular phase of the menstrual cycle. These values may be near the threshold value for a breast cancer promoting effect, so that some women will manifest increased risk and some will not; this will depend on their varying genetic susceptibility factors, including family history of breast cancer, abnormal BRCA1, BRCA2, or p53 genes [36]. The intake of alcohol by a postmenopausal woman receiving estrogens raises estradiol levels in the blood to values characteristic of the periovulatory peak in the menstrual cycle. These values may be well above the threshold associated with a breast cancer promoting effect for all women, regardless of genetic background. The alcohol-breast cancer hypothesis is intriguing; however, more refined temporal, quantitative, and qualitative indicators of alcohol exposure in future studies are needed, together with further exploration of endocrine and metabolic effects of moderate alcohol intake.

Oral Contraceptives (OC)

With the widespread diffusion of hormonal therapies, there is an increasing number of women who have used both oral contraceptives (OC) and HRT during their lifetime. Furthermore, women who have used OC are more likely than non-OC users to take estrogens once they reach the menopause. Prior use of OC before HRT has been indicated as a possible modifier of breast cancer risk, though data in support of this hypothesis are quite scant.

The Nurses' Health Study shows that women who used OC in the past and who are classified as current users of HRT, have an increased risk of breast cancer (RR 1.49) and that among never-users of OC, current HRT use is associated with a RR of 1.31 compared to never-users [23]. Mills et al. did find a nonsignificantly increased breast cancer risk for women who used both OC and HRT (RR 1.42) compared with women who never used OC or HRT, but this result was based on just 10 cases [5].

In the Dutch cohort study for women who used both OC and HRT, no increased breast cancer risk was found; the RR for combined use in the multivariate analysis was 1.0 (95% CI 0.51-1.94) [37].

The CGHFBC reanalysis did not find any evidence of a substantial interactive effect of combined OC and HRT exposure; the RR of breast cancer associated with current or recent use of HRT for ≥ 5 years was 1.35 among women who had never used OC and 1.35 in previous users of OC [18]. In any case, the interactive effect of CO and HRT in promoting breast cancer is not biologically plausible considering that women using these compounds belong to different age groups. A recent meta-analysis indicates that breast cancer risk among CO users returns to baseline levels after 10 years of therapy discontinuation [38]. Consequently, there is just a short overlap between women at increased risk of breast cancer due to OC use (women < 50 years) and those at risk for HRT (postmenopausal women).

Conclusion

Epidemiological data on breast cancer risk among HRT users are reassuring: treatments shorter than 5 years are safe, whereas some studies suggest the existence a 20-30% increase of breast cancer risk for longer treatments [39]. Nevertheless, there are groups of women at higher risk of developing breast cancer due to their family and reproductive history, habits, and due to a previous diagnosis of proliferative BBD [17]. A critical issue is how to advise these patients about postmenopausal use of HRT, as the presence of such risk factors could be associated with an unusual susceptibility to the stimulatory effect of estrogens.

A previous diagnosis of BBD, including typical hyperplasia, is not a contraindication to HRT as in most studies conducted on estrogens users this variable does not modify breast cancer risk. As far as family history is concerned data are confusing; some authors report a nonstatistically significant increase of breast cancer risk associated with estrogen use in this group of women. However, it has been consistently reported that the overall mortality of the women with a positive family history on HRT is reduced as compared to non-HRT users. No recommendations can be made with confidence in carriers of mutations of the BRCA genes regarding the safety of HRT due to the complete lack of data.

The modest increase of breast cancer risk associated with long-term HRT is more evident in thin women, whereas in obese women, whose baseline breast cancer risk is higher, the potential unfavorable effect of HRT appears quite limited. Alcohol intake has been consistently reported to increase breast cancer risk by acting synergistically with estrogens; it is then advisable that women on HRT avoid excessive drinking, particularly in case of long-term treatments. On the contrary, there is no evidence that OC use may influence breast cancer risk in menopause, even if HRT is a frequent choice of past OC users. There are no sufficient data to ascertain the risk associated with HRT in elderly women; as some studies show that the risk may increase after 60 years of age, the major value of a judicious evaluation of benefits and harms has to be emphasized.

References

1. Steinberg KK, Thacker SB, Smith SJ, et al. A meta-analysis of the effect of estrogen

replacement therapy on the risk of breast cancer. JAMA 1991;265:1985-90.
2. Colditz GA, Egan KM, Stampfer MJ, et al. Hormone replacement therapy and risk of breast cancer: Results from epidemiologic studies. Am J Obstet Gynecol 1993;168:1473-80.
3. Brinton LA, Hoover R, Fraumeni JF Jr. Menopausal oestrogens and breast cancer risk: an expanded case-control study. Br J Cancer 1986;54:825-32.
4. Newcomb PA, Longnecker MP, Storer BE, et al. Long-term hormone replacement therapy and risk of breast cancer in post-menopausal women. Am J Epidemiol 1995;142:788-95.
5. Mills PK, Beeson WL, Phillips RL, et al. Prospective study of exogenous hormone use and breast cancer in Seventh-day Adventist. Cancer 1989;64:591-97.
6. Palmer JL, Rosenberg L, Clarke EA, et al. Breast cancer risk after estrogen replacement therapy: Result from the Toronto breast cancer study. Am J Epidemiol 1991;134:1386-1401.
7. Colditz GA, Rosner BA, Speizer FE. Risk factors for breast cancer according to family history of breast cancer. J Natl Cancer Inst 1996;88:365-71.
8. Sellers TA, Mink PJ, Cerhan JR, et al. The role of hormone replacement therapy in the risk for breast cancer and total mortality in women with a family history of breast cancer. Ann Int Med 1997;127:973-80.
9. Grodstein F, Stampfer MJ, Colditz GA, et al. Postmenopausal hormone therapy and mortality. N Engl J Med 1997;336:1769-75.
10. Dupont WD, Page DL, Rogers LW, et al. Influence of exogenous estrogens, proliferative breast disease, and other variables on breast cancer risk. Cancer 1989;63:948-57.
11. Hoover R, Glass A, Finkle WD, et al. Conjugated estrogens and breast cancer risk in women. J Natl Cancer Inst 1981;67:815-20.
12. Thomas DB, Persing JP, Hutchinson WB. Exogenous estrogens and other risk factors for breast cancer in women with benign breast disease. J Natl Cancer Inst 1982;69:1017- 25.
13. Dupont WD, Page DL. Menopausal estrogen replacement therapy and breast cancer. Arch Intern Med 1991;151:67-72.
14. Stanford J, Weiss N, Voigt L, et al. Combined estrogen and progestin hormone replacement therapy in relation to risk of breast cancer in middle-aged women. JAMA 1995;274:137-42.
15. La Vecchia C, Negri E, Franceschi S, et al. Hormone replacement treatment and breast cancer risk: A cooperative Italian study. Br J Cancer 1995;72:244-48.
16. Foth D, Cline JM. Effects of estrogen monotherapy on breast tissue in the primate model. Zentralbl Gynacol 1997;119:607-10.
17. Brinton LA. Hormone replacement therapy and risk for breast cancer. Endocr Metab Clin North Am 1997;26:361-78.
18. Collaborative Group of Hormonal Factors in Breast Cancer. Breast cancer and hormone replacement therapy: Collaborative reanalysis of data from 51 epidemiological studies of 52705 women without breast cancer. Lancet 1997;350:1047-59.
19. Wingo PA, Layde PM, Lee NC, et al. The risk of breast cancer in postmenopausal women who have used estrogen replacement therapy. JAMA 1987;257:209-15.
20. Colditz GA, Hankinson SE Hunter DJ, et al. The use of estrogens and progestins and the risk of breast cancer in postmenopausal women. New Engl J Med 1995;332:1589-93.
21. Tavani A, Braga C, La Vecchia C, et al. Hormone replacement treatment and breast cancer risk: An age-specific analysis. Cancer Epidemiol Biomark Prev 1997;6:11-14.
22. Pujol P, Galtier-Dereure F, Bringer J. Obesity and breast cancer risk. Hum Reprod 1997;12 Suppl:116-25.
23. Colditz GA, Stampfer MJ, Willet WC, et al. Type of postmenopausal hormone use and risk

of breast cancer: 12-year follow-up from the Nurses' Health Study. Cancer Causes Control 1992;3:433-39.

24. Harris RE, Namboodiri KK, Winder EL. Breast cancer risk: Effects of estrogen replacement therapy and body mass. J Natl Cancer Inst 1992;84:1575-82.
25. Kaufman DW, Palmer JR, de Mouzon J, et al. Estrogen replacement therapy and the risk of breast cancer: Results from the case-control surveillance study. Am J Epidemiol 1991;134: 1375-85.
26. Weinstein AL, Mahoney MC, Nasca PC, et al. Oestrogen replacement therapy and breast cancer risk: A case-control study. Int J Epidemiol 1993;22:781-89.
27. Campagnoli C, Biglia N, Belforte P, et al. Post-menopausal breast cancer risk: Oral estrogen treatment and abdominal obesity induce opposite changes in possibly important biological variables. Eur J Gynaec Oncol 1992;13:139-54.
28. Campagnoli C, Biglia N, Peris C, Sismondi P. Potential impact on breast cancer risk of circulating insulin-like growth factor I modifications induced by oral HRT in menopause. Gynecol Endocrinol 1995;9:67-74.
29. Campagnoli C, Biglia N, Cantamessa C, et al. Insulin-like growth factor I (IGF-I) serum level modifications during transdermal estradiol treatment in postmenopausal women: A possible bimodal effect depending on basal IGF-I values. Gynecol Endocrinol 1997; in press.
30. Longnecker MP. Alcoholic beverage consumption in relation to risk of breast cancer. Meta-analysis and review. Cancer Causes Control 1994;5:73-82.
31. Smith-Warner SA, Spiegelman D, Yaun SS, et al. Alcohol and breast cancer in women: A pooled analysis of cohort studies. JAMA 1998;279:535-40.
32. Thun MJ, Peto R, Lopez AD, et al. Alcohol consumption and mortality among middle-aged and elderly U.S. adults. N Engl J Med 1997;337:1705-14.
33. Gapstur SM, Potter JD, Sellers TA, et al. Increased risk of breast cancer with alcohol consumption in postmenopausal women. Am J Epidemiol 1992;136:1221-31.
34. Colditz GA, Stampfer MJ, Willet WC, et al. Prospective study of estrogen replacement therapy and risk of breast cancer in postmenopausal women. JAMA 1990;264:2648-53.
35. Ginsburg EL, Mello NK, Mendelson JH, et al. Effects of alcohol ingestion on estrogens in postmenopausal women. JAMA 1996;276:1747-51.
36. Zumoff B. Editorial: The critical role of alcohol consumption in determining the risk of breast cancer with postmenopausal estrogen administration. J Clin Endocrinol Metab 1997;82: 1656-58.
37. Schuurman AG, van den Brandt PA, Goldbohm RA. Exogenous hormone use and the risk of postmenopausal breast cancer: Results from the Netherlands cohort study. Cancer Causes and Control 1995;6:416-24.
38. Collaborative Group on Hormonal Factors in Breast Cancer. Breast cancer and hormonal contraceptives: Collaborative reanalysis of individual data on 53297 women with breast cancer and 100239 women without breast cancer from 54 epidemiological studies. Lancet 1996;347:1713-27.
39. Sismondi P, Biglia N, Giai M, Campagnoli C. Hormone replacement therapy and breast cancer. Eur Menopause J 1996;3:227-31.

Breast Cancer and HRT: Collaborative Reanalysis of Data

Martin P. Vessey

Introduction

Large numbers of epidemiological studies of the relationship between use of oral contraceptives and hormone replacement therapy (HRT) on the one hand and the risk of breast cancer on the other have been published over the past two decades. The results have generally been regarded as conflicting and difficult to interpret. Some years ago, Professor Valerie Beral (Director of the Imperial Cancer Research Fund Cancer Epidemiology Unit, Oxford, UK) decided that the best way forward would be to undertake a collaborative re-analysis of the studies. She and her team succeeded in gaining the support of almost all the investigators worldwide and in 1996 a collaborative re-analysis of the data on breast cancer and oral contraceptive use was published [1]. This was followed by a similar reanalysis concerning the data on breast cancer and hormone replacement therapy which appeared in 1997 [2]. The latter analysis is the subject of the present report.

Methods

A detailed description of the methodology will be found in the original publication [2] and only a few general points can be made here.

First, no effort was spared in trying to identify all relevant studies, published and unpublished. In total, 63 eligible studies were found. Original data were contributed by the authors of 51 of these, 49 published and 2 unpublished. These studies had been conducted in 21 different countries and encompassed 52,705 women with breast cancer (17,949 postmenopausal) and 108,411 controls (35,919 postmenopausal). Original data could not be retrieved for 10 studies and one research group declined to participate. The reanalysis thus covered about 90% of the worldwide epidemiological evidence up to 1996 and it was possible to determine from the available results of the missing studies that their inclusion would not have substantially altered the findings in the reanalysis.

Secondly, it must be stressed that the reanalysis involved access to the original data for individual women so that, as far as possible, all the data could be handled in the same way. This is a vastly more satisfactory approach than that commonly used in meta-analyses where reliance is placed on the content of published scientific papers.

Thirdly, the data were checked with great care and were analyzed by well-

R. Paoletti et al. (eds.), Women's Health and Menopause, 243–248.

established methods using the Mantel-Haenszel stratification technique. Each individual study formed a separate stratum in every analysis thus ensuring that comparisons were always made on a within study basis.

Results

Menopause and Breast Cancer Risk

It was found that one-third of the postmenopausal women in the analysis had used HRT at some time and that one-third of these had used HRT for five years or longer. The median age at first use was 48 years and there was very little information beyond age 70 years.

It was recognized at the outset that any effect of HRT on breast cancer risk would be confounded by the effect of age at menopause. Thus early age at menopause is known to decrease the risk of breast cancer but to increase the likelihood of HRT use (at any given age, say 45 years). Failure to allow for the effect of age at menopause might, therefore, lead to underestimation of any adverse effect of HRT (or even to failure to detect the presence of such an effect). To examine the effect of the menopause on breast cancer risk independently of the effect of HRT, initial analyses were restricted to women who had never used HRT. It was found that the relationship between age at menopause and breast cancer risk was similar for women whose menopause was natural and for those whose menopause was the result of bilateral oophorectomy; the relative risk increased by 2.9% (SE 0.3%) and 2.4% (SE 1.0%), respectively for each year of increase in age at menopause. Accordingly, all analyses concerning HRT use were stratified by age at menopause (or time since menopause which amounts to the same thing) as well as for other confounding variables such as age at cancer diagnosis, body mass index, parity, and age at first term birth.

Ever-Use of HRT and Breast Cancer Risk

The studies were divided into 3 categories (prospective studies, case-control studies with population controls, and case-control studies with hospital controls). The results for each group of studies were closely similar. Overall, there was a significant increase in the relative risk of breast cancer associated with ever-use of HRT (relative risk 1.14, SE 0.03, $2p = 0.00001$).

Timing of HRT Use and Breast Cancer Risk

Ever-use is clearly a very crude measure of HRT exposure, so further analysis focused on total duration of use, time since first use, and time since last use. These three measures of exposure are correlated and it was only because of the large numbers of subjects in the reanalysis that their independent effects could be estimated. It was found that after duration of use and time since last use had been taken into account, there was no residual effect of time since first use. The key results are given in Table 1. For those whose last use was less than five years before diagnosis (including current users), there was strong evidence of a trend of increasing relative risk of breast cancer with increasing duration of HRT use; the risk increased by 2.3% (SE 0.6%) for each year of use. By contrast, those women who had stopped HRT use five or more years before diagnosis showed no significant overall increase in the relative risk of breast cancer.

Table 1. Relative risk of breast cancer by duration of use within categories of time since last use of HRT. Floated standard errors in parentheses.

Duration of use and time since last use	Cases/Controls	Relative risk
Never user	12467/23568	1.00 (0.021)
Last use < 5 yrs before diagnosis		
Duration < 1 yr	368/866	0.99 (0.085)
" 1-4 yrs	891/2037	1.08 (0.060)
" 5-9 yrs	588/1279	1.31 (0.079)
" 10-14 yrs	304/633	1.24 (0.108)
" ≥ 15 yrs	294/514	1.56 (0.128)
Last use ≥ 5 yrs before diagnosis		
Duration < 1 yr	437/896	1.12 (0.079)
" 1-4 yrs	566/1256	1.12 (0.068)
" 5-9 yrs	151/374	0.90 (0.115)
" ≥ 10 yrs	93/233	0.95 (0.145)

Adapted from [2].

Consistency of Main Findings

A comprehensive search was undertaken to try to identify factors modifying the findings shown in the table. Taken into consideration were age at diagnosis, family history, ethnic group, education, height, weight, body mass index, age at menarche, parity, age at first term birth, past oral contraceptive use, alcohol intake, smoking history, and type of menopause. Only two closely related factors had a significant effect; thus the relative risk associated with recent (within five years) long-term (five or more years) use of HRT was greater for light women than for heavy women and for women with a low body mass index than for women with a high body mass index.

Tumor Spread

Relevant information was available for only 54% of the postmenopausal women with breast cancer. Among recent users (within five years), the excess risk of breast cancer was confined to localized disease (but there was still a duration of use effect in those with cancer spread beyond the breast).

Hormonal Constituents

Information about the hormonal constituents of the HRT preparations used most was available for only 39% of eligible women. Of these, 80% had mostly used preparations containing oestrogen alone. There was no significant evidence of any heterogeneity in the effects of the different preparations.

Key Points

Combination of Studies

This seems valid as there was no significant heterogeneity of findings across the 51 studies. In addition the data from the missing studies are compatible with the overall results of the reanalysis.

Confounding and Bias

The usual confounding factors were considered together with age at menopause (or interval since menopause) and body mass index. The control of confounding would thus appear to be adequate. It is possible that women recently using HRT had increased "breast awareness," increased diagnostic surveillance, or increased rates of mammography in comparison with never users or past users of HRT. If so, this might have contributed to the reported findings.

Risk in Recent (Including Current) Users

The relative risk of breast cancer increased by 2.3% for each year of HRT use. The only risk modifiers were body weight and body mass index (slim women being at greater risk). The data were insufficient to study HRT constituents adequately; in particular, there were few data about combination products. The findings with respect to tumor spread are of uncertain meaning.

Limitations of Data

On average, the cases were diagnosed in 1985 and there is a dearth of data on modern HRT preparations. There is little information about women over the age of 70 years. The reanalysis cannot address mortality from breast cancer, only the incidence of the disease. There is thus a need for continuing large-scale research. Key findings are likely to emerge before long from the British Million Women Study (based on the women taking part in the NHS Breast Screening Programme in the UK).

POSSIBLE BIOLOGICAL EXPLANATIONS FOR RESULTS

It is tempting to think of the administration of HRT as essentially "delaying the menopause." Thus, for recent HRT users, the relative risk associated with each year of use is 1.023. The relative risk associated with delay of the menopause is closely similar, 1.028 per year of delay.

MAGNITUDE OF EFFECT OF HRT USE IN ABSOLUTE TERMS

Illustrative figures are provided in Table 2 which helps to put the risk in proportion. In addition, of course, it is important to think of breast cancer risk in relation to the complete array of benefits and risks of HRT.

Table 2. Estimated Cumulative Incidence of Breast Cancer in 1,000 Women (In UK or USA) Associated with Use of HRT

	Cumulative incidence/1000 women			
Up to age (yrs)	Never users	Use beginning at age 50		
		Use for 5 yrs	Use for 10 yrs	Use for 15 yrs
50	18	18	18	18
55	27	28	28	28
60	38	40	41	41
65	50	52	56	57
70	63	65	69	75
75	77	79	83	89

Adapted from [2].

Conclusion

The collaborative reanalysis has provided important information indicating a modest increase in the risk of breast cancer in current and recent users of HRT which is dependent on duration of use. Further information is required to assess the effects of modern HRT preparations and this is likely to be available from the Million Women Study before too long. In the meantime, in assessing the overall impact of HRT benefits must be considered as well as risks.

References

1. Collaborative Group on Hormonal Factors in Breast Cancer. Breast cancer and hormonal contraceptives: Collaborative reanalysis of individual data on 53,297 women with breast cancer and 100,239 women without breast cancer from 54 epidemiological studies. Lancet 1996;347:1713-27.
2. Collaborative Group on Hormonal Factors in Breast Cancer. Breast cancer and hormone replacement therapy: Collaborative reanalysis of data from 51 epidemiological studies of 52,705 women with breast cancer and 108,411 women without breast cancer. Lancet 1997; 350:1047-59.

A Summary of the Evidence Relating Postmenopausal Hormone Use and Large Bowel Cancer Risk

Polly A. Newcomb

Introduction

Colorectal cancer is second only to lung cancer as the leading cause of cancer mortality and remains a significant cause of morbidity. Increasing evidence indicates that postmenopausal hormone replacement (HRT) use may significantly decrease the risk of colorectal cancer in women. This report summarizes the epidemiologic and biologic support for a link between estrogens and large bowel cancer.

Descriptive Epidemiology

Historically, colorectal cancer occurred with approximately equal frequency in men and in women. However, in countries like the U.S. the rates for men now exceed those for women by over one third, with rates diverging after age 50 [1]. Over the past 30 years, large bowel cancer mortality declined 29% in women, but only 7% in men. As with most cancers, colorectal cancer rates increase with increasing age. For colorectal cancer, the slope of the rise in incidence is particularly steep—age to the power of 4 or 5. Thus, nearly 95% of female colorectal cancer occurs in postmenopausal women.. There are some suggestions of gender differences in occurrence in subsites within the colon [2]; for example, the incidence of proximal (cecum and ascending) colon cancer is higher for women than men. Thus, the descriptive epidemiology of large bowel cancer suggests that women have adopted some behavior quite different from men, and these changes in risk around the time of menopause appear to suggest a role for hormones [3].

Reproductive and Other Factors and Colorectal Cancer

Several observations appear to support a role for reproductive and hormonal factors in the etiology of colorectal cancer [4]. The incidence and mortality rates of colon and breast cancers are positively correlated internationally and within populations, consistent with common etiologies. Further, nuns have been observed to have elevated mortality from colon cancer, as well as breast and ovary cancers. The higher incidence of colon cancer in single women compared to married women also appear consistent with a reproductive role.

R. Paoletti et al. (eds.), Women's Health and Menopause, 249–255.

In 1980, McMichael and Potter considered these and other observations, including the gender differentials in occurrence and gut metabolism and postulated that the long-term consequences of the hormonal changes associated with pregnancy would result in a reduction in bile acid synthesis and a consequent lower incidence of colon cancer [5]. They hypothesized further that hormonal manipulation (at that time, limited primarily to the use of oral contraceptives) would modify hepatic bile acid production and thus exert a beneficial effect on colon cancer incidence.

Tests of this thesis have yielded somewhat inconsistent results. The available data on reproduction and colon cancer risk has not demonstrated clear associations with parity, age at first birth or age at menopause [4]. In studies of oral contraceptives and colorectal cancer, the results have been conflicting. While a slight majority suggest an inverse association–the large Nurse's Health Study found a relative risk of 0.84 (95% CI 0.69-1.02) [6]–only a few show a significant inverse effect [6-8]. However, eight others have not found an association. It is notable that the inverse effect was more likely to be observed in younger study populations where the use of oral contraceptives may have been more recent. While these evaluations of reproduction and oral contraceptives have not strongly supported a role for estrogens, recent evidence from studies of postmenopausal hormones in relation to colorectal cancer risk have been more compelling.

Exogenous Hormone Use and Large Bowel Cancer

In earlier epidemiologic studies of HRT and colorectal cancer, no relation was suggested [9]. However most recent studies have demonstrated an inverse association, particularly for current HRT users. Of the nine case-control studies, three published prior to 1990 found little or no association between HRT and colorectal cancer [10-12]. Subsequent case-control studies have reported an inverse association between HRT and colorectal cancer incidence [6,13-18], as have several cohort studies [19-23]. In these studies (including the large Nurse's Health Study cohort and the Cancer Prevention Study cohort) reductions are similar and range from 0.6 to 0.9. Three smaller studies found no association [10,24-26]; differences in the reference group likely explain some of this disparity. When recency of use is examined, the reduction in risk appears greater for current use—30-50%. Because of limited study sizes, it is not clear whether the inverse association is stronger among long-term users independent of the reduction associated with recent use; long duration of use does not appear to afford further protection when recency is considered. Few studies have reported on the relation between specific preparations and colorectal cancer risk; it appears risk estimates were similarly reduced for use of estrogen only and estrogen and progestin use [13,21]. Studies of adenomatous polyps, precursor lesions to colorectal cancer [27], have found reductions in risk as well, with relative risk estimates ranging from 0.8 to 0.9 [19,28,29]. It is notable that Grodstein et al. observed a decreased risk for large adenoma in current users, but no association for small adenoma [19], suggesting that the action of estrogen is a late effect.

Among HRT users, certain subgroups of women appear to be at decreased risk, in particular, women of lean body mass (e.g., RR:0.28 for women using HRT within the past 10 years in the lowest quartile in the study of Newcomb et al. [13]). This effect modification was also observed in the adenomatous polyp study of Potter et al. [29]. Fewer studies have examined the relation between cancer of the rectum and postmenopausal hormone use. In these small case groups (most < 100), relative risks have ranged from 0.2-1.5 with six of the studies finding risks below one. However only one study observed a significant reduction in rectal cancer among HRT users [15].

The inconsistencies in the direction and magnitude of effect of HRT on risk of large bowel cancer may be attributable to 1) secular differences in the use of replacement hormone; 2) confounding, specifically the "healthy user" profile; and 3) at least theoretically, early detection practices. In the earlier studies, those published prior to 1990, data were presented without respect to recency, a potentially important characteristic of use. When these older studies were conducted, use of HRT was generally short term, prescribed for the acute symptoms of menopause. While these null studies do not generally provide any information on recency, we can assume the use of HRT was very likely historical, since the women in these studies were usually older and past menopause. Thus the salient exposure—recent use—may have been absent in these study populations. There is some evidence to suggest that HRT users represent a population that is healthier than women who do not use HRT [30,31] and that HRT users tend to have decreased mortality from a range of conditions [32]. Thus, it may be attributes of the women who choose to take HRT and not the hormone itself which is beneficial. However, adjustment for a large number of health-related characteristics in full multivariate models have not demonstrated significant bias in previous studies of HRT and colon cancer [13,14,19,22]. Finally, HRT users may have more intensive surveillance than nonusers. The use of sigmoidoscopy and perhaps fecal occult blood testing followed by colonoscopy decreases large bowel cancer mortality by as much as 50-70% [33]. This reduction occurs both through the earlier detection and treatment of cancer as well as by identifying premalignant lesions for removal. Thus, if HRT users regularly participate in colorectal screening, as suggested by Barrett-Connor [31], the effect would be a reduction in incidence that would tend to overestimate the benefits of HRT. However, a reduced risk of cancer among current hormone users remained after adjustment for history of screening sigmoidoscopy [7,13]. In the study of Calle et al., it was estimated that the impact of increased surveillance among HRT users would result in an overall reduced mortality among current users of 8% and an RR of 0.92—clearly insufficient to explain the observed magnitude of HRT risk reduction [22].

Evidence from Other Exogenous Hormone Use

While the data are preliminary, the adjuvant use of the antiestrogen tamoxifen may be associated with an increased risk of large bowel cancer. In a pooled analysis of data from three randomized controlled trials, gastrointestinal tumors were increased 2-3 fold among tamoxifen users [34]. While this finding, based on only 57 cases (in 32,091 person years) could be spurious, it appears consistent with a beneficial role for estrogens in the large bowel. We recently completed an analysis of hormonal therapy, presumably tamoxifen, in a population-based cohort of breast cancer patients reported to SEER [35]. In this analysis of 14,984 tamoxifen users and 70,427 nonusers, the adjusted relative risk of colorectal cancer was 1.39 (95% confidence interval 1.09-1.77), with increasing time since use associated with greater risk.

Hormones and the Biology of Colorectal Carcinogenesis

While some aspects of colon carcinogenesis are related to DNA damaging agents and processes, emerging evidence suggests that hormones—specifically estrogens—are also relevant. Given the range of estrogen's actions and the multi-stage nature of colon carcinogenesis [27], several hormone-related mechanisms, both indirect and direct, could be germane to the development of large bowel cancer.

As an indirect mechanism, first posited by Potter and McMichael [5], bile acid secretion may be favorably influenced by hormones (both endogenous and exogenous), leading to reduced risk of colorectal cancer. In support of this theory, bile acids appear to cause proliferation and promote colon cancer in laboratory animals [36]. There is also some human data that support this relationship [37]. It is not clear whether exogenous hormones necessarily result in decreased secretion, however this does suggest that the bile acid profiles may change in exogenous hormone users, towards decreased synthesis.

More importantly, hormones, including exogenous hormones, may have a direct effect on the colonic epithelium. Steroid receptors are a prerequisite for hormone responsiveness and there is substantial data that demonstrates the presence of receptors for estrogen (ER) and progesterone (PR) in the normal and malignant colon [38], with some suggestion that levels may be higher in normal compared to malignant tumor. The clinical significance of these receptors is not yet clear, although one study suggested that survival was prolonged among patients with ER positive tumors consistent with the prognostic benefit of positive hormone receptors in the breast [38]. An association between exogenous hormones and ER/PR status in the breast has been observed in several studies. In our recent analysis of current HRT and breast cancer risk, we observed a doubling of risk of ER positive tumors among HRT users [39]. The effect of the antiestrogen tamoxifen in ER positive breast tumors suggests the plausibility of the association in the bowel.

It is usually thought that estrogen, in the presence of ER, acts as a proliferative signal; however, estrogen-dependent antiproliferative effects in the normal breast in other tissues are well established [40]. Recent experimental studies may provide evidence of a direct antiproliferative role of estrogen in inhibiting colon carcinogeneses. The addition of ethyl estradiol to cultures of a colon cancer line resulted in significant decreases in cell proliferation [41]. It is difficult however to directly relate this experiment to concentrations of estrone and estradiol in women using HRT. Reduction in the size of colonic crypts has been shown to be induced by estradiol in both male and female mice [42]. Tutton and Barkla [43] have shown that oophorectomy and the consequent loss of estrogen results in an increase in proliferation in the normal rat colon. Thus, while aspects of the mechanisms through which estrogens stimulate growth are understood to a degree, the mechanism of estrogen receptor-mediated inhibition is unknown and deserves investigation. Recent evidence provided by Issa et al. indicates that, with age, ER gene expression declines and is ultimately silenced [44]. If this silencing can be reversed, as is suggested by work in breast cell lines [45], exogenous hormones might reactivate ER gene expression. The expression of downstream estrogen-regulated genes appears critical to colon carcinogenesis [46]. These gene products regulate cell-cell interaction, cell proliferation, and apoptosis. A reversal of this hypermethylation by estrogen (provided by exogenous sources) is a plausible explanation for the epidemiologic observation of a reduced risk of colon cancer among users of HRT.

Summary

Hormone replacement therapy is used by nearly half of middle-aged women in the U.S. Epidemiologic evidence strongly and consistently supports a substantial reduction in colorectal cancer risk among women taking postmenopausal estrogens. Biologic data provide further evidence for a link between estrogen and colorectal cancer, perhaps through expression of the estrogen receptor gene. Women and their physicians should consider this evidence along with estrogen's other benefits and risks.

References

1. Ries LAG, Kosary CL, Hankey BF, Harras A, Miller BA, Edwards BK, editors. SEER Cancer Statistics Review, 1973-1993: Tables and Graphs. Bethesda, MD: National Cancer Institute, 1996.
2. McMichael AJ, Potter JD. Do intrinsic sex differences in lower alimentary tract physiology influence the sex-specific risks for bowel cancer and other biliary and intestinal diseases? Am J Epidemiol 1983;118:620-27.
3. Potter JD. Editorial: Hormones and Colon Cancer. J Natl Cancer Inst 1995;87:1039-40.
4. Potter JD, Slattery ML, Bostick RM, Gapstur SM. Colon cancer: A review of the epidemiology. Epi Rev 1993;15:499-545.
5. McMichael AJ, Potter JD. Reproduction, endogenous and exogenous sex hormones, and colon cancer: A review and hypothesis. J Natl Cancer Inst 1980;65:1201-7.
6. Martinez ME, Grodstein F, Giovannucci E, et al. A prospective study of reproductive factors, oral contraceptive use and risk of colorectal cancer. Cancer Epidemiol Biomarkers 1977;6:1-5.
7. Potter JD, McMichael AJ. Large bowel cancer in women in relation to reproductive and hormonal factors: A case-control study. J Nat Cancer Inst 1983;71:703-9.
8. Fernandez E, LaVecchia C, Franceschi S, et al. Oral contraceptive use and risk of colorectal cancer. Epidemiology 1998;9:295-300.
9. Calle EE. Hormone replacement therapy and colorectal cancer: Interpreting the evidence. Cancer Causes Control 1997;8:127-29.
10. Weiss NS, Daling JR, Chow WH. Incidence of cancer of the large bowel in women in relation to reproductive and hormonal factors. J Natl Cancer Inst 1981;67:57-60.

11. Peters RK, Pike MC, Chang WWL, Mack TM. Reproductive factors and colon cancers. Br J Cancer 1990;61:741-48.
12. Davis FG, Furner SE, Persky V, Koch M. The influence of parity and exogenous female hormones on the risk of colorectal cancer. Int J Cancer 1989;43:587-90.
13. Newcomb PA, Storer BE. Postmenopausal hormone use and risk of large bowel cancer. J Natl Cancer Inst 1995;87:1067-71.
14. Jacobs EJ, White E, Weiss NS. Exogenous hormones, reproductive history, and colon cancer (Seattle, Washington, USA). Cancer Causes Control 1994;5:359-66.
15. Furner SE, Davis FG, Nelson RL, Haenszel W. A case-control study of large bowel cancer and hormone exposure in women. Cancer Res 1989;49:4936-40.
16. Kampman E, Potter JD, Slattery ML, et al. Hormone replacement therapy, reproductive history and colon cancer. Cancer Causes Control 1997;8:146-58.
17. Gerhardsson de Verdier M, London S. Reproductive factors, exogenous female hormones, and colorectal cancer by subsite. Cancer Causes Control 1992;3:355-60.
18. Fernandez E, La Vecchia C, D'Avanzo B, et al. Oral contraceptives, hormone replacement therapy and risk of colorectal cancer. Br J Cancer 1996;73:1431-35.
19. Grodstein F, Martinez ME, Giovannucci EL, et al. Postmenopausal hormone use and colorectal cancer in the Nurses' Health Study. Ann Int Med 1998;128:705-12.
20. Folsom AR, Mink JJ, Sellers TA, et al. Hormone replacement therapy and morbidity and mortality in a prospective study of postmenopausal women. Am J Public Health 1995;85: 1128-32.
21. Persson I, Yven J, Bergkvist L, Schairer C. Cancer incidence and mortality in women receiving estrogen and progestin therapy. Int J Cancer 1996;67:327-32.
22. Calle EE, Miracle-McMahill HL, Thun MJ, Heath CW. Estrogen replacement therapy and risk of fatal colon cancer in a prospective cohort of postmenopausal women. J Natl Cancer Inst 1995; 87:517-23.
23. Sturgeon SK, Schairer C, Brinton LA, et al. Evidence of a healthy estrogen user survivor effect. Epidemiology 1995;6:227-31.
24. Risch HA, Howe GR. Menopausal hormone use and colorectal cancer in Saskatchewan: A record linkage cohort study. Cancer Epidemiol Biomarkers Prev 1995;4:21-28.
25. Troisi R, Schairer C, Chow WH, et al. Reproductive factors, oral contraceptive use and risk of colorectal cancer in the United States. Epidemiology 1997;8:75-79.
26. Wu AH, Paganini-Hill A, Ross RK, Henderson BE. Alcohol, physical activity and other risk factors for colorectal cancer: A prospective study. Br J Cancer 1987;55:687-94.
27. Fearon ER, Vogelstein B. A genetic model for colorectal tumorigenesis. Cell 1990;61:759-67.
28. Pepins LA, Newman B, Sandher RS. Reproductive history, use of exogenous hormones and risk of adenomatous polyps. Cancer Epidemiology Biomarkers Prev 1997;6:671-75.
29. Potter JD, Bostick RM, Grandits GH, et al. Hormone replacement therapy is associated with a lower risk of adenometous polyps of the large bowel. Cancer Epidemiol Biomarkers Prev 1996;5:779-84.
30. Matthews KA, Kuller LH, Wing RR, et al. Prior to use of estrogen replacement therapy are users healthier than non-users? Am J Epidemiol 1996;143:971-78.
31. Barrett-Connor E. Postmenopausal estrogen and prevention bias. Ann Intern Med 1991; 115:455-56.
32. Schairer C, Adami HO, Hoover R, Persson I. Case-specific mortality in women receiving

hormone replacement therapy. Epidemiology 1997;8:59-65.

33. U.S. Preventive Services Task Force. Report. Guide to Clinical Preventive Services Screening for Colorectal Cancer. Baltimore:Williams and Wilkins, 1996:89-104.

34. Rutqvist LE, Johansson H, Signomklao T, et al. Adjuvant tamoxifen therapy for early stage breast cancer and second primary malignancies. Stockholm Breast Cancer Study. J Natl Cancer 1995;87:645-51.

35. Newcomb PA, Solomon C, White E. Tamoxifin and risk of large bowel cancer in women with breast cancer. Breast Cancer Res Treat, in press.

36. Reddy BS, Watanabe K, Weisburger JH, Wynder EL. Promoting effect of bile acids in colon carcinogenesis in germ-free and conventional F344 rats. Cancer Res 1977;37:2533-39.

37. McKeown-Eyssen G. Epidemiology of colorectal cancer revisited: Are serum triglycerides and/or plasma glucose associated with risk? Cancer Epidemiol Biomarkers Prev 1994;3: 687-95.

38. Di Leo A, Messa C, Russo F, et al. Prognostic value of cytosolic estrogen receptors in human colorectal carcinoma and surrounding mucosa,. Digestive Diseases and Science 1994;39:2038-42.

39. Trentham-Dietz A, Newcomb PA, Gilchrist K. Hormone replacement therapy and risk of postmenopausal breast cancer according to estrogen and progesterone receptor status. Proc Amer Assoc Cancer Res 1998;39:101.

40. Yager JD, Liehr JG. Molecular mechanisms of estrogen carcinogenesis. Annu Rev Pharmacol Toxicol 1996;36:203-32.

41. Lointer P, Wildrick DU, Boman N. The effects of steroid hormones on human cancer cell lining *in vitro*. Anti Cancer Research 1992;12:1327-30.

42. Hoff MB, Chang WWL. The effect of estrogen on epithelial cell proliferation and differentiation in the crypts of the descending colon of the mouse: A radioautographic study. Am J Anat 1979;155:507-16.

43. Tutton PJM, Barkla DH. Differential effects of estrogenic hormones on cell proliferation in the epithelium and colonic carcinomata of rats. Anticancer Res 1982;2:199-202.

44. Issa J-P, Ottaviano YL, Celano P, Hamilton SR, Davidson NE, Baylin SB. Methylation of the estrogen receptor CpG island links aging and neoplasia in human colon. Nature Genetics 1994;7:536-40.

45. Ferguson AT, Lapidus RG, Baylin SB, Davidson NE. Demethylation of the estrogen receptor gene in estrogen receptor-negative breast cells can reactivate estrogen receptor gene expression. Cancer Res 1995;55:2279-83.

46. Gee JM, Robertson JF, Ellis IO, et al. Immunocytochemical localization of BCL-2 protein in human breast cancers and its relationship to a series of prognostic markers and response to endocrine therapy. Int J Cancer 1994;59:619-28.

Aims, Methods, and Results of the Progetto Menopausa Italia

Giovan Battista Serra, Stefania Ricci, Carlo Piscicelli, Paola Manna, Cesare Pasquinucci, Antonio Chiantera, and participants of the Progetto Menopausa Italia

Introduction

The aim of this study, named Progetto Menopausa Italia (PMI), is to collect epidemiological data on Italian postmenopausal women to ascertain how risk factors for cardiovascular diseases (CVD) and other pathologies correlate with menopause, what is the incidence of such risk factors in Italy, to verify the appropriateness of medical strategies, and to evaluate differences across northern, central, and southern regions. The common epidemiological tool of this study, involving 132 menopause centers all over the country, is a standard clinical file sheet, automatically scanned by means of original software, allowing easy, comparable, and complete archiving and analysis of collected data by a national central server. These data should pursue educational as well as cultural goals for women and health care providers, both public and private, with an objective of indicating several areas where more effort should be made to improve health care. Women enrolled in the PMI study (although they cannot be taken as a representative sample of the Italian population since they voluntarily consult a gynecologist) can help to describe the characteristics of menopausal women in Italian society, since the medical centers involved in the PMI study are very well distributed throughout the country. This sample includes a large and growing number of women, who once enrolled, are continually monitored (8,869 women as of the fall, 1998).

Material and Methods

To facilitate collection, archiving, and extrapolation of clinical data from each physician, innovative tools have been developed. Two standard clinical file sheets, one for the first examination, the other for the follow-up visits, have been used by all the participants, thus providing a common clinical method to investigate the patient's health. Each clinical file sheet has several sections: demographic, anamnestic, with lifestyle questions, climacteric syndrome score, usually associated with multiple choice answers. There are other sections for pelvic examination evaluation, laboratory and imaging data, medical risk assessment, and therapeutic prescriptions. Particular attention has been placed on the therapeutic management and compliance: one section is dedicated to past or current therapies, while

R. Paoletti et al. (eds.), Women's Health and Menopause, 257–266.

a second section is dedicated to prescriptions. In this context it is possible to file information on several specific treatments, using numerical codes, reported in the file sheet, regarding preparations, dosages, routes of administration, and therapeutic schemes (sequential or combined, cyclic or continuous).

This clinical file sheet is available for every physician wherever he or she works (private and/or public out-patient services) if participating in the study. Each out-patient service, with one or more doctors, represents an operative unit (OU). Data archiving may be possible in larger centers where a computer, a scanner, and specifically dedicated software have been installed. These centers are called SPAC (Stazione Periferica di Archiviazione Computerizzata: peripheral computerized archiving station). Specifically dedicated software, using a scanner permits alpha-numeric symbols to be read, as well as black marks placed on the multiple choices spaces. Ranges with function of filter are foreseen in each field for the numeric data input. For the data source recognition there are, in the card, proper compartments for 3 code numbers. The first one is the SPAC code, the second one is the OU code (each SPAC can consist in maximum 999 operative units), the third one is the code number of the physicians working inside each OU Every OU sends the cards to its own SPAC. The archiving of a single file sheet takes about 1.5 minutes. In each SPAC, local statistical elaborations are available. SPACs are connected, via modem or Internet, with a national central server able to produce national epidemiologic statistics. At this moment 132 SPAC have been installed in Italy and 78 of them are already operative. SPACs are spread throughout the country, with good geographic representation. Italy has been divided in 3 macro-regions: north (Valle D'Aosta, Piemonte, Liguria, Lombardia, Trentino Alto Adige, Friuli Venezia Giulia, Veneto, and Emilia Romagna), central (Toscana, Marche, Umbria, Lazio, Abruzzo, Molise, and Sardegna), and south (Campania, Puglia, Basilicata, Calabria, and Sicilia).

Results

All women attending one of these PMI study out-patient clinics are enrolled. Preliminary data from the first 8,869 patients enrolled are presented in this paper.

The demographic data indicated a mean age of spontaneous menopause in Italy at 48.9 $\pm$ 4.1 (mean $\pm$ SD) years. Mean surgical menopausal age was 46.6 $\pm$ 4.8 years. Mean menopausal age, considering both surgical and spontaneous menopause, was 48.0 $\pm$ 4.9. The mean age of patients was 53.1 $\pm$ 6.8 years. Among all the women evaluated, 22.2% had undergone hysterectomies. Data comparison among the northern, central, and southern areas does not indicate regional significant differences.

Table 1 shows some demographic and clinical characteristics of the PMI study women taken as a whole (national) and divided by region (north, central, and south).

Among all patients, 25.1% indicate positive familial history for CVD, 23.5% for diabetes, 17.1% for osteoporosis, and 10.3% for breast cancer. Personal anamnestic data show that 20.0% of all patients are hypertensive, 17.5% suffer from varicose veins, 9.3% from thyroid disease, 7.9% have had cholecystectomy, and 5.0% suffer from diabetes.

Table 1. Some demographic and clinical characteristics of postmenopausal Italian women enrolled in the PMI study

	National (n = 8,869)	North (n = 1,554)	Central (n = 4,285)	South (n = 3,030
Married	85.5%	84.2%	86.5%	84.9%
Parity				
• 0	15.0%	12.9%	16.9%	13.7%
• 1	13.1%	23.3%	13.5%	7.3%
• 2	40.5%	44.5%	44.4%	32.9%
• >2	31.4%	19.3%	25.2%	46.1%
Education				
• Primary School	42.2%	41.0%	42.5%	42.3%
• Secondary School	27.6%	33.9%	27.6%	24.5%
• High School	21.9%	21.6%	20.5%	23.7%
• University	8.2%	3.5%	9.3%	9.4%
OC past users	26.4%	31.0%	28.8%	21.6%
Smokers	26.1%	21.7%	25.7%	28.6%
Alcohol drinkers	22.3%	29.1%	20.3%	21.7%
Sedative users	19.2%	25.0%	21.6%	13.3%
Regular exercisers	19.6%	28.4%	21.5%	13.4%

The national mean body mass index (BMI) (weight/height^2 [Kg/m^2]) value was 25.7 $\pm$ 4.3; similar mean values were found in women from the north (25.3 $\pm$ 4.2) and from the central region (25.5 $\pm$ 4.1), however, the mean BMI was higher in women from the south (26.3 $\pm$ 4.4) ($p < 0.001$). In the whole national sample of postmenopausal women, 41.6% were overweight (BMI > 25 and < 30) and 17.7% were obese (BMI $\geq$ 30).The overweight and obese women were more prevalent in the south (43.8% and 24.3%, respectively).

Apart from the differences in normal, overweight, and obese women across the country, in all 3 regions there also were significant differences regarding education, lifestyle, and lipemic profile between normal-weighted women (BMI < 25) and overweighted women (BMI ≥ 25) (Table 2).

Table 2. Education level, parity, smoking, physical activity, and lipemic values in normal-weighted (BMI < 25) and in overweighted (BMI ≥ 25) women in different Italian regions

	Normal-weighted women	Overweighted women	p
North			
High school or more	30.8% (202/655)	18.3% (123/672)	< 0.001
Multiparous (> 2)	15.5% (106/682)	22.3% (155/695)	< 0.001
Cigarette smokers	24.8% (164/661)	17.2% (113/660)	< 0.001
Regular exercisers	32.4% (211/653)	23.7% (156/659)	< 0.001
Total Chol, mg/dl	222 ± 41 (309)	230 ± 41 (360)	< 0.01
HDL-Chol, mg/dl	64 ± 14 (282)	58 ± 14 (329)	< 0.001
LDL-Chol, mg/dl	160 ± 55 (57)	161 ± 41 (64)	ns
Triglycerides, mg/dl	93 ± 50 (288)	119 ± 57 (335)	< 0.001
Central			
High school or more	39.5% (487/1231)	23.3% (398/1708)	< 0.001
Multiparous (> 2)	19.0% (304/1596)	30.0% (612/2041)	<0.001
Cigarette smokers	31.0% (440/1421)	20.5% (455/2211)	< 0.001
Regular exercisers	29.5% (406/1373)	16.2% (294/1816)	< 0.001
Total Chol, mg/dl	228 ± 41 (552)	234 ± 42 (619)	< 0.01
HDL-Chol, mg/dl	59 ± 14 (391)	54 ± 12 (391)	< 0.001
LDL-Chol, mg/dl	156 ± 37	158 ± 36 (214)	ns
Triglycerides, mg/dl	106 ± 52 (467)	133 ± 68 (531)	< 0.001
South			
High school or more	48.4% (400/826)	25.2% (444/1761)	< 0.001
Multiparous (> 2)	36.5% (319/873)	51.2% (952/1860)	< 0.001
Cigarette smokers	37.4% (320/858)	25.0% (458/1831)	< 0.001
Regular exercisers	18.5% (156/843)	10.6% (192/1808)	< 0.001
Total Chol, mg/dL	216 ± 40 (236)	220 ± 40 (546)	ns
HDL-Chol, mg/dL	54 ± 11 (136)	51 ± 12 (317)	< 0.01
LDL-Chol, mg/dL	148 ± 30 (58)	151 ± 32 (105)	NS
Triglycerides, mg/dL	93 ± 45 (206)	119 ± 65 (501)	< 0.001

Regarding weight gain and menopause, in postmenopausal women enrolled in the

PMI study, increases in mean BMI values were independent of the years (1 to 10) elapsed since menopause. BMI records from 4,970 women aged 20-70 years, who came to our hospital in Rome for different reasons, indicated that the steeper weight gain occurred between 41 and 47 years of age.

About 22.2% (1,649/7,408) of Italian women enrolled in the PMI study presented systolic blood pressure ≥ 140 mmHg and diastolic blood pressure ≥ 90 mmHg at the first visit, with regional differences: 27.4% (390/1,423) in the north, 22.4% (699/3,116) in the central area, and 19.5% (560/2,869) in the south.

Comparing a group of 337 premenopausal women (mean age 48.8 ± 3.3 years) with another group of 1,865 postmenopausal women (mean age 53.7 ± 5.8 years), it was noted that 24.3% of premenopausal women were hypertensive (anamnestic data or blood pressure ≥ 140/90 mmHg at the first visit) versus 31.6% of postmenopausal women ($p < 0.001$). Considering 2 subgroups of premenopausal and postmenopausal women of the same age (48-50 years old), 28.8% of premenopausal women (n = 118) were hypertensive versus 28.2% of postmenopausal women (n = 340) (p = ns). Dividing postmenopausal women of the same age (48-50 years) by years elapsed since menopause (from 0 to 10 years, 3-year-intervals for each group), mean systolic and diastolic blood pressure values were not significantly different among the different groups.

Table 3 shows cholesterol (total, high density lipoprotein [HDL], low density lipoprotein [LDL]) and triglycerides (TG) levels of the PMI study women taken as a whole (national) and divided by region (north, central, south).

Table 3. Lipemic values of postmenopausal Italian women enrolled in the PMI study

	National	North	Central	South
Total Chol, mg/dL	226 ± 41 (2,916)	227 ± 41 (728)	231 ± 42 (1,310)	219 ± 40* (872)
HDL-Chol, mg/dL	57 ± 14 (2,064)	1 61 ± 4 (670)	56 ± 13 (887)	52 ± 12*(507)
LDL-Chol, mg/dL	156 ± 38 (808)	159 ± 47 (133)	157 ± 37 (484)	151 ± 33 (189)
TG, mg/dl	114 ± 61 (2,587)	108 ± 55 (682)	120 ± 62** (1,111)	112 ± 63 (788)

* $p < 0.001$ south versus central versus north; ** $p < 0.01$ central versus north and versus south.

Total and HDL cholesterol levels were significantly higher in the northern and in the central areas than in the south ($p < 0.001$).

Among all postmenopausal women, 16.2% and 19.2% presented, respectively, moderate (≥ 240 mg/dl) and high hypercholesterolemia (≥ 260 mg/dl). Regional analysis found a lower percentage incidence of moderate (13.9%) and high hypercholesterolemic women (13.9%) in the south.

At least one cardiovascular risk factor (such as hypercholesterolemia [≥ 240 mg/dl], smoking, hypertension [≥ 140/90 mmHg], and obesity) was found in 60% of all postmenopausal women, one risk factor in 39%, two risk factors in 17%.

Considering the data from the first visit, and, where available, from the following visits, it appears that 53.6% of all women have had mammograms, with differences across the country: 70.2% underwent this examination in the north, 53.1% in the central area, and 46.1% in the south. Of all women, 29.1% underwent bone mineral density examination: 47.0% in the north, 27.7% in the central region, and 22.0% in the south.

Regarding hormonal replacement therapy (HRT), at the first visit 14% of women enrolled in the PMI study (n = 8,869) were current or past-HRT users and 27% received prescriptions for HRT. At the second visit (patients = 1,590), about 10 months after the first visit, 35% of women were current or past-HRT users and 45% received prescriptions for HRT. At the third visit (patients = 509), about 19 months after the first visit, 50% of women were current or past users and 50% received prescriptions for HRT. We found similar percentages at following visits.

To ascertain HRT use in postmenopausal women in our country (in addition to the use by women enrolled in PMI study), we evaluated the region-wide use of estrogen blisters purchased in Rome and Umbria under National Health System prescriptions presented in pharmacies and national sales from 1989 to 1997. The region-wide estrogen use has been evaluated in a casual population sample (83,840 women aged 40-74) in Rome from 1990 to 1992 and in the whole female population in Umbria (193,142 women aged 40-74) from 1994 to 1997 (Table 4).

Considering the population sample in Rome, among women aged 45-74, HRT users increased from 1.0% in 1990 to 1.9% in 1992, with the highest prevalence of use among women aged 50-54, increasing from 2.1% in 1990 to 4.2% in 1992.

Table 4. Prevalence of estrogen use in Umbria in women aged 40-74 (n = 193,142) from 1994 to 1997

Years old	40-44	45-49	50-54	55-59	60-64	65-69	70-74	45-74
n. of women	27,030	27,520	26,776	27,846	28,226	28,945	26,799	166,112
HRT users 1994	0.9%	3.1%	4.6%	2.7%	0.7%	0.3%	0.1%	1.8%
HRT users 1995	1.5%	4.7%	4.7%	2.0%	0.6%	0.2%	0.1%	2.0%
HRT users 1996	1.4%	6.2%	6.3%	3.3%	0.9%	0.3%	0.1%	2.7%
HRT users 1997	1.8%	7.4%	7.4%	3.3%	0.9%	0.3%	0.1%	3.2%

Considering the entire female population in Umbria in 1994, among women aged 45-74, HRT users represented 1.8%. This prevalence of use increased progressively during the study period: 2.0% in 1995, 2.7% in 1996, and 3.2% in 1997. The highest prevalence of HRT use resulted in 1994 among women aged 50-54, with 4.6% of HRT users, followed by 3.1% of HRT users among women aged 45-49. The prevalence of estrogen use increased progressively in the following years: 4.7% in 1995, 6.3% in 1996, and 7.4% in 1997 among women aged 50-54; among women aged 45-49 use increased to 4.7%, 6.2%, and 7.4%, respectively, in 1995, 1996, and 1997. During the study period, age percent distribution of women purchasing estrogen showed a prevalence of HRT use less than 1% in the age groups over 60 years.

National sales indicated a constant increase in the estrogen sales from 358,000 blisters sold in 1989 to 6,092,000 in 1997.

Discussion

The realization and use of the new clinical file sheet for the postmenopausal patients represents a valid tool to obtain comparable data from patients across the country.

This easy, computerized data gathering enables doctors involved primarily in general gynecology to participate in this large epidemiological study and also check national data and compare them with those obtained in other countries. According to the PMI study the mean age of spontaneous menopause occurs earlier in Italy than in the United States [1]. Several factors may influence this: diet, geographic localization, lifestyles, and selective sampling.

Data on the educational levels of our patients indicated that the majority of these women had only a first degree diploma, although their educational level was higher than that reported in the overall national data. This should indicate the need to provide medical information simple enough to be understood by women with this educational background and the need for wider access to medical care. Moreover, among women who visited the gynecologist, women with university educations were present in higher numbers in the central and southern regions than in the north. This suggests that, in the north, women visit gynecologists regardless of their educational level.

The different incidence of multiparous women across the country indicated how significant the variation between the northern and southern regions can be. This reflects different cultural and economical backgrounds. Another interesting observation was that past users of oral contraceptives (considered as one of the determinants of HRT) were more representative in the northern than in the central and southern regions, suggesting that there may be a greater propensity to use HRT among women from the north.

Cigarette smoking, an important risk factor for CVD and osteoporosis, appeared to be a wide-spread habit in the Italian postmenopausal population; alcohol is an important risk factor both for osteoporosis and breast cancer [2]. Appropriate information of the increased risk for breast cancer and CVD in these different modifiable conditions should also be available from the central health agencies.

Mean BMI is quite similar to that found by Bush in the United States [1]. However, since higher BMI results in an increased risk for breast neoplasia [3], as well as for CVD, a first recommendation can be to promote X-ray mammography in southern regions where we found the highest prevalence of obese women and the least use of this examination. Apart from incidental differences in normal, overweighted, and obese women across the country, in all regions, north, central, and south, there were significant differences regarding education, lifestyle, and lipemic values between women with BMI < 25 and women with BMI ≥ 25. Normal-weighted women seemed to be more educated, have fewer children, be more predisposed to be regular exercisers, but also smoke, and have better lipemic values than overweighted and obese women across the country, independent of their location, indicating the importance of cultural background in controlling weight gain. Regarding weight gain, our data showed that it seemed not to occur at the moment of menopause but in the late years of the fertile period, in particular from 41 to 47 years of age, and that BMI remained constant after menopause.

Concerning hypertension, an important cardiovascular risk factor, about 22.2% of postmenopausal Italian women enrolled in the PMI study were hypertensive in accordance with data collected in the United States [4]. Data on hypertension and menopause are not conclusive. We found a significant difference in the prevalence of hypertensive women when considering premenopause versus postmenopause, but when we eliminated the confounding age factor, this difference disappeared. Moreover, considering years since the onset of menopause in women of the same age, we did not find significant differences in the mean blood pressure values. Therefore our data seem to stress that blood pressure increase is correlated to aging rather than menopause.

Mean national and regional cholesterol, HDL, LDL, and triglycerides values showed to be between normal ranges, as in postmenopausal American women [1]. But 35.4% of all Italian postmenopausal women, as do their American counterparts [4], presented levels ≥ 240 mg/dl, thus being at risk for CVD [5].

Considering the incidence of hypertension, smoking, hypercholesterolemia, and obesity, 60% of the women presented at least one risk factor for CVD. Italian postmenopausal women seem to have risk factors similar to those of their American counterparts. In Italy, the mortality trend for coronary heart disease decreases less rapidly than in the United States, yet, on the whole, the mortality rate for coronary heart disease is lower in Italy than in the U.S.A. [6]. It is probably due to the fact that in Italy fewer women engage in physical activity, more are smokers, and therefore they seem to be less concerned with early prevention; additionally, there are fewer HRT users than in the United States [7].

National sales of estrogens for HRT clearly show that before 1989 in Italy HRT users were virtually nonexistent. In Rome from 1990 to 1992, HRT users increased from 1.0% to 1.9% among women aged 45-74, with the highest prevalence among women aged 50-54, which showed an increase from 2.1% in 1990 to 4.2% in 1992. Among women of the same age (50-54 years), considering all female population in Umbria, HRT users increased from 4.6% in 1994 to 7.4% in 1997. An increase also occurred among women aged 45-49 from 3.1% in 1994 to 7.4%, but in the over-60-year age groups the HRT use remained very low, less than 1%, in women more susceptible to coronary heart disease and fractures. Thus, the prevalence of HRT users among all women (45-74 years old) increased through these years from 2.5% in 1994 to 3.2% in 1997. This low use prevalence found in Umbria also in 1997 is due to the fact that we have considered the entire female population of the region rather than only those women who visit the gynecologist. Therefore, this low HRT use is closer to the reality of the country.

Data from PMI study indicated that the prevalence of HRT use was greater among women who visit their gynecologists than among the entire female population, as data from Umbria showed, and also that this higher prevalence increased from 14% and stabilized at approximately 50% after the third visit, which stresses the importance of continued follow up with the same physician or team.

Data collected from the PMI study and the subsequent results help to present a more complete picture of postmenopausal Italian women, allow us to recognize differences across the country, and determine those areas in which intervention (preventive or therapeutic) is necessary.

The Participants in the *Progetto Menopausa Italia*:

G. Accinelli, G. Adinolfi, D. Agostinelli, G. Angeloni, E. Ansaldi, R. Arienzo, V. Azzini, A. Bacchi Modena, E. Baldaccini, G. Balsotti, P. Barbacini, G. Barese, A. Becorpi, U. Bellati, S. Bennici, F. Bertelli, B. Bersellini, P. Breveglieri, E. Bocchin, E. Bonifante, P. Bongiovanni, S. Bottino, G. Brigato, P. Bruno, D. Bucari, G. Buonanno, M. Buonerba, G. Buoso, C. Campagnoli, A. Campanella, P. Cancellieri, L. Cantiello, P. Ceccarelli, A. Cetera, G. Cicchetti, E. Cirese, E. Cittadini, A. Cisternino, A. Coco, F. Corazza, N. Cornopatri, N. Corrado, P. Cristiani, P. Curiel, A. D'Aandrea, N. D'Antona, P. D'Aquino, S. D'Asta, D. De Aloysio, V. De Leo, M.G. De Silvio, G. Del Frate, P. Di Donato, L. Di Prisco, R. Dibitonto, G. Dolfin, G. Donini, C. Donati Sarti, V.B. Ercolano, M. Fabiani, C. Ferruccio, S. Filinceri, G. Finocchiaro, S. Filinceri, G. Finocchiaro, G. Galbiati, M. Gallo, H. Gamper, S. Garofalo, S. Garzarelli, Gentile, V. Giambanco, C. Giannice, G. Giannola, G. Giannone, C. Gigli, Giulini, F. Goisis, R. Graziano, G.L. Grismondi, L. Labate, G. Linsalata, M. Lombardo, R. Lombardo, R. Lorefice, M. Luerti, C. Malanetto, GP Mandruzzato, V. Marsoni, G. Masciari, L. Massaccesi, M. Massobrio, G. Meli, C. Minervini, A. Mondo, M. Monti, D. Mossotto, N. Natale, O. Negro, F. Nobili, F. Nocera, M. Obassi, F. Ognissanti, U. Omodei, G. Palumbo, A. Pascarella, A.R. Pastore, P. Pesando, E.R. Poddi, G. Polizzotti, P. Pomini, L. Provenzano, C. Pullè, P. Pupita, S. Quaranta, RV Raso, S. Rastelli, F. Repetti, L. Ripamonti, C. Roccasalva, C. Romanini, G. Santeufemia, C. Sbiroli, S. Schönauer, C. Simonato, G. Sordi, P. Stelleri, C.M. Stigliano, O. Sorrentino, Tartaglino, F.G. Tinelli, F. Tirozzi, PA Todaro, F. Tomacelli, S. Tramontana, P.F. Tropea, P. Vadalà, G. Vegna, M. Viani, M. Vitale, S. Votano, V. Zacchi, G. Zandonini

References

1. Bush TL, Barrett-Connor E, Cowan LD, et al. Cardiovascular mortality and noncontraceptive use of estrogen in women: Results from the lipid research clinics program follow-up study. Circulation 1987;75:1102-9.
2. van den Brandt PA, Goldbohm RA, van't Veer P. Alcohol and breast cancer: Results from the Netherlands Cohort Study. Am J Epidemiol 1995;141:907-15.
3. La Vecchia C, Negri E, Franceschi S, et al. Body mass index and postmenopausal breast cancer: An age-specific analysis. Br J Cancer 1997;75(3):441-44.
4. Perlman JA, Wolf PH, Ray R, Lieberknecht G. Cardiovascular risk factors, premature heart disease, and all-cause mortality in a cohort of Northern California women. Am J Obstet Gynecol 1988;158:1568-74.
5. National Institute of Health Consensus Development Panel on Lowering Blood Cholesterol to Prevent Heart Disease. JAMA 1985;253:2080-86.
6. World Health Organization. World Health Statistics Annual, 1990. Geneva: WHO, 1990.
7. Brett KM, Madans JH. Use of postmenopausal hormone replacement therapy: Estimates from a nationally representative cohort study. Am J Epidemiol 1997;145:536-45.

Frequency of Cardiovascular Risks Factor in Women Attending Menopause Clinics: Findings from the ICARUS Data Base

P. Affinito, D. Agostinelli, C. Albanese, F. Bartiromo, G. Bonaccorsi, A. Cagnacci, S. Guaschino, F. Mangino, B. Mauro, B. Molteni, D. Morano, U. Omodei, G.L. Radici, G. Ranchet, G. Sciacchitano, M. Milani, on behalf of ICARUS Cardiovascular Risk Study Subgroup

Introduction and Study Objective

Coronary heart disease is generally considered a "man's disease," but it also the greatest killer of women older than 50 [1]. Cardiovascular diseases remain the main cause of morbidity and mortality in postmenopausal women [1]. Coronary heart disease is more dependent on age in women than in men; women are usually 10 years older than men when any coronary manifestations first appear [2]. Hypertension, smoking, and hyper-cholesterolemia are frequently observed in women [3]. Thus preventive measures are important to lower cardiovascular morbidity and mortality.

Since the frequency of cardiovascular disease increases after menopause, menopause clinics may have an important role in evaluating risk factors for cardiovascular disease. Given this perspective, the ICARUS study started in 1995 in menopause clinics in Italy. In this brief paper we present the frequency of the main risk factors for cardiovascular diseases in this selected population and compare it with information from the Italian population.

Patients and Methods

The ICARUS project is an epidemiological study conducted since 1995 in 79 menopause clinics across Italy. Women who attended the participating centers for the first time during the study period were eligible for the study. The protocol did not foresee any exclusion criteria. A total of 8,498 postmenopausal women (mean age 55 $\pm$ 5 years), who never received hormone replacement therapy (HRT) were enrolled in the study between January 1995 and December 1997. During the first visit, information is collected by trained interviewers using a standardized information system, based on a computerized medical record shared by all participating centers. Quality and consistency of the entered data were checked by the study monitor during on-site visits before data were sent to the central database. Women were defined as postmenopausal if menses had naturally ceased for 12

R. Paoletti et al. (eds.), Women's Health and Menopause, 267–269.

or more months. The following cardiovascular risk factors were assessed: arterial hypertension, hypercholesterolemia, hyperglycemia, obesity, smoking, and a family history of cardiovascular disease. Hypertensive women were women currently taking drugs for hypertension or with a blood pressure > 140/90 mmHg at study entry. Hypercholesterolemic women were women currently taking drugs for hypercholesterolemia or with a serum total cholesterol > 240 mg/dL. Patients taking drugs for hyperglycemia or with a plasma glucose > 140 mg/dL were considered hyperglycemic. Patients were considered obese when the body mass index (BMI, kg/m^{-2}) was > 28.6.

Results

Prevalence of cardiovascular risk factors in this population was as follow: 42% of women were hypercholesterolemic. Hyperglycemia was observed in 2% of women; hypertension was found in 22%. A positive family history was found in 16% of the patients. Obesity was detected in the 19%. Smokers represented 16%. Two or more risk factors were observed in 31% of the population. Eighteen percent of these postmenopausal women had three or more cardiovascular risk factors.

To determine if the women attending menopause clinics in Italy possibly represented a specific group at risk of cardiovascular disease on which to focus a preventive campaign, the frequency of main risk factors for cardiovascular disease in this population, stratified by age, has been compared with the frequency in a random sample of Italian population [4] (see Table 1). Never-HRT users attending the menopause clinics, who participated in the ICARUS study, reported hypertension and diabetes more frequently and were more frequently overweight, suggesting that they represent a population at risk of cardiovascular diseases.

Discussion

Our results show that main cardiovascular risk factors are frequently observed in HRT-free postmenopausal women attending menopause clinics in Italy. A considerable percentage of women presented a high cardiovascular risk profile (two or more risk factors).

Women who entered this study were identified in 79 menopause clinics in Italy. These clinics are not formally representative of all Italian menopause clinics; however, participating centers represent about 30% of all such centers in Italy. These centers were well distributed in the main regions of the country. In consideration of these limitations, any inferences must be made cautiously.

Hypercholesterolemia was the most common risk factor observed in our population. It is well known that cholesterol concentration continues to predict coronary risk in postmenopausal women [5]. Lowering cholesterol levels in postmyocardial infarction women is beneficial, as is in men [6]. Cigarette smoking triples the risk for myocardial infarction, with the greatest risk in women with other coronary risk factors and in older women; a relatively high prevalence of smokers was found in our population. Several risk factors for cardiovascular disease and overweight (ie hypertension, diabetes) were more frequently observed in women attending menopause clinics than in the general Italian population (ISTAT). These data suggest that women attending menopausal clinics represent a population at risk of cardiovascular disease. Specific efforts should be made to focus preventive campaigns for cardiovascular diseases in never-HRT users attending menopause clinics in Italy.

Table 1. Comparison of frequency (in %) of selected risk factors for cardiovascular disease in women never-HRT users attending menopause clinics who entered the ICARUS Study and in a random sample of Italian population (ISTAT Study [4]) in strata of age*.

	ICARUS			ISTAT		
Class age	35-44	45-54	55-61	35-44	45-54	55-64
Smoking (Yes)	24.7	17.8	11.0	27.6	21.2	12.6
Hypertension (Yes)	9.7	16.6	26.5	3.8	12.5	26.6
Diabetes (Yes)	2.8	3.7	5.7	0.8	2.7	8.8
Body mass index (Kg/m^2)						
< 18	2.0	0.6	0.6	2.6	1.2	0.8
18-19.9	5.3	4.3	4.3	12.3	6.7	4.7
20-24.9	49.9	48.8	46.4	57.3	50.9	42.9
25-26.9	15.3	17.2	19.2	14.1	17.5	20.4
27-29.9	15.0	15.7	15.9	8.1	13.5	17.9
> 30	12.4	13.4	13.6	5.0	9.8	12.7

* In some cases the sum does not add up to 100% because of missing values.

References

1. Thomas J, Braus P. Coronary artery disease in women. Arch Intern Med 1998;158:333-37.
2. Hisia J. Cardiovascular diseases in women. Medical Clinics of North America, 1998;82:1-19.
3. Wenger N. Coronary heart disease: An older woman's major health risk. BMJ 1997;315: 1085-90.
4. ISTAT. Condizioni di salute e ricorso ai servizi sanitari (Anno 1994). Indagine Multiscopo sulle famiglie (A cura di: A. Urbano, R. Vivio). Informaizoni n. 54 - 1997.
5. Kannel WB. Nutrition and the occurrence and prevention of cardiovascular disease in the elderly. Nutr Rev 1988;46:68-78.
6. Scandinavian Simvastatin Survival Study Group. Randomised trial of cholesterol lowering in 4444 patients with coronary artery disease: The 4S Study. Lancet 1994;344:1383-89.

Determinants of Age at Menopause and Symptomatological Profile in Italian Women Attending Menopause Clinics in Italy

A. Cordopatri, A. Becorpi, M. Bottari, G. Bonaccorsi, A. Cagnacci, E. Cevasco, P. Masi, D. Morano, S. Ottanelli, G. Sciacchitano, A. Valerio, and S. Viglino for the ICARUS Sociodemographic Study Subgroup

Introduction

Age at menopause is population-specific, with a range from the early 40s in Yucatan Mayan women to the early 50s for the Ede population in the Netherlands. Even within the same population we may observe some differences. For instance, the mean age at menopause for Indonesians in Central Java is 50.2 years in the educated population and 46.0 years in the rural population. Also typical and atypical climacteric symptoms are different among different populations, e.g. the vasomotor symptoms cause virtually no difficulties in Asian women and the psychological impact of the menopause is also much less of a problem in Asian than in Western women [1-3].

Furthermore, there are some interesting cultural differences as regards the effects of social status on climacteric symptoms [4]. In Europe and in the United States, the women most bothered by the climacteric symptoms are those in the lower socioeconomic classes. In Asia the situation is exactly the opposite, severe symptoms being reported frequently by women in the higher socioeconomic classes [5].

These aspects, i.e. sociodemographic descriptors, symptomatological and clinical correlates of menopause, will be investigated in the ICARUS study.

Subjects and Methods

The ICARUS study is a prospective study on the effects of menopause on women's health conducted since 1995 in menopause clinics through Italy. Women who attended a participating center for the first time during the study period were eligible for the study. The protocol did not foresee any exclusion criteria.

During first visit, information regarding sociodemographic data, reproductive and medical history, menopausal symptoms, use of female hormones, and a selected medical history were collected by trained interviewers using a standardized information system, based on a computerized medical record shared by all participating centers. Quality and consistency of the entered data were checked by study monitor during on-site visits, before

R. Paoletti et al. (eds.), Women's Health and Menopause, 271–274.

the data were sent to the central database. The computerized medical report form allows the uniform and standardized collection of clinical data and their transfer to a central database for statistical analysis.

The study started in 20 menopause clinics in 1995. The number of centers increased to 79 during the study. Data collected until December 1997 are considered in the present paper.

Body mass index (BMI) was calculated as weight in Kg/height in m^2.

Climacteric symptoms were assessed according to the Greene Climacteric Scale, which includes: a vasomotor score, a psychological score (anxiety plus depression), a somatic score, and a score on sexuality. Vaginal dryness was also evaluated. For all the scores we considered only women who were not hormone replacement therapy (HRT) users at the time of data collection.

Results

A total of 15,464 women (mean age 53.7, SD 5.4) entered the study. Of these, 9,351 were in postmenopause (7,415 spontaneous menopause, 1,896 surgical, and 40 pharmacological menopause).

The mean age for the onset of spontaneous menopause was 49.16 ± 4.43 years, with no variation according to the geographical area of residence. No difference in mean age at menopause emerged when the analysis was conducted in strata of education (data not shown).

The symptomatological picture of the climacterium, assessed only in HRT-free women (11,699 subjects) according to the Greene Climacteric Scale, showed a rapid increase of the vasomotor score from the premenopause (10.4 ± 9.3, mean ± SD based on 2,385 women) to the peri- (12.4 ± 9.3 based on 3,728 women) and postmenopause (12.8 ± 9.2). The psychological score increased from the premenopause (5.9 ± 6.1) to the peri- (6.5 ± 5.9) and postmenopause (6.8 ± 5.9). By contrast no clear trend emerged for somatic score: (2.0 ± 3.0) in premenopause (2.9 ± 3.0) in peri- and (3.1±3.1) in postmenopause.

Vaginal dryness progressively increased from premenopausal values (0.3 ± 0.6) to maximal observed values reached more than 60 months after the menopause (1.2 ± 1.0). No marked difference emerged in Green Climatic Scale in strata of education and geographic areas of the country (data not shown).

Discussion

It should be stressed that the study population consisted of women attending menopause clinics for the first time. Thus, they represent a selected group of women sensitized towards menopause. In Italy, access to menopause clinics is free of charge and does not require a prescription from general practitioners. Thus, the study population can not be considered representative of the general one. For example, women who entered the ICARUS study were more educated, although no marked differences emerged in prevalence of medical conditions such as hypertension and overweight. (Table 1 shows a comparison of characteristics of women who entered ICARUS study and of a random sample of Italian population which entered a national research study of the health status of the Italian population conducted in 1994 by the Italian Central Institute of Statistics for two selected strata of age [6]).

Table 1. Comparison of frequency of selected characteristics (in %) of women who entered the ICARUS study and a random sample of 29,166 Italian women interviewed in a national survey of the Italian population conducted in 1994 by the National Institute of Statistics (ISTAT, 1997) in strata of age.

	ICARUS		ISTAT	
Class age	45-54	55-64	45-54	55-64
Education				
None/Elementary school/Intermediate school	71.1	77.6	71.4	82.8
High school/University degree	28.9	22.4	28.6	17.2
Body mass index (kg/m^2)				
< 18	0.7	0.4	1.2	0.8
18-19.9	4.2	4.8	6.7	4.7
20-24.9	48.5	45.7	50.9	42.9
25-26.9	17.1	19.4	17.5	20.4
27-29.9	15.7	16.3	13.5	17.9
>30	13.8	13.4	9.8	12.7
Frequency of hypertension	15.9	20.5	10.6	20.2

Further, women were identified in 79 menopausal clinics. They represent about 30% of all such centers in Italy. These centers were well distributed in the main areas of the country.

In consideration of these limitations, any inference must be made in strictly comparative terms.

Among the strengths of the analysis is the large sample size, the nationwide coverage of participating centers, and the use of standardized criteria of definition of menopause and other clinical variables. The data moreover were collected by the physician. Thus, information regarding medical data should be particularly accurate.

Age at spontaneous menopause was not related in the study with education and area of the country. By stratifying the data for menopausal status, we were able to document a great effect of menopause on both vasomotor symptoms, which dramatically increase in perimenopause, and vaginal dryness which progressively increases with time since the menopause. By contrast, psychological symptoms did not seem to be greatly influenced by the menopausal status, although the Greene Scale for climacteric symptoms may not be as sensitive as other scales in detecting psychological disturbances. The present data, obtained in a very large number of subjects, support the possibility that psychological discomfort is not one of the main consequences of the menopausal transition.

References

1. Beyene Y. From Menarche to Menopause. Albany, NY: State University of New York Press, 1989: 115-26.
2. Jaszmann L, Van Lith ND, Zaat JCA. The age at menopause in the Netherlands: The statistical analysis of a survey. Int J Fertil 1969;14:106-17.
3. Rekers H. Mastering the menopause. In: Burger H, Boulet M (editors). A Portrait of the menopause. London: Parthenon Publishing, 1991: 23-43.
4. Payer L. The menopause in various cultures. In: Burger H, Boulet M (editors). A Portrait of the menopause. London: Parthenon Publishing, 1991: 3-22.
5. Flint M. and Samil R.S. Cultural and subcultural meanings of the menopause. Ann NY Acad Sci, 1990;592:134-48.
6. ISTAT. Condizioni di salute e ricorso ai servizi sanitari (Anno 1994). Indagine Multiscopo sulle famiglie. (A cura di: A. Urbano, R. Vivio). Informazioni n. 54 - 1997.

Osteoporosis Risk Profile for Peri- and Postmenopausal Women Attending Menopause Clinics in Italy

M. Massobrio, R. Chionna, M. Ciammella, V. De Leo, G. De Luigi, M. Gallo, A. La Marca, R. Lombardo, M. Mauloni, U. Omodei, G. Sciacchitano for the ICARUS Osteoporosis Study Subgroup

Introduction

Bone mineral density (BMD) is the best predictor of osteoporotic fractures since each standard deviation decline is associated with a 2- or 3-fold increase in the risk of fracture, depending on the skeletal site evaluated. It would be of interest to look for relations between BMD and risk factors for osteoporosis (such as body mass index (BMI), menopausal status, menarcheal age). For example, in 1995 in a large epidemiological study of more than 9,000 women over 65, it was possible to see correlations between BMD and specific single factors, many of which could be ameliorated by prevention and treatment [1].

The ICARUS study is an Italian epidemiological study on the effect of menopause on women's health. The study involved 79 menopause clinics throughout Italy and between 1995 and 1997 enrolled a total population of 15,464 women. One subproject of the ICARUS Study is the analysis of parameters relating to the diagnosis of osteoporosis.

Subjects and Methods

The ICARUS study is a prospective study conducted since 1995 in menopause clinics through Italy. Women who attended the participating centers for the first time during the study period were eligible for the study. The protocol did not foresee any exclusion criteria.

During first visit, information regarding sociodemographic data, reproductive and medical history, menopausal symptoms, use of female hormones, and a selected medical history were collected by trained interviewers using a standardized information system, based on a computerized medical record shared by all participating centers. Quality and consistency of the entered data were checked by the study monitor during on-site visits, before the data were sent to the central database. The computerized medical report form allowed the uniform and standardized collection of clinical data and their transfer to a central database for statistical analysis.

The study started in 20 menopause clinics in 1995. The number of centers increased

R. Paoletti et al. (eds.), Women's Health and Menopause, 275–278.

to 79 during the study. Data collected until December 1997 are considered in the present paper.

For the purpose of this analysis, participants who had never used hormone replacement therapy (HRT) were observed at the 10 centers which participated in the osteoporosis subproject and performed densitometry with DXA instruments.

Of the total eligible population of 15,464 women, who entered the ICARUS study, 11,699 were never-HRT users at study entry. Of these, 4,416 women were observed in the 10 centers which performed DXA evaluation. Among the 4,416 subjects, DXA was performed in 2,621 (59.4%). Table 1 shows the characteristics of 4,416 never-HRT users observed in the 10 participating centers according to the collection of DXA information. No marked differences emerged between women on whom DXA was performed and those on whom it was not with regard to age, age at menopause, duration of fertile period, and frequency of smoking.

Table 1. Distribution of never-HRT users observed in the 10 centers considered in this analysis according to DXA.

	Densitometry DXA					
	No			Yes		
	No.	Mean	(SD)	No.	Mean	(SD
Age	2621	52.1	(5.0)	1795	52.4	(4.
BMI	2621	25.2	(4.1)	1795	25.1	(4.
Age at menopause	2621	49.1	(4.2)	1795	49.3	(4.
Duration of fertile period (years)	2621	36.5	(4.5)	1793	36.6	(4.
Smoking (yes, %)	2604	17.5%		1778	17.2%	

Women were classified as normal, osteopenic, or osteoporotic according to the WHO criteria (1994) (normal bone mass = a value for T score that is within 1 SD of the young adult reference mean; low bone mass (osteopenia) = a value for T score more than 1 SD below the young adult mean, but less than 2.5 SD below this value; osteoporosis = a value for T score 2.5 SD or more below the young adult mean) [2]. Women were divided according to menopausal status: 9.9 of subjects were in premenopause, 19.4 in perimenopause, and 67.7% in postmenopause, either spontaneous (60.1%) or surgical (7.6%) (in the 3.0% of subjects menopausal status was not defined).

The risk profile for osteoporosis was examined, considering the presence of the following factors: smoking, age at menarche, menopausal status, age at menopause, BMI, and duration of fertile period.

A multiple regression analysis was carried out in order to identify factors significantly influencing the development of osteoporosis. It considered the following variables: age, menarcheal age, menopausal status, duration of fertile period, BMI, smoking.

Results

Analyzing distribution of T score values according to WHO classification, osteoporosis was diagnosed in 2.2% of premenopausal women, in 4.2% of perimenopause subjects, and in 12.2% of postmenopause patients, while osteopenia was observed in 22.9%, 25.6%, and 30.8% of subjects, respectively.

The occurrence of late menarche ($\geq$ 15 years) and short fertile period (< 30 years), were in the multivariate analysis significantly ($p < 0.01$ and < 0.05, respectively) more frequently in women with osteoporosis (15.9% and 12.3%, respectively) than in women with either normal bone mass (9.7% and 6.0%, respectively) or osteopenia (7.9% and 7.6%, respectively). Moreover, obesity (BMI $\geq$ 28.6) was more frequently ($p < 0.05$) observed in subjects with normal bone mass or with osteopenia (21.7% and 18.8%, respectively) than in subjects with osteoporosis (11.4%).

Discussion

The present results underline the influence of menopausal status, BMI, late menarche, and duration of fertile period in the development of osteoporosis in this group of middle-aged women attending menopause clinic in Italy.

This population represents a selected population. Women considered in this analysis were observed in 10 Italian menopausal clinics. Thus, the study population can not be considered representative of the general one.

For example, women who entered the ICARUS study were more educated, but no marked differences emerged in prevalence of medical conditions such as hypertension and overweight [3]. Further, only about 60% of never-HRT users attending menopause clinics participating to the study had DXA. However, no marked difference was observed with reference to main determinants of osteoporosis between women who had and did not have DXA. The population considered in this analysis is a nonrandom sample of a selected population, thus any inference can be made only in strict comparative terms (i.e. characteristics of women of this population with osteoporosis/osteopenia or normal DXA values).

Densitometry assay was conducted in different centers using the centers' DXA instruments.

In our study BMI was significantly related to T score, confirming previous observations of an increased risk of vertebral fractures associated with low body weight/BMI [4]. Furthermore, EVOS Study showed an inverse relation between hip BMD and body weight. The same relationship was observed at the femoral neck and at the trochanter [5]. There are clearly established hormonal influences on the susceptibility to osteoporosis. Our results showed that late menarche ($\geq$ 15 years) and short fertile period (< 30 years) were significant predictors of osteoporosis. The inverse relation between BMD and age at menarche has been reported by several studies. Delayed onset of menarche, sometimes associated with disordered growth and delayed development, probably interferes with the normal accretion of peak bone mass [6]. If an increased rate of bone loss was then superimposed on a lower peak bone mass, the risk for osteoporosis would be markedly increased. Cross-sectional data suggest that bone density is negatively related to the age at menarche in both normal premenopausal and anoretic women [6]. Older age at menarche, early menopause, and consequently a decreased fertile period are associated with an increased risk of vertebral deformities [7]. Several studies showed that the number of fertile years is positively associated with lumbar BMD [8,9] and a reduced risk of hip fracture [10]. EVOS Study reported a risk of vertebral deformity increased by 48% in late menarche women, whereas a risk increased by 15 % was seen in women with a fertile period shorter than 34.6 years [11].

In conclusion this study confirms that late menarche, a short fertile period, and low BMI are risk factors for osteoporosis in a group of women attending the Italian menopause clinics.

References

1. Cummings SR, Nevitt MC, Browner WS, et al. Risk factors for hip fracture in white women. N Engl J Med 1995;332:767-72.
2. World Health Organization. Assessment of fracture risk and its application to screening for postmenopausal osteoporosis. WHO Technical Report Series, 843, Geneva, 1994.
3. ISTAT. Condizioni di salute e ricorso ai servizi sanitari (Anno 1994). Indagine Multiscopo sulle famiglie. (A cura di: A. Urbano, R. Vivio). Informazioni n. 54 - 1997.
4. Melton LJ III, Lane AW, Cooper C, Eastell R, O'Fallon WM, Riggs BL. Prevalence and incidence of vertebral deformities. Osteoporosis Int 1993;3:113-19.
5. Johnell O, Nilsson BE, O'Neill T, et al. Height loss and weight changes and vertebral deformities-the EVOS Study. Osteoporosis Int. 1996;6(Suppl.1):144.
6. Rosenthal DI, Mayo-Smith W, Hayes CW, et al. Age and bone mass in premenopausal women. J Bone Miner Res 1989;4:533-8.
7. Silman AJ, O'Neill TW, Varlow JR, et al. Hormonal and gynaecological factors and risk of vertebral deformity (EVOS Study). J Bone Miner Res 1995;10(Suppl.1):S261.
8. Kritz-Silverstein D, Barrett-Connor E. Early menopause, number of reproductive years and bone mineral density in postmenopausal women. Am J Public Health 1993;83:983-88.
9. Vico L, Prallet B, Chappard D, Pallot-Prades B, Pupier R, Alexandre C. Contributions of chronological age, age at menarche and menopause and of anthropometric parameters to axial and peripheral bone densities. Osteoporosis Int 1992;2:153-58.
10. Johnell O, Gullberg B, Kanis JA, et al. Risk factors for hip fracture in European women: the MEDOS Study. J Bone Miner Res 1995;10:1802-15.
11. OíNeill TW, Silman AJ, Naves Diaz M, Cooper C, Kanis J, Felsenberg D and the European Vertebral Osteoporosis Study Group. Influence of hormonal and reproductive factors on the risk of vertebral deformity in European women. Osteoporosis Int 1997;7:72-8.

MENOPAUSE: WHAT IS THE ROLE OF THE GENERAL PRACTITIONER?

G. Tresoldi for the Italian Society of General Practitioners

The complexity and the variety of problems to be managed in the climacteric period strongly suggests that one doctor act as a primary "point of reference" for the patient offering advice based on her individual history, family, and lifestyle [1-3].

The general practitioner (GP) possesses both the information and technical ability to assume this crucial role for the postmenopausal patient [4,5]. Specifically, he or she can perform the following duties:

- To provide women of all ages with accurate information about the climacteric and its problems; to provide appropriate health education; and to play an active part in any initiatives related to the climacteric;
- To recognize the symptoms of estrogen deficiency and to identify any such signs during the clinical examination;
- To identify any signs or symptoms that require a specialist approach;
- To diagnose, with case history assistance, or any examinations of a technical, hemato-chemical, or other nature, the physiological, early, pathological, or iatrogenic menopausal states;
- To carry out a clinical examination of the patient with a view to any eventual hormone replacement therapy (HRT), and with a view to the prevention of menopausal complications, or complications arising from the HRT itself;
- To discuss all aspects of HRT with the patient, so that she may make an informed decision regarding the management of her health during the climacteric. If, on the basis of the patient interview, her expectations, and the results of the clinical values measured during the examination, it seems appropriate, then replacement therapy or an alternative should be prescribed. In all cases, such factors as behavioral, dietary, life-style, and contraception, must be considered [6];
- To carry out methodical follow-up of women undergoing replacement or other therapies [7-11]; and
- To offer psychological support to the patient during all phases of the climacteric, through patient and family counseling. This counseling will focus on the psychological, interpersonal, and sexual changes associated with the climacteric [12,13].

The GP's role in the climacteric is characterized by the fact that all aspects (risk factors, family history, desire, and ability to comply with treatment, etc.) of clinical,

R. Paoletti et al. (eds.), Women's Health and Menopause, 279–284.

emotional, and interpersonal circumstances must be determined for each individual. On the basis of this, the physician must then establish with the patient whether to carry out HRT immediately after the last menstruation period (because of a higher risk of cardiovascular disease or because of particularly intense symptoms) [14-20]; whether to carry out HRT later (because of risk of osteoporotic fracture) [15,21-23]; whether to carry out a therapy alternative to HRT (statin, biphosphonate, etc); or whether to do anything at all.

The conscious choice of the accurately informed patient who may place more value on nonmedical considerations should be the major criterion in making the decision. This will only be possible within the context of a good, long-standing relationship with one's doctor.

The management of women in the climacteric by Italian general practitioners is currently not without its problems. The introduction in the 1970s of family consultants led the Italian Society of General Practitioners to delegate all problems associated with obstetrics and gynecology to the Italian National Health Service and to neglect updating, training, and education in these areas. Thus GP, who previously were responsible for the management of the menopause, no longer are on the cutting edge of the field.

At the same time, the increase in life expectancy has notably broadened the issue of the climacteric [24-27]. Every general practitioner with a patient load of 1,500 can expect to have (depending on the pediatric population) approximately 100-150 women between 45 and 55 years of age, i.e. 100-150 women presenting climacteric symptoms. This GP may also expect to have 250 to 350 or more women aged over 50, presenting the signs and symptoms of estrogen deficiency. Moreover, the approach to, and management of, replacement therapy in menopause has changed remarkably in recent years as a result of rapid developments in estrogen drug delivery technology, the introduction of echography, transvaginal sounding, and hysteroscopy, all of which have widened the safety margins regarding HRT's major deterrent, endometrial cancer.

Thus HRT no longer falls within the realm of specialist medicine, but is a more common, wide-ranging, therapy that should fall within the range of general practitioner abilities.

It is therefore necessary to define the changes that general practice must face to effectively carry out the important social and health tasks involved. One initial problem is the authority of the doctor in the eyes of the patient. After 20 years of referring all menopausal conditions to the gynecologist, GP now need to present themselves to the patient in such a way that the patient realizes that they are knowledgeable about, and are willing and able to deal with, the problems surrounding the menopause and HRT. This can be done via informal contacts, specially arranged patient meetings, and by the use of appropriate audio-visual material in the waiting room.

In fact, women in this age range are the most frequent visitors to the doctor's office. The reasons for the visits are varied and often concern family problems, as well as those related to the menopause (vasomotor symptoms, menstrual problems, sometimes requests for clarification on what they may have heard or read about the menopause). This can present an ideal opportunity to discuss breast or uterine cancer prevention and the possibilities of HRT.

Health education measures the GP may propose to the menopausal woman may be summarized as follows:

1) Programs directed at both large and small groups (town, district, one's own patients, etc.);
2) Articles in local publications according to legal decrees DL 175/92 and 657/94; and
3) Personal counseling, leaflets, presentations, and videos for the waiting room.

Such means allow the patient to receive and select information and knowledge which can then be discussed again individually with the doctor.

The availability of technology such as a personal computer or video-recorder can make a great difference in the delivery of health education.

For comprehensive health education on the climacteric, it is essential to cover the following information: What the menopause is; what causes it; what symptoms/disturbances are most frequently associated with it; what can be done about them; what are the possible health consequences of the menopause; what preventive and behavioral measures can be taken; whether it is worth taking hormones; if one does take hormones, will the risk of cancer be increased; and menopause as a natural state of health The messages imparted must be positive. The overall approach to the content should be reassuring and underline the following: the menopause is a normal stage of life; it does not mean the end of femininity; there is no reason to worry; check that the information you have is correct–don't be unduly influenced by friends or the newspapers; see your doctor if you experience loss of blood after entering the menopause; regularly examine your breasts; undergo a Pap test and any examinations agreed upon with your doctor; the doctor, knowing your case history and that of your relatives, is in the best position to help; talk with him about your hopes and fears; together you can find the best solution.

Another problem relevant to the daily management of such patients is that posed by the rapport existing between the general practitioner and the gynecologist/menopause center. The existence of menopause centers seems to be justified for consultation of particular cases referred by the GP or for research and study. The criteria distinguishing those situations requiring specialized skills as opposed to those handled by the GP may be as follows: if the climacteric shows all signs of normality, a good general practitioner can manage perfectly well. In pathological situations (early menopause, endocrinopathy, previous breast or ovarian disease, etc.) or problems during therapy (atypical bleeding), the patient should be referred to the gynecologist. The specialist may also be consulted if other clinical problems associated with the climacteric are present (previous cardiac or vascular disease, pathologies, uncompensated diabetes, serious osteoporosis).

It is anachronistic in this phase of the history of the National Health Service to consider establishing specialized services at high running costs aimed at a large percentage of women in order to effect a service already in existence. Furthermore, many of these services use study protocols requiring computerized bone mineralometry, endometrial biopsy, and hysteroscopies whose value in routine work, according to the most recent literature, is debatable.

The gynecological specialists and menopause centers could undertake work in the following areas:

- Early menopause;
- Patients with menopause-related disease (hirsutism, endometriosis);
- Patients with previous or current genital or mammary neoplasia;
- Patient consultation for those with serious risk factors (hypertension, dyslipidemia, etc.);
- Bleeding in, or due to, therapy and complications arising from therapy;
- Abnormal results in follow-up examinations;
- Research;
- Health information and education in group settings and educational meetings for the public;
- Rapid and convenient service; and
- Collaboration with general practitioners.

However, it is important to remember that GP have abdicated their gynecological skills; training, education, and updating in the clinical and personal issues surrounding the menopause and HRT are necessary. The first step in this process would be drawing up common guidelines and agreements between the Scientific Society of General Practitioners and the Society of Obstetrics and Gynecology. This could enable the respective roles in managing the climactic to be clearly defined and could also lead to the establishment of a framework of education and training courses for GP.

The general educational objective to accomplish this goal may be: "The GP should be able to meet and deal with the problems that the woman presents in the climacteric with regard to information, diagnosis of the actual menopausal condition, the prescription of hormone replacement therapy and its alternatives, and the prevention of complications due to hormone deficiency, thereby encouraging informed decision-making on the part of the patient while offering counseling throughout this period. Furthermore, the GP should offer appropriate health education for the woman at this time of life using everyday means (e.g. leaflets, presentations) or by educating small patient groups or larger public groups."

This general objective can be broken down into specific learning subobjectives : for example, to recognize the signs and symptoms of under-production of estrogen; to diagnose the menopausal state; to prescribe hormone replacement therapy or alternatives; to agree, with woman in the menopause or under HRT, upon a program of periodic check-ups over time. An analysis of these educational subobjectives leads to the establishment of a full course. The first step in constructing such a course is that of deciding on the choice of teacher to impart the above-mentioned objectives. After agreement with the teacher has been reached and any necessary amendments to the course content have been made, means of testing need to be devised that will ensure that the course objectives have been met.

The methodology normally used involves the Small Interactive Group Learning technique as a way of achieving teaching aims, i.e. those behavioral models in keeping with learning objectives. This teaching method usually requires one or two teachers and is employed with a limited number of learners (around 40). It is based on the possibility for interaction between participants and generates learning based on direct comparison of ideas, notions, and real-life experiences. As the subject of the "female climacteric" comprises cognitive, behavioral, and interpersonal components, teaching methods can be varied to a great extent, involving written clinical cases, homework projects, questionnaires, test grids, observation and evaluation, and role-playing. Evaluation of learning may be carried out in many ways, according to the type of testing and should, in any case, result in a positive mark reflecting a correct answer and a negative mark reflecting an incorrect answer [28].

References

1. Guidelines for counseling postmenopausal women about preventive hormone therapy. American College of Physicians Ann Intern Med 1992;117(12):1038-41.
2. Genazzani AR, Zichella L (editors). La terapia ormonale in climaterio o postmenopausa. Atti della 1 conferenza Nazionale di consenso in scienze ginecologoche e ostetriche, Madonna do Campiglio 17-24 marzo 1996 - Churchill Livingstone.
3. "Women's health" supplement to The Lancet. Lancet 1997;349(Suppl.):1-26.
4. Traynor V, Britt H, Sayer GP. Menopause: Its management in general practice. Aus Family Physician 1995;24:407-11.
5. O'Connor V, et al. The menopause and hormone replacement therapy: Australian general practitioners' self-reported opinions, attitudes and behaviour. Fam Pract 1996;13:421-26.
6. Hammond CB. Management of menopause. Am Fam Phys 1997;55:1667-73.
7. Beresford SA, et al. Risk of endometrial cancer in relation to use of oestrogen combined with cyclic progestogen therapy in postmenopausal women. Lancet 1997;349:458-61.
8. Colditz GA, et al. The use of estrogens and progestins and the risk of breast cancer in postmenopausal women. New Engl J Med 1995;332:1589-93.
9. Wise J. Hormone replacement therapy increases risk of breast cancer. BMJ 1997;315:969.
10. Sellers TA, et al. The role of hormone replacement therapy in the risk for breast cancer and total mortality in women with a family history of breast cancer. Ann Intern Med 1997; 127:973-80.
11. Collaborative Group on Hormonal Factors in Breast Cancer. Breast cancer and hormone replacement therapy: Collaborative reanalysis. Lancet 1997;350:1047-43. See also commentaries pages 1042, 1043.
12. Nicol-Smith L. Causality, menopause, and depression: A critical review of the literature. BMJ 1996;313:1229-32.
13. Hunter MS. Depression and the menopause. BMJ 1996;313:1217-18.
14. AHA Science Advisory. Guide to primary prevention of cardiovascular diseases. Circulation 1997;95:2329-31.
15. The Writing Group for the PEPI Trial. Effects of estrogen or estrogen/progestogen regimens on heart disease risk factors in postmenopausal women. The Postmenopausal Estrogen/Progestogen Interventions (PEPI) Trial. JAMA 1995;273:199-208.
16. Wenger NK, Speroff L, Packard B. Cardiovascular health and disease in women. New Engl J Med 1993;329:247-53.

17. Grodstein F, et al. Postmenopausal estrogen and progestin use and the risk of cardiovascular disease. New Engl J Med 1996;335:453-61.
18. Rossouw JE. Estrogens for prevention of coronary heart disease. Putting the brakes on the bandwagon. Circulation 1996;94:2982-85.
19. Wenger NK. Coronary heart disease: An older woman's major health risk. BMJ 1997;315:1085-90.
20. Pedersen AT, et al. Hormone replacement therapy and risk of non-fatal stroke. Lancet 1997;350:1277-83.
21. Consensus development statement: Who are candidates for prevention and treatment for osteoporosis? Osteoporosis Int 1997;7:1-6.
22. Schneider DL, et al. Timing of postmenopausal estrogen for optimal bone mineral density. The Rancho Bernardo Study. JAMA 1997;277:543-47 (JAMA ed. italiana vol. 9 n.6:264-70).
23. Borroni M, et al. Idee guida per la prevenzione e la diagnosi di osteporosi e per la prevenzione delle fratture su base osteoporotica. Ricerca e Practica 1997;13:158-67.
24. Grodstein F, Stampfer MJ, et al. Postmenopausal hormone therapy and mortality. New Engl J Med 1997;336:1769-75.
25. Grady D, et al. Hormone therapy to prevent disease and prolong life in postmenopausal women. Ann Intern Med 1992;117:1016-37.
26. Ettinger B, et al. Reduced mortality associated with long-term postmenopausal estrogen therapy. Obstet Gynecol 1996;87:6-12.
27. Crosignani PG, Paoletti R (editors). Hormone replacement and the menopause. Eur Men J 1996;3(3):195-256.
28. Guilbert JJ. Guida pedagogica. Armando editore 1981.

BIOETHICS, MENOPAUSE, AND AGING: A VIEW FROM THE ITALIAN MENOPAUSE SOCIETY

I. Baldaro Verde, A. Bompiani, F. D'Agostino, C. Nappi, A. Spinelli, A.G. Spagnolo, and L. Zichella for the Italian Society of Menopause

The phenomenon of "menopause" has become more frequent in this century because of the prolongation of women's lives. Therefore, an evaluation of the bioethical issues associated with menopause is timely and advisable.

In Italy the overall picture seems unfocused, notwithstanding the efforts being made by those concerned: the woman, the physician, the pharmaceutical companies, the mass media, and politicians. These attitudes have caused many physicians to underestimate the symptoms and risks of hypoestrogenism and support the resistance of those women who already have accepted their physical decline, while the general public, in particular those in government responsible for health care, is disinterested in the subject, slowed by bureaucracy, or both.

Today correct bioethical evaluation of the management of menopause must take into account at least three aspects of this phenomenon: 1) the appropriate evaluation of the symptoms and of the efficacy of the therapies; 2) the sociological values affecting those involved; and 3) the rules that guarantee these values be identified.

Bioethically, the issues associated with menopause are not unlike those addressed in gender differences. Gender or sexual differences are at the same time both factual and a symbolic phenomenon. It is also one of the most serious normative problems of modern consciousness. Thus it may not be considered simply on its factual or its psychological front. The advances in biomedical knowledge have provided new opportunities to view menopause both as a physiological process as well as in terms of symbolic dynamics, thereby altering society's perception of sexual differences.

Using demographic data, it has been possible to evaluate the population at risk. There has been an increase worldwide in the number of women aged 50 and over (generally considered postmenopausal women). For instance, in Italy in 1953 there were 5,684,890 women in this age group (23.5% of the total female population); in 1996 this number almost doubled to 10,966,892 (37.2%).

Analytical epidemiological studies (cohort and case-control designs) have been used to identify associations between risk factors and undesirable signs or symptoms (e.g. the relationship between breast cancer and the use of hormonal replacement therapy). The results of these studies are sometimes contradictory. Further risk-benefit and cost-

R. Paoletti et al. (eds.), Women's Health and Menopause. 285–286.

effectiveness research should be carried out, giving greater attention to the differences among countries with regard to the meaning of menopause and the incidence of disease associated with menopause.

Analysis of mortality rates by age may provide some indication of the health consequences of the hormonal changes that occur at the time of menopause. The ratios of female-to-male mortality rates by age and broad cause show very different patterns for cancers and cardiovascular diseases. In general, men have higher mortality rates than women (ratio < 1). However, for malignant neoplasms, the ratio is greater than 1 in the age groups 25-34 and 35-44, declines through the menopausal period, and remains subsequently low. In contrast, adjusted rates for cardiovascular diseases are lower in premenopausal women than in men, but increase postmenopausally. Thus, in terms of mortality, menopause seems to have a beneficial effect on cancer and a negative impact on cardiovascular diseases.

Menopause, as pregnancy, should not be considered as a disease but as a physiological phase of women's lives. As such, both excessive medication for menopause and hormone replacement therapy (HRT) as a "fountain of youth" are inappropriate responses to the situation. Additionally both approaches to menopausal women, as well as trial protocols in the field, run the risk of being influenced by the rules of the pharmaceutical marketplace.

One of the most frequent trial protocols submitted to the Ethical Review Committees (ECs) concerns new methods of administration or new combination of estrogen and progestin as part of HRT. According to the recent European Good Clinical Practice, ECs have the responsibility to ensure the protection of the rights, safety, and well-being of women involved in the trial and provide public assurance of that protection. Another main topic that ECs must deal with is the risk/benefit ratio.

ECs reviewing protocols for replacement therapy are asked to consider, but not be limited to, the following ethical points: 1) justification of predictable risks and inconveniences weighed against the benefits for the women; 2) adequacy and completeness of written information to be given to the women, including the uncertainty of benefits and the severity of possible risks; and 3) adequacy of provisions made for monitoring and auditing the conduct of research.

Furthermore, it must be noted that, according to the International Ethical Guidelines for Biomedical Research Involving Human Subjects (CIOMS, 1993), scientific review and ethical review committees must consider both the scientific and the ethical aspects of proposed research.

In conclusion, the most adequate bioethical method is to bring to light the supposed neutrality of the scientific method and reach a nondescriptive, nondefensive bioethical modality, both legally and medically, especially in those areas affecting the field of menopause.

INDEX

Medical Science Symposia Series

1. A.M. Gotto, C. Lenfant, R. Paoletti (eds.) and M. Soma (ass.ed.): *Multiple Risk Factors in Cardiovascular Disease.* 1992 ISBN 0-7923-1938-9
2. A.L. Catapano, A.M. Gotto, Jr., L.C. Smith and R. Paoletti (eds.): *Drugs Affecting Lipid Metabolism.* 1993 ISBN 0-7923-2232-0
3. T. Godfraind, S. Govoni, R. Paoletti and P.M. Vanhoutte (eds.): *Calcium Antagonists. Pharmacology and Clinical Research.* 1993 ISBN 0-7923-2259-2
4. D. Galmarini, L.R. Fassati, R. Paoletti and S. Sherlock (eds.): *Drugs and the Liver: High Risk Patients and Transplantation.* 1993 ISBN 0-7923-2307-6
5. P.M. Vanhoutte, P.R. Saxena, R. Paoletti, N. Brunello (eds.) and A.S. Jackson (ass.ed.): *Serotonin. From Cell Biology to Pharmacology and Therapeutics.* 1993 ISBN 0-7923-2518-4
6. A.G. Dalgleish, A. Albertini and R. Paoletti (eds.): *The Impact of Biotechnology on Autoimmunity.* 1994 ISBN 0-7923-2724-1
7. P.G. Crosignani, R. Paoletti, P.M. Sarrel, N.K. Wenger (eds.), M. Meschia and M. Soma (ass.eds.): *Women's Health in Menopause. Behaviour, Cancer, Cardiovascular Disease, Hormone Replacement Therapy.* 1994 ISBN 0-7923-3068-4
8. A.M. Gotto Jr., C. Lenfant, A.L. Catapano and R. Paoletti (eds.): *Multiple Risk Factors in Cardiovascular Disease.* 1995 ISBN 0-7923-3503-1
9. T. Godfraind, G. Mancia, M.P. Abbracchio, L. Aguilar-Bryan and S. Govoni (eds.): *Pharmacological Control of Calcium and Potassium Homeostasis. Biological, Therapeutical, and Clinical Aspects.* 1995 ISBN 0-7923-3604-6
10. A.M. Gotto Jr., R. Paoletti, L.C. Smith, A.L. Catapano and A.S. Jackson (eds.): *Drugs Affecting Lipid Metabolism. Risk Factors and Future Directions.* 1996 ISBN 0-7923-4167-8
11. R. Paoletti, P.G. Crosignani, P. Kenemans, G. Samsioe, M. Soma and A.S. Jackson (eds.): *Women's Health and Menopause.* Risk Reduction Strategies. 1997 ISBN 0-7923-4697-1
12. A.M. Gotto Jr., C. Lenfant, R. Paoletti, A.L. Catapano and A.S. Jackson (eds.): *Multiple Risk Factors in Cardiovascular Disease.* Strategies of Prevention of Coronary Heart Disease, Cardiac Failure, and Stroke. 1998 ISBN 0-7923-5023-5
13. R. Paoletti, P.G. Crosignani, P. Kenemans, N.K. Wenger and A.S. Jackson (eds.): *Women's Health and Menopause.* Risk Reduction Strategies – Improved Quality of Health. 1999 ISBN 0-7923-5906-2

KLUWER ACADEMIC PUBLISHERS – DORDRECHT / BOSTON / LONDON